OPPORTUNITIES FOR BIOLOGICAL NITROGEN FIXATION IN RICE AN

Developments in Plant and Soil Sciences

VOLUME 75

Opportunities for Biological Nitrogen Fixation in Rice and Other Non-Legumes

Papers presented at the Second Working Group Meeting of the Frontier Project on Nitrogen Fixation in Rice held at the National Institute for Biotechnology and Genetic Engineering (NIBGE), Faisalabad, Pakistan, 13–15 October 1996

Edited by

J. K. LADHA, F. J. de BRUIJN and K. A. MALIK

Reprinted from *Plant and Soil*, Volume 194, Nos. 1–2 (1997)

KLUWER ACADEMIC PUBLISHERS
DORDRECHT / BOSTON / LONDON

A C.I.P. Catalogue record for this book is available from
the Library of Congress.

ISBN 0-7923-4514-2 (HB)
ISBN 0-7923-4748-X (PB)

Published by Kluwer Academic Publishers,
P.O. Box 17, 3300 AA Dordrecht, The Netherlands

Sold and distributed in the U.S.A. and Canada
by Kluwer Academic Publishers,
101 Philip Drive, Norwell, MA 02061, U.S.A.

In all other countries, sold and distributed
by Kluwer Academic Publishers,
P.O. Box 322, 3300 AH Dordrecht, The Netherlands

Cover photo: Transmission electron micrograph showing *Azorhizobium caulinodans* within a dead rice (IR42) root cell with an altered wall adjacent to an intact uninfected root cell. *Plant and Soil* Vol. 194, pp. 81–98.

Printed on acid-free paper

Printed in the Netherlands

Contents

Preface

During the next 30 years, farmers must produce 70% more rice than the 550 millions tons produced today to feed the increasing population. Nitrogen (N) is the nutrient that most frequently limits rice production. At current levels of N use efficiency, we will require at least double the 10 million tons of N fertilizer that are currently used each year for rice production. Global agriculture now relies heavily on N fertilizers derived from petroleum, which, in turn, is vulnerable to political and economic fluctuations in the oil markets. N fertilizers, therefore, are expensive inputs, costing agriculture more than US$45 billion annually.

Rice suffers from a mismatch of its N demand and N supplied as fertilizer, resulting in a 50-70% loss of applied N fertilizer. Two basic approaches may be used to solve this problem. One is to regulate the timing of N application based on needs of the plants, thus partly increasing the efficiency of the plants' use of applied N. The other is to increase the ability of the rice system to fix its own N. The latter approach is a long-term strategy, but it would have enormous environmental benefits while helping resource-poor farmers. Furthermore, farmers more easily adopt a genotype or variety with useful traits than they do crop and soil management practices that may be associated with additional costs.

New frontiers of science offer exciting opportunities to stretch rice research horizons. Recent advances in understanding symbiotic Rhizobium-legume interactions at the molecular level and the ability to introduce new genes into rice through transformation have created an excellent opportunity to investigate the possibilities for incorporating N fixation capability in rice. During a think-tank workshop organized by IRRI in 1992, the participants reaffirmed that such opportunities do exist for cereals and recommended that rice be used as a model system. Subsequently, IRRI developed a *New Frontier Project* to coordinate the worldwide collaborative efforts among research centers committed to reducing dependency of rice on mineral N resources. An international Rice Biological Nitrogen Fixation (BNF) working group was established to review, share research results/materials, and to catalyze research.

This volume contains 20 papers presented at the second working group meeting, held 13-15 October 1996 at the National Institute of Biotechnology and Genetic Engineering (NIBGE) in Faislabad, Pakistan. It took place in conjunction with the 7th International Symposium on Nitrogen Fixation with Nonlegumes held 16-21 October 1996 at Faislabad. Dr. J.K. Ladha, IRRI, and Dr. K.A. Malik, NIBGE, were the convenors and Dr. Christina Kennedy, University of Arizona, chaired the meeting. Drs. J.K. Ladha, F.J. deBruijn, and K.A. Malik served as technical editors.

George Rothschild
Director General

Plant and Soil **194**: 1–10, 1997.

1

Introduction: Assessing opportunities for nitrogen fixation in rice - a frontier project

J.K. Ladha[1], F.J. de Bruijn[2] and K.A. Malik[3]

[1]*International Rice Research Institute, P.O. Box 933, Manila, Philippines** [2]*DOE Plant Research Laboratory, Michigan State University, East Lansing, Michigan, U.S.A. and* [3]*National Institute for Biotechnology and Genetic Engineering, Faisalabad, Pakistan*

Key words: endophyte, legume-*Rhizobium*, *nif* gene, nitrogen fixation, rice

Abstract

Recent advances in understanding symbiotic *Rhizobium*-legume interactions at the molecular level, the discovery of endophytic interactions of nitrogen-fixing organisms with non-legumes, and the ability to introduce genes into rice by transformation have stimulated researchers world wide to harness opportunities for nitrogen fixation and improved N nutrition in rice. In a think-tank workshop organized by IRRI in 1992, the participants reaffirmed that such opportunities do exist for cereals and recommended that rice be used as a model system. Subsequently, IRRI developed a *New Frontier Project* to coordinate the worldwide collaborative efforts among research centers committed to reducing dependency of rice on mineral N resources. An international Biological Nitrogen Fixation (BNF) working group was established to review, share research results/materials and to catalyze research.

The strategies of enabling rice to fix its own N are complex and of a long-term nature. However, if achieved, they could enhance rice productivity, resource conservation, and environmental security. The rate of obtaining success would, of course, benefit tremendously from concerted efforts from a critical mass of committed scientists around the world, as well as a constant and continued funding support from the "donor" community.

Population growth and increasing demand for rice

Rice is the most important staple food for over two billion people in Asia and for hundreds of millions in Africa and Latin America. To feed the ever-increasing population of these regions, the world's annual rice production must increase from the present 460 million to 560 million by the year 2000 and to 760 million by 2020 (IRRI, 1993). If future increase in rice production has to come from the same or even a reduced land area, rice productivity (yield ha^{-1}) must be greatly enhanced to meet these goals.

Inorganic N fertilizer, a key input for rice production

Nitrogen is the nutrient that most frequently limits agricultural production. Global agriculture now relies heavily on N fertilizers derived at the expense of

petroleum, which in turn is vulnerable to political and economic fluctuations in the oil markets. Nitrogen fertilizers, therefore, are expensive inputs, costing agriculture more than $45 billion (US) per year.

In the tropics, lowland rice yields 2–3.5 t ha^{-1} utilizing naturally available N derived from biological nitrogen fixation (BNF) by free-living and plant-associated diazotrophs (Watanabe and Roger, 1984; Ladha et al., 1993) and from mineralization of soil N (Bouldin, 1986; Kundu and Ladha, 1994). For higher yields, additional N must be applied. Achieving the 50% higher rice yields needed by 2020 will require at least double the 10 million t of N fertilizer that is currently used each year for rice production (IFA-IFDC-FAO, 1992; IRRI, 1993). Manufacturing the fertilizer for today's needs requires 544×10^9 MJ of fossil fuel energy annually (Mudahar, 1987a,b). Industrially produced N fertilizer depletes non-renewable resource and poses human and environmental hazards. In spite of an unlimited supply of N_2 in the air, manufacturing 1 kg of N fertilizer requires 6 times more energy than that

---
* FAX No: +6328711292. E-mail: J.K.Ladha@cgnet.com

needed to produce either P or K fertilizers (Da Silva et al., 1978). Nevertheless, over the past two and a half decades, farmers have become increasingly dependent on chemical sources of N for obtaining higher grain yields to meet the demands of enlarging population (see Figure 1).

Although the use of N fertilizer has increased substantially, a large number of farmers still use little or no N fertilizer because of several factors: its non-availability at times, lack of cash to buy it, and poor yield response due to adverse conditions. Furthermore, more than half the applied fertilizer N is lost (through denitrification, ammonia volatilization, leaching and runoff) because rice is grown in an environment conducive to N losses (see Figure 2). This not only represents a cash loss to the farmer but may lead to considerable environmental pollution. In addition, large denitrification losses during the transition from aerobic to anaerobic soil conditions may represent an important source of nitrous oxide, a gas linked with the greenhouse effect and the destruction of the stratospheric ozone layer. It is in this context BNF-derived N assumes importance in the lowland soils that provide about 86% of the world's rice.

Alternative sources of nitrogen

Free-living and associative systems

Diverse N_2-fixing microorganisms (aerobes, facultative anaerobes, heterotrophs, phototrophs) grow in wetland rice fields and contribute to soil N pools. The major BNF systems known include cyanobacteria and photosynthetic bacteria that inhabit floodwaters and the soil surface, and heterotrophic bacteria in the root zone (rhizosphere), or in the bulk soil.

The contributions of cyanobacterial BNF are estimated to be 10–80 kg N ha^{-1} crop^{-1}, averaging about 30 kg N ha^{-1} crop^{-1} (Roger and Watanabe, 1986). Since the discovery of the importance of cyanobacteria in N gain under flooded conditions, many inoculation experiments have been conducted using cultured cyanobacteria to improve soil fertility and grain yields of rice. Based on an extensive review of the literature in which gains in grain yields ranged between 0–3.7 t, Roger and Watanabe (1986) calculated that cyanobacterial inoculation increased rice yields only by an average of 337 kg grain ha^{-1} crop^{-1}.

Heterotrophic bacterial BNF averages 7 kg N ha^{-1} (App et al., 1986), ranging from 11–16 kg N ha^{-1} and

contributing to 16–21% of total rice N need (Zhu et al., 1984; Shrestha and Ladha, 1996).

Some free-living heterotrophic bacteria form associations with roots and other submerged portions of the rice plant. Varietal differences in supporting rhizospheric N_2 fixation in rice have been shown (Ladha et al., 1993). The latter raises the possibility to select and breed rice genotypes with higher BNF potential, high soil N uptake, and high N use efficiency to obtain high yields (Ladha et al., 1988, 1993, 1997). Recently Wu et al. (1995) mapped several rice loci mediating the variety dependent ability to stimulate N_2 fixation. Their results indicate that the varietal ability of rice to enhance N_2 fixation in the rhizosphere is controlled by multiple genes. The identification of quantitative trait loci underlying this trait provides the first real evidence for the presence of genetic factors which mediate the interaction with diazotrophs in the rice rhizosphere and lays the foundation for increasing N_2 fixation through genetic manipulation of rice plants and their partner microbes.

Green-manure systems

Aquatic plant like the water fern *Azolla*, and the semi-aquatic legumes such as *Sesbania*, *Aeschynomene* or *Astragalus*, are recommended green-manure plants for rice since they fix N_2 symbiotically. In a continuous long-term experiment at IRRI, an average of 16 crops gave an estimate of about 60 kg N ha^{-1} crop-1 of *Azolla* and *Sesbania* BNF in 50-60 days (Ladha et al., 1993). Yields with *Azolla* and *Sesbania* were equivalent to those with 60 kg N ha^{-1} as urea fertilizer in the wet season. In the dry season, the yields were 6.5 t ha^{-1} with *Sesbania* and 7.7 t ha^{-1} with *Azolla* as compared to 5.8 t ha^{-1} in a treatment of 60 kg N ha^{-1} urea. These data demonstrate the potential of *Azolla* and *Sesbania* to produce yields of 6–8 t ha^{-1}, roughly equivalent to an application of 100-200 kg N ha^{-1} as urea. Farmers, however, usually have little economic incentive to chose *Azolla* or *Sesbania* over N fertilizer since additional costs such as labor, land opportunity, irrigation, seed/inoculum, P and pesticides are involved.

Conventional versus novel systems

Estimates of the N supply potential of different BNF systems are provided in Table 1. Among the conventional systems, the free-living/associative diazotrophs have low to moderate potential to supply N to rice because the N_2 fixed outside the plant is subject to

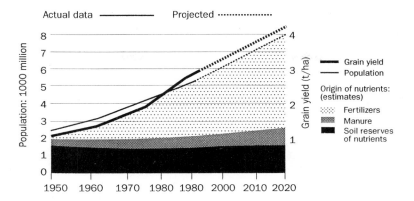

Figure 1. Global trends in population growth, cereal grain yield and origin of plant nutrients (source: Bøckman et al., 1990).

Table 1. Conventional and future biological nitrogen fixation systems for sustainable rice production (from Ladha and Reddy, 1995)

BNF system	N supply potential (kg ha^{-1})	Rice grain yield potential (t ha^{-1})	Essential trait in rice genotype[a]	Technology availability	Economic feasibility and farmers adoption
CONVENTIONAL BNF SYSTEMS					
Free-living/Associative (phototrophs and heterotrophs)	50–100	3–6	ANFS NAE NUE	3–5 yrs	High
Green-manure (*Azolla* and *Sesbania*)	100–200	5–8	NAE NUE	Available	Low
FUTURE BNF SYSTEMS					
Endophytic	?	?	Endo$^+$ *fix*$^+$ NUE	3–5 yrs	High
Induced Symbiosis (Rhizobia, *Frankia*, etc)	>200	>8	*nod*$^+$ *fix*$^+$ NUE	>5 yrs	High
nif gene transfer	>200	>8	*nif*$^+$ *fix*$^+$ NUE	>5 yrs	High

[a] ANFS = associative N_2 fixation stimulation; NAE = nitrogen acquisition efficiency; NUE = nitrogen utilization efficiency; Endo = Endophytic; *fix* = N_2 fixation ability; *nif* = N_2 fixation genes; *nod* = nodulation ability.

losses due to various loss processes as described earlier. Therefore, only genotypes having appropriate N use efficiency traits (i.e., large associative N_2 fixation and N acquisition, and efficient N utilization traits) can produce grain yields of 3–5 t ha^{-1} when depending on N supply from free-living/associative diazotrophs (Ladha et al., 1993). However, these yield levels cannot meet the growing demand for rice. Green-manures have high N supply potential to support rice grain yields of 5–8 t ha^{-1}, but due to the reasons discussed above, they do not present an attractive option for the farmer. Green-manures are also known to enhance the emission of methane which contributes to greenhouse effect (see Figure 2). If symbiotic or other nitrogen fixation systems could be incorporated/assembled in rice then the N supply potential could be higher because fixed N

Table 2. Rice as a model plant

- Excellent knowledge base on plant morphology, anatomy, agronomy, physiology and genetics is available;
- Rice has a small genome (415 million bp/1C) size: 3 x >*Arabidopsis*; 5.5 x < maize; 39 x < wheat);
- A High density restriction fragment length polymorphism map is available;
- Transposition of Ac/Ds systems have been documented;
- Co-linearity of genes of rice with genomes of other cereal has been established (syntony);
- Transformation and regeneration through protoplast, biolistic and *Agrobacterium* in both indica (i.e., IR43; 58; 64; 74) and japonica (i.e., Taipei 309; Nipponbare; Kinokihari; Yamabiko) rice has been achieved;
- Rice is amenable to heterologous gene expression;
- Several agronomically important genes (disease and pest resistance) have been integrated and expressed;
- Inheritance of foreign gene (i.e., *Bt, CryIa(b)*; Chitanase, *Chi11*) in both indicas and japonicas has been studied;
- Rice has been extensively utilized in studying the biochemistry and molecular genetics of host-pathogen (bacterial blight, blast, sheath blight, virus) interaction. Disease resistance (bacterial blight and blast) genes have been cloned and their expression is being studied in rice.

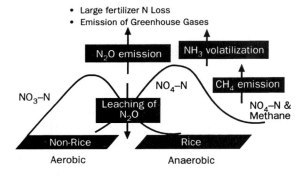

Figure 2. Rice is grown in an environment conducive to significant loss of N.

would be available directly to the plant with little or no loss. This in turn, would promote higher grain yield in rice varieties with efficient N utilization traits.

Assessing opportunities for nitrogen fixation in rice - a frontier project

As pointed out above, more rice must be produced from less land with a minimum of cost. Moreover, environmentally adverse inputs must be reduced to meet the challenge of feeding the world's growing population of rice consumers on a sustainable and equitable basis. Nowhere is this challenge more intense than in sup-

plying N to the crop. IRRI's objective of increasing the yield plateau of rice grown under tropical conditions from 10 to 15 t ha^{-1} for the next 30 yr demands that attention be paid to sustainable methods of N supply. Rice suffers from a mismatch of its N demand and N supplied as fertilizer, resulting in 50–70% loss of applied N fertilizer. Two basic approaches may be used to solve this problem. One is to regulate the timing of N application based on the plant's needs, thus increasing the efficiency of the plant's use of applied N (Cassman et al., 1997). The other is to increase the ability of the rice system to fix its own N (Bennett and Ladha, 1992; Ladha and Reddy, 1995; Reddy and Ladha, 1995; de Bruijn et al., 1995). The latter approach is a long-term strategy, but it has large environmental benefits while helping specifically resource poor farmers. Furthermore, farmers more easily adopt a genotype or variety with useful traits than they do crop and soil management practices that are associated with additional costs. If a BNF system could be assembled in the rice plant itself, it would enhance the N supply potential with little or no loss, besides ensuring no additional economic burden on farmers.

As mentioned above, in 1992, IRRI organized a Think-Tank Workshop to assess the feasibility of (symbiotic) nitrogen fixation in rice (Khush and Bennett, 1992). The experts attending the Think-Tank Meeting agreed to work toward the goal of achieving (symbiotic) associations/nodulation and nitrogen fixation in

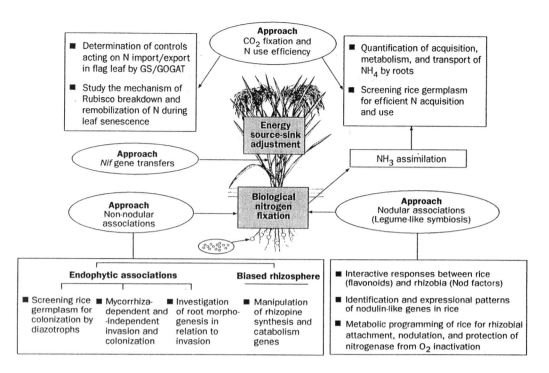

Figure 3. Potential research approaches for achieving improved nitrogen fixation and N use efficiency in rice.

rice. However, they also concluded that exploratory research would clearly be needed to assess the feasibility of novel approaches, that rice was an excellent plant model system (see Table 2), and identified four major short- and long-term approaches (see Figure 3):

- Non-nodular associations - Improve the associations between rice and nitrogen-fixing soil bacteria. This includes achieving colonization and invasion of rice roots by suitable diazotrophs.

- Nodular associations (legume-like symbiosis) - Lay the foundation for the engineering of rice plants capable of "nodulation". This approach includes identifying compatible rhizobia and varieties of rice, examining the defense response of rice to find ways to avoid responses that would inhibit symbiosis or the nitrogen fixation process.

- Transferring N_2 fixation (*nif*) genes - Transform rice with *nif* genes to ensure expression of nitrogenase, protection of nitrogenase from inactivation by oxygen, and an energy supply for N_2 fixation without compromising yield.

- CO_2 fixation and N use efficiency - Increase the understanding of nitrogen metabolism in rice and impact of N_2 fixation on carbon and energy budgets.

Considerable interest and support for this project was generated since the Think-Tank Workshop. A New Frontier Project, Assessing Opportunities for Nitrogen Fixation in Rice, was developed in 1994 and included in IRRI's 1994-1997 and 1998-2006 Medium-Term Plans. The long term objective of this project is to enable rice plants to fix their own nitrogen. This research project involves a committed group of scientists from research disciplines and several institutes around the world. The project has a Working Group (WG), through which IRRI facilitates communication among scientists all over the world with active research interests in nitrogen fixation in rice and other cereals (see appendix for a recent WG meeting report on current status and recommendations for future research). The second BNF working group meeting was organized. As a satellite to the meeting on BNF in non-legumes, held in Faisalabad in November of 1996 (see also accompanying volume by Malik et al., 1997). The exciting papers presented at this satellite meeting are included in this volume, and represent a comprehensive and current picture of our knowledge in the area of assessing opportunities for N_2 fixation in rice and other non-legumes.

References

App A A, Watanabe I, Ventura T S, Bravo M and Jurey C D 1986 The effect of cultivated and wild rice varieties on the nitrogen balance of flooded soil. Soil Sci. 141, 448–452.

Bennett J, Ladha J K 1992 Introduction: feasibility of nodulation and nitrogen fixation in rice. *In* Nodulation and nitrogen fixation in rice. Eds. G S Khush and J Bennett. pp 1–14. International Rice Research Institute, Philippines.

Bøckman O C, Kaarstad O, Lie O H and Richards I 1990 Agriculture and Fertilizers. Norsk Hydro, Oslo.

Bouldin D R 1986 The chemistry and biology of flooded soils in relation to the nitrogen economy in rice fields. *In* Nitrogen economy of flooded rice soils. Eds. S K De Datta and W H Patrick Jr. pp 1–14. Martinus Nijhoff Publishers, The Netherlands.

Cassman K G, Peng S, Olk D C, Ladha J K, Reichardt W, Dobermann A and Singh U 1997 Opportunities for increased nitrogen use efficiency from improved resource management in irrigated rice systems. Field Crops Res. *(In press)*.

Da Silva J G, Serra G E, Moreira J R, Goncalves J C, Goldenberg J 1978 Energy balance for ethyl alcohol production from crops. Science 210, 903–906.

de Bruijn F J, Jing Y and Dazzo F B 1995 Potential and pitfalls of trying to extend symbiotic interactions of nitrogen-fixing organisms to presently non-nodulated plants, such as rice. Plant Soil 172, 207–219.

IFA-IFDC-FAO 1992 Fertilizer use by crop. FAO, Rome.

IRRI 1993 Rice Research in a time of change. International Rice Research Institute's medium-term plan for 1994–1998.

Khush G S and Bennett J (eds) 1992 Nodulation and nitrogen fixation in rice: Potential and prospects. International Rice Research Institute, Philippines.

Kundu D K and Ladha J K 1994 Efficient management of soil and biologically fixed nitrogen in intensively cultivated rice fields. Soil Biol. Biochem. *(In press)*.

Ladha J K, Kirk G J D, Bennett J, Peng S, Reddy C K, Reddy P M and Singh U 1997 Opportunities for increased nitrogen use efficiency from improved lowland rice germplasm. Field Crops Res. *(In press)*.

Ladha J K and Reddy P M 1995 Extension of nitrogen fixation to rice - Necessity and possibilities. Geojournal 35, 363–372.

Ladha J K, Tirol-Padre A, Punzalan G C, Watanabe I and De Datta S K 1988 Ability of wetland rice to stimulate biological nitrogen fixation and utilize soil nitrogen. *In* Nitrogen fixation: Hundred years after. Eds. H Bothe, F J de Bruijn and W E Newton. pp 747–752. Gustav Fischer, Stuttgart, New York.

Ladha J K, Tirol-Padre A, Reddy C K and Ventura W 1993 Prospects and problems of biological nitrogen fixation in rice production: A critical assessment. *In* New horizons in nitrogen fixation. Eds. R Palacios, J Mora and W E Newton. pp 677–682. Kluwer Academic Publishers, The Netherlands.

Malik K A, Sajjad Mirza M and Ladha J K 1997 Nitrogen fixation with non-legumes. Proceedings of the 7th International Symposium on nitrogen fixation with non-legumes. Kluwer Academic Publishers, Dordrecht. *(In press)*.

Mudahar M S 1987a Energy requirements, technology and resources in fertilizer sector. *In* Energy in plant nutrition and pest control. Ed. Z R Helsel. pp 25–62. Elsevier Publications, New York.

Mudahar M S 1987b Energy, efficiency, economics and policy in the fertilizer sector. *In* Energy in plant nutrition and pest control. Ed. Z R Helsel. pp 133–163. Elsevier Publications, New York.

Reddy P M and Ladha J K 1995 Can symbiotic nitrogen fixation be extended to rice? *In* Nitrogen fixation: Fundamentals and applications. Eds. I A Tikhonovich, N A Provorov, V I Romanov and W E Newton. pp 629–633. Kluwer Academic Publishers, Dordrecht, The Netherlands.

Roger P A and Watanabe I 1986 Technologies for utilizing biological nitrogen fixation in wetland rice: Potentialities current usage and limiting factors. Fert. Res. 9, 39–77.

Shrestha R K and Ladha J K 1996 Genotypic variation in promotion of rice nitrogen fixation as determined by nitrogen-^{15}N dilution. Soil Sci. Soc. Am. J. 60, 1815–1821.

Watanabe I and Roger P A 1984 Nitrogen fixation in wetland rice fields. *In* Current developments in biological nitrogen fixation. Ed. N S Subba Rao. pp 237–276. Oxford-IBM Publications, New Delhi.

Wu P, Zhang G, Huang N and Ladha J K 1995 Non-allelic interaction conditioning spikelet sterility in an F_2 population of indica/japonica cross in rice. Theor. Appl. Genet. 91, 825–829.

Zhu Zhao-liang, Liu Chong-qun and Jiang Bai-fan 1984 Mineralization of organic nitrogen, phosphorus and sulfur in some paddy soils of China. *In* Organic matter and rice. pp 259–272. International Rice Research Institute, Philippines.

Appendix: Report of recent Working Group meeting on current status and recommendations for future research.

The Second BNF Working Group meeting was held 13–15 October 1996 at the Institute of Biotechnology and Genetic Engineering (NIBGE) in Faisalabad, Pakistan. It took place in conjunction with the 7th International Symposium on Nitrogen Fixation with Non-legumes on 16–21 October at Faisalabad. Participants shared their results and prioritised the areas of research in rice endophytic/rhizobial interactions and nitrogen metabolism in rice. Following is the summary of the group discussions.

Rice-endophyte interactions

Three approaches to obtain additional N inputs from biological nitrogen fixation in rice were adopted in the 1992 think-tank meeting at IRRI. One of these approaches was to force a tight association between rice and diazotroph to which the sugarcane-*Acetobacter diazotrophicus* interaction could serve as an example. Due to their strategic and advantageous location and the possibility of rapid exchange of nutrients including fixed N, endophytic diazotrophs such as *A. diazotrophicus* are held responsible for the substantial supply of biologically fixed N to the plant.

In the First Working Group meeting held at IRRI in 1995, participating scientists made recommendations for future research in rice-endophyte interaction. Although more than half of these recommendations have been satisfied as shown by the data presented at the Second Working Group meeting held at NIBGE,

Faisalabad, much work remains to be done. While some of the studies presently underway may need to be continued, results currently available will allow us to move forward to other important aspects of the project. The following were reported at the meeting:

- Method for surface sterilization of rice tissues to enable proper enumeration and isolation of putative endophytes had been standardized.
- Different rice genotypes had been screened to find a potential diazotrophic endophyte-genotype couple and a predominant nitrogen-fixing endophyte.
- Genetic diversity of putative endophytic population had been studied using polymerase chain reaction-based techniques.
- Colonization studies using mostly non-indigenous endophytes marked with *gusA* had been performed.

The discussion at the meeting revolved around many different aspects related to the rice-endophyte interactions although more discussion was focused on the endophyte side than on the plant side. In summary, following are the important suggestions for future research:

- A need to extend molecular tools in comparison to microbiological methods in trying to find the "right diazotrophic endophyte" and examine the genetic diversity of the diazotrophic endophyte population.
- The *Azoarcus*-Kallar grass system is presently the most studied diazotrophic endophyte-plant system. With this in mind, there is a need to continue finding an endophyte-rice genotype couple, and examine the host genes that are specifically expressed during the development of association with endophytes. Suggestion was made to examine perennial rices to find a "right diazotrophic endophyte" because these rices would provide a greater chance for selecting a more beneficial association than annual rices. Studies should also concentrate on a few chosen endophytes and rice genotypes to establish facts regarding infection, distribution, location, growth stimulation, and nitrogen fixation within the plant. About 200 diazotrophic putative endophytes that have been isolated from rice so far could be used for this purpose to screen for the most potential, stable, and aggressive colonizer.
- There is a need to examine and correlate in planta numbers of endophytes and the level of fixed nitrogen required (aimed at about 50% of the N required by the plant) to produce a real impact on plant growth and yield. It has been calculated that the maximum demand for N is 150 kg ha^{-1} for 7–8 t rice yield levels. If the endophyte system is not able to provide this amount of N, then an integrated approach involving a combination of soil N, added fertilizer N, and N from endophyte was suggested. A three-point test for assessing the benefits an endophyte could give to the plant was suggested. The test includes ^{15}N feeding experiments, growth stimulation to the plant under N-limiting conditions (no potential nitrogen-fixing endophyte has been conclusively shown to benefit plant via fixed nitrogen alone), and use of *nif* negative mutant strains to check the growth responses of the system.

- On the plant side, a need for addressing the question of which rice cultivar to be used in the test experiments was raised. Proposed varieties are IR42 (not tested yet for transformation), Lamont (a hybrid japonica variety), IR72 and Taipei 309 (responsive to transformation), and Nipponbare (small seed, small biomass, photoperiod sensitive, transformable). It was stressed that a rice cultivar which is to be used in the studies should be easily transformable as it is helpful for future molecular and genetic studies. It was, however, pointed out that, at present, the use of cultivar is more important regardless of whether the variety is transformable.

Rice-rhizobial interactions and genetic manipulation of rice for nitrogen fixation/symbiosis

Of the various naturally occurring nitrogen-fixing plant-microbe associations, the symbiosis between rhizobia and legume plants is the best understood scientifically. Moreover, this association is also the most successful agriculturally, providing fixed nitrogen for many of the world's major crop species (e.g., soybean). Therefore, rhizobia-legume symbioses form natural models for efforts to develop a sustainable nitrogen-fixing system in rice. However, it is recognized that this is likely a long-term approach.

The first working group meeting in 1995 itemized in its report a number of research areas likely to yield important information. Most of these research areas have been addressed since 1995 with some notable success. Following are some of the highlights of the research results reported:

- As outlined in 1995, efforts have been made to understand how rhizobia colonize rice, wheat, and *Arabidopsis* roots using *Azorhizobium caulinodans*. These experiments have revealed a consistent picture that rhizobia colonize non-legume roots at sites of lateral root emergence.

■ New technology has been developed [e.g., use of GUS gene (β-glucuronidase) fusions] to study bacterial root colonization.

■ The first bonafide rhizobial endophytes of rice have been identified. These strains may be well adapted for rice colonization and should provide better research material for future experiments.

■ The most surprising finding is that rice possesses some of the same genes found in legume nodules, so-called nodulin genes. These results suggest that rice may indeed have some of the functions necessary for the formation of a symbiosis with rhizobia.

■ Studies of rice plants expressing a GUS fusion to the early nodulin gene *ENOD12* has, for the first time, shown that rice can respond to inoculation by rhizobia. In this case, the pattern of *ENOD12* expression within rice roots was noticeably changed upon the addition of rhizobial Nod factors. These exciting results suggest that at least a portion of the signal transduction machinery important for legume nodulation likely exists in rice.

New research information has also led to the abandonment of a few approaches that appeared promising in 1995 but now are obviously less so. For example, it now appears that efforts to induce preferred artificial sites of rhizobial colonization on rice roots are unnecessary as colonization naturally occurs at the sites of lateral root emergence, and rhizobial inoculation appears to increase the formation of lateral roots. The rapid progress using whole plants has also made unnecessary to pursue studies based on previous suggestions employing tissue culture to develop rhizobial-rice interactions.

Work in the two and a half years has provided valuable new information on how rhizobia and rice roots interact. This new information has helped to better focus current research efforts. Therefore, the suggestions of the WG for areas of future research can now be written in a more focused manner than those arising from the first meeting of the WG in 1995. These suggestions are as follows:

■ Efforts should be made to isolate rhizobial strains better adapted for rice colonization. Recent research has identified the existence of true rhizobial endophytes of rice plants. These endophytes can be best identified in areas where rice and legumes are grown in rotation. Efforts should be made to isolate additional strains that may be better adapted for rice colonization than strains isolated from legumes. Such strains that are currently available will be distributed to researchers interested in basic and applied studies of rhizobial colonization of rice. Another source for such strains may be perennial rice varieties. Sites should be sought where such varieties are grown in rotation with legume crops.

■ Efforts should be made to improve colonization of rice roots by rhizobia. Researchers should seek to improve the growth conditions of rice in their experiments since it is likely that this will improve colonization. Basic research is needed to identify the various environmental, physiological, and nutritional factors limiting colonization. Co-inoculation of rhizobia with other bacterial strains may enhance colonization. Efforts should be made to understand how rice plants respond to rhizobial colonization. Do rice plants elicit a defense response to such colonization? Do factors such as ethylene production limit colonization? These are questions that need to be addressed.

■ Aggressive efforts are needed to build upon past results suggesting that rice possesses the genetic predisposition to respond to rhizobial colonization. Current efforts using endogenous rice homologues of legume nodulin genes, as well as transgenic rice plants expressing legume nodulins, should be used to analyze the ability of rice to respond to rhizobial colonization and rhizobial factors essential for nodulation (e.g., the lipo-chitin nodulation signals).

■ Current information clearly shows that hormonal responses in legume roots are essential for nodule formation. Studies are needed that compare the hormonal responses of rice roots and legume roots, specifically targeting those events known to occur in legume roots upon rhizobial inoculation.

■ The WG strongly encourages long-term research approaches (e.g., engineering nitrogenase expression in plants) that seek to develop a sustainable nitrogen-fixing system in rice. Such efforts should involve increased support for basic research, such as studies of plastid gene expression, that will be essential for developing the tools needed for such endeavors.

■ Future research should not ignore the potential to improve rice production through rhizobial inoculation via mechanisms that do not involve biological nitrogen fixation. Endophytic and rhizosphere colonization of rice roots by rhizobia can be beneficial through biocontrol of pathogens and due to plant growth promotion effects. Additional research is needed to understand the mechanism(s) of such effects.

Past research has shown some notable successes but much work remains. As more information has become available, it has also become clear that an understanding of the ability of rhizobial colonization to improve rice production will likely involve other areas besides biological nitrogen fixation. Because of this, it is recommended that the composition of the WG be expanded to include members with expertise in areas such as biocontrol, plant pathogens, and plant growth promotion by microorganisms.

Nitrogen metabolism in rice

The WG on N metabolism, a component of Frontier Project on Assessing Opportunities of Nitrogen Fixation in Rice, met 9-10 Sep 1996 at IRRI. Dr J B Bennett presented the report of this meeting during the 13-15 Oct meeting in Faislabad. The work carried out since the first working group meeting in 1995 was reported. The highlights of research work presented are as follows:

- High capacity of rice roots for NH_4^+ uptake is associated with tolerance of high intracellular contents of NH_4^+. Ammonium uptake is stimulated by the anaerobic conditions of the rice field .
- Nitrogen transport to shoots is tightly coupled with N uptake.
- Rice roots contain two forms of glutamine synthetase 1 (GS1), one constitutive (20%) and another induced by NO_3^- or NH_4^+ (80%). Rice leaves concentrate GS1 in phloem companion cells to drive export of N from leaves.
- Rubisco is rate-limiting for photosynthesis under high light conditions. Rubisco breakdown in vivo may occur by a mechanism similar to that caused by Cu^{2+} and mediated by H_2O_2. Rice genes that code for the enzymes catalase and ascorbate peroxidase involved in H_2O_2 metabolism have been isolated.
- Reactivation of the *gus* gene using 5-azacytidine requires DNA replication. This phenomenon provides a convenient *gus* reporter gene system for studying the control of DNA replication in apical, cortical, and pericycle cells of rice roots.

Future research activities include

- *Uptake and transport.* Dissection of regulation of NH_4^+ uptake in roots; dissection of interaction between N uptake and N transport; studies on uptake of N compounds other than NH_4^+; role of glutamine synthetase in NH_4^+ uptake; and increasing C efficiency of N transport from root to shoot.
- *Rubisco and photosynthesis under high-light stress.* Mechanism of Rubisco breakdown in vivo and in vitro; oxidative stress resistance in rice leaves; rate-limiting factors in photosynthesis in wet and dry seasons; development of simple phenotypic screens for delayed senescence for use in breeding programs.
- *Source-sink relations.* Limitations by source strength, sink strength and source-sink transport rate during grain filling; enhancement of source strength and sink strength through genetic engineering of cytokinin levels using *ipt* gene.
- *Regulation of cell division in roots.* Activation of cell division in rice cortex by hormones and Nod factors.

The group identified the following aspects of N metabolism having linkages with BNF research.

- Over the last three years, research on the timing of N fertilizer supply has reduced the total N required per hectare per season to support high-yielding rice cultivars. This finding increases the probability that BNF will be able to supply a significant proportion of total N requirement and may be able to reverse the trend of increased reliance on chemical fertilizers.
- Research on ^{13}N-NH_4^+ uptake is providing new information on the rates and sites of N uptake by rice roots. This information will be valuable in relation to the sites of colonization of rice roots by diazotrophs and the efficiency of N transfer from bacteria to plant cells.
- Investigations on glutamine and asparagine metabolism in rice roots and their transport to the leaves and the grain are focusing on the regulation and the efficiency of the processes, including minimizing the C requirement for N transport. In this respect, the research is being guided by results on nodulated legumes where the C-efficiency of N transport is maximized through transport of N as asparagine or ureides.
- Research on cell cycle control of root development is providing tools for the study of the cell cycle in relation to nodulation in rice roots.

Several additional issues linking the two research areas emerged from the meeting. They include

- Identification of niches inside the rice root for the most effective and mutually beneficial exchange of C and N between diazotroph and plant.
- Broadening the scope of research on N uptake beyond ammonium and nitrate to include amino acids and other nitrogenous molecules secreted by diazotrophs.
- Inclusion of diazotrophs in system modeling of C/N interactions in plants.
- Focusing both N metabolism research and BNF research on the rice plant at flowering rather than at the seedling stage.

Guest editors: J K Ladha, F J de Bruijn and K A Malik

Plant and Soil **194**: 11–14, 1997.

Fertilizers and biological nitrogen fixation as sources of plant nutrients: Perspectives for future agriculture

O.C. Bøckman
Norsk Hydro Research Centre, P.O. Box 2560, N-3901 Porsgrunn, Norway

Key words: biological nitrogen fixation, fertilizer nitrogen, sustainable agriculture

Abstract

Biological nitrogen fixation (BNF) has an assured place in agriculture, mainly as a source of nitrogen for legumes. Legumes are currently grown mostly as a source of vegetable oil and as food for humans and animals, but not as nitrogen source.

Other crops with BNF capability may be eventually be developed eventually. Such crops will also need mineral fertilizers to maintain a good status of soil nutrients, but their possible effects to the environment is also a concern. Fertilizers, however, will remain a necessary and sustainable input to agriculture to feed the present and increasing human population. It is not a case of whether BNF is better or worse than mineral fertilizers because both plays an important role in agriculture.

Introduction

One of the triumphs of human endeavour is the increase in agricultural productivity that has taken place in the last 50 years. Global population has grown from 2.5 billion in 1950 to to 5.7 billion in 1995. During this time, yields of cereals have increased in harmony with the population boom, providing enough food for the majority of the people and which even exceeded the demand (Bøckman et al., 1990). While malnutrition and hunger are still widespread, they are due to political factors (e.g., wars, poverty), and not because agriculture has already reached an absolute biological or resource ceiling for food production.

World human population is still increasing. It is expected to reach about 8.3 billion by 2025 before attaining a stable level later in the next century. This spells out the need for increasing rice yields further and expanding the areas for intensive rice production.

Nitrogen from biological and industrial nitrogen fixation in agriculture

The increase in food production is due to the interaction of factors such as plant breeding, irrigation, use of multiple cropping, crop protection with agrochemicals, and increased availability of plant nutrients. The most common nutrient that affects the yield is N.

Biological nitrogen fixation (BNF) and mineral fertilizers provide the principal input to agricultural soils which is N. The atmospheric deposition of nitrogen oxides and ammonia also contributes to the supply of N. These depositions originate mainly from pollution.

On the global scale, BNF provides the largest input of N to soils. Various estimates of this contribution have been published, ranging from 44 to 200 Tg N/year (Søderlund and Rosswall, 1982). In general, about 140 Tg N/year is the estimate mostly used.

In 1991-92 about 11% of all arable land was used for producing legume oil seeds (soybeans and groundnuts) and pulses (Peoples et al., 1995). In addition grazed grass-clover swards are commonly rotated with cereals in regions where yields are uncertain and generally low due to water deficiency, such as in some parts of Australia. Other sources of BNF for agriculture are *Azollae* and *Sesbania* sp. which are used as green manure in rice production (Ladha and Garrity, 1994).

Past and present, this proves that BNF has an assured place in agriculture. However, the steady increase in agricultural productivity would not have been possible without mineral fertilizers. Fertilizer N use has increased from 9.6 Tg N in 1960 to 77 Tg N

Table 1. Fossile fuel consumption in 1995 (in 10^9 t oil equivalents)

Oil	3.2
Gas	1.9
Coal	2.3
Total consumption	7.3

Source: BP (1996)

Table 2. Phosphate reserves(in 10^{12} t phosphate rock)

Rock price: 40 USD/t	11
Rock price: 100 USD/t	37
Present annual consumption	0,13

Source: US Bureau of mines (1995).

in 1995. Fertilizer dependence has given rise to the following concerns:

- The production of fertilizers depends on the use of nonrenewable resources. Is this sustainable ? Will agriculture face a resource crisis in the future ?

- Intensive agriculture implies enhanced availability of N with increased risk of N losses that are of environmental concern, such as NO_3^- leaching and gaseous emissions of NH_3, N_2O and NO from the soil.

These concerns form part of the desire to complement and eventually substitute mineral fertilizers with BNF to increase the sustainability of future agriculture. It is the purpose of this paper to examine these concerns.

Sustainability of fertilizer production

Nitrogen

The preferred raw material for industrial nitrogen fixation is natural gas because it is readily available and cheaper than other raw materials. Modern ammonia plants using natural gas are very efficient with emissions and energy wastage at a level approaching the theoretical minimum.

All forms of energy can be used as basis for hydrogen production to feed Haber-Bosch reactors for ammonia production. Coal, oils, and electricity (for water electrolysis) are all being used, or have been used in the past.

Industrial process efficiency varies considerably, but it takes on average about 1.3 t oil (or equivalent amount of energy) to fix 1 t of NH_3 (Ladha and Reddy 1995). The annual production of about 77×10^6 t NH_3-N requires about 0.1×10^9 t oil equivalents. This is a large amount of fossil fuel, but is it is only 1,4% of the present total consumption (Table 1).

The vast use of fossil fuel is a major concern. Present known reserves of oil and natural gas will only last for about 50 years. In the last 50 years, discoveries of new resources have kept in pace with reserve depletions and this is expected to continue although not forever. Sometime in the future, concerns about reserve availability and sustainability will become more urgent, and prices will rise. This will result in increased use of coal and energy sources such as solar power. The pattern of energy use will shift. Since industrial nitrogen fixation can utilize all forms of energy, the process pattern within the industry will follow any stable changes in energy prices and availability.

Energy price increases will cause strains in society, as has been seen in short-time energy crises in the past. However, it seems unlikely that such strains will result in food shortages due to inability of the industry to produce fertilizers.

Phosphate

The other fertilizer nutrient that is dependent on apparently limited reserves is phosphorous. But what is a reserve depends on both geological and economic factors. At present only large rich deposits, located so that shipping costs are low, can be worked on. If prices increase, known but presently uneconomic resources are likely to be utilized (Table 2).

Known phosphate resources are so large that there is little incentive at present for prospecting. Also, phosphorus is not a rare element, i.e., it stands at no. 11 in the list of the most abundant elements in the earth's crust. A real scarcity of phosphate thus seems very remote. Minerals rich in other nutrients such as potassium, sulfur, magnesium, and calcium are also so abundantly available that a global resource crisis is impossible despite possible increases in production prices and regional needs for import.

It is thus unlikely that a future resource crisis will prevent the fertilizer industry from producing the needed fertilizers to supply a highly productive agriculture, although prices of agricultural inputs may increase.

Environmental issues: NO_3^- leaching and gaseous losses of NH_3, N_2O and NO

Leaching of NO_3^- to ground and surface waters from cultivated land is a serious issue. Restrictions are placed on land use over large areas to prevent NO_3^- concentration in ground waters rising above 50 mg $NO_3^- \ L^{-1}$.

This problem did not exist before WWII when yields were less than 50% of present (Bøckman et al., 1990) and N availability to crops were correspondingly less. High yields depend on high levels of N available to the plants. N residues in roots and in easily mineralized soil organic matter have increased with the intensification of agriculture. It is this left-over N that forms the principal source of leached N, while direct losses of fertilizer N by leaching mostly takes place on a modest scale (Addiscott et al., 1991).

Legumes leave N in root residues mainly as organic N. This can be of use for subsequent crops after mineralization, but some may also be lost by leaching. It is therefore unlikely that substitution of mineral fertilizer with BNF will substantially decrease the problem of with NO_3^- leaching in intensive agricultural production. A probable exception is where frequent heavy rains make precipitation larger than evapotranspiration. Such conditions favor loss of fertilizer N by leaching.

The same can said about gaseous emissions of N_2O and NO. They originate from with soil microbial processes. Limited evidence indicates that, in general, more N_2O is emitted from fields with legumes than from fertilized crops except where excessive amounts of N are applied, (Eichner, 1990; Granli and Bøckman, 1994).

However, losses of N with vaporization of NH_3 should be less from crops with BNF than from fields fertilized with ammonium compounds especially where urea or ammonium carbonate is applied to the surface.

Crops exchange NH_3 with the atmosphere, but the net emission is modest, usually only 1 to 2 kg N ha[1] per year. Inspection of a list of all published emission measurements up to 1994 gives no indication that legumes emit more NH_3 than other crops (Holtan-Hartvig and Bøckman, 1994), and a field study of actinorhizal

Table 3. Fertilizer recommendations for legumes and rice

	N	P	K
Field bean (*Vicia Faba*)	0–40	35–50	80–170
Pea (*Pisum sativum*)	0	35–50	80–170
Soybean (*Glycine max.*)	0–60	0–25	0–150
Groundnut (*Arachis hypogaea*)	0–30	5–35	0–65
Tropical pulses	10–30	15–25	25–35
Rice (*Oryza sativa*)	75–125	10–20	0-40

Source: IFA (1992)

Alnus glutinosa gave no evidence for NH_3 emissions (Bøckman, unpublished observation).

NH_3 losses through vaporization are usually only a few percent where ammonium nitrate or NP fertilizers are applied, but where urea is broadcast, 15 to 20% is usually lost (ECETOC, 1994).

Emissions from paddy rice can be even larger. De Datta et al. (1991) reported 46–54% losses through NH_3 volatilization during 8 days after urea application to irrigated lowland rice in the Philippines.

The emissions of NH_3 from fertilizer application can be greatly reduced by the use of appropriate technology, e.g., in rice cultivation, deep placements of urea granules will reduce vaporization (Savant and Stangel, 1990).

Emissions of NH_3 are a cause of concern in Northern Europe, because what is emitted is eventually deposited and the increased input to pristine ecosystems may cause undesirable changes, e.g., in the biodiversity (Hornung et al., 1995). It is not known if rice-growing areas face similar problems.

Maintenance of soil productivity

Crops also need nutrients other than N. When yields are increased by N application alone, other nutrients are slowly consumed resulting in declining soil productivity. Giving common crops the ability to fix nitrogen is equivalent to fertilization with N alone, which will not eliminate the need for fertilizer with other nutrients. Table 3 gives the ranges of fertilizer recommendations for various grain legumes compared with rice.

It would be pure speculation to predict what the fertilizer recommendation would be for a rice with BNF capability, but it will not be nil.

Economic considerations

Purchased inputs can put small farmers at economic risk, as crops may fail, or bumper crops may glut the market so the produce can not be sold with profit. Under such circumstances, subsistence farming may be the best policy for poor farmers. Developing crops with BNF capability may help such farmers to overcome N limitations and increase yield without the need for purchasing N fertilizers such as urea.

N fixation requires energy, so crops with BNF capability may yield less than crops given optimal fertilization, but this concern should not be important for farmers looking for the highest possible yields with few purchased inputs.

However, in commercial high-yielding agriculture, the merits of BNF over N fertilizers N depends on (1) relative yields, (2) relative demand for labor, and (3) relative prices of inputs and produce. These factors vary with time and "state of the art". They cannot be predicted. However, it seems likely that both sources of N will be utilized in the future as in the past, depending on local conditions.

References

Addiscott T M, Whitmore A P and Powlson D S 1991 Farming, fertilizers and the nitrate problem. CAB International, Wallingford.

[BP] British Petroleum 1996 Statistical review of world energy. British Petroleum, London.

Bøckman, O C, Kaarstad O, Lie O H and Richards I 1990 Agriculture and Fertilizers. Norsk Hydro, Oslo.

De Datta S K, Buresh R J, Samson M I, Obcemea W N and Real J G 1991 Direct measurement of ammonia and denitrification fluxes from urea applied to rice. Soil. Sci. Soc. Am. J. 55, 543–548.

ECETOC 1994 Ammonia emissions to air in Western Europe. Technical Report no. 62. European Centre for Ecotoxicology and Toxicology of Chemicals, Brussels.

Eichner M J 1990 Nitrous oxide emissions from fertilized soils: Summary of available data. J. Environ. Qual. 19, 272–280.

Granli T and Bøckman O C 1994 Nitrous oxide from agriculture. Norw. J. Agric. Sci. Suppl. no 12.

Holtan-Hartvig L and Bøckman O C 1994 Ammonia exchange between crops and air. Norw. J. Agric. Sci. Suppl. no 14.

Hornung M, Sutton M A and Wilson R B (Eds) 1995 Mapping and modelling of critical loads for nitrogen: a workshop report. Institute of Terrestrial Ecology, Edinburgh.

IFA 1992 World Fertilizer use manual. International Fertilizer Industry Association, Paris.

Ladha J K and Garrity (Eds) 1994 Green manure production systems for Asian ricelands. IRRI, Manila.

Ladha J K and Reddy P M 1995 Extension of nitrogen fixation to rice - necessity and possibilities. Geojournal 35, 363–372.

Peoples M B, Herridge D F and Ladha J K 1995 Biological nitrogen fixation: An efficient source of nitrogen for sustainable agricultural production. Plant Soil 174, 3–28.

Savant N K and Stangel P J 1990 Deep placement of urea supergranules in transplanted rice: principles and practices. Fert Res 25, 1–83.

Søderlund R and Rosswall T 1982 The nitrogen cycle. In Ed. O Hutzinger. The handbook of environmental chemistry. Vol 1B. The natural environment and the biogeochemical cycles pp 60-81. Springer Verlag, Berlin/Heidelberg.

US Bureau of Mines 1995 Mineral Commodity Summaries 1995. Phosphate Rock. Washington DC pp 124–125.

Guest editors: J K Ladha, F J de Bruijn and K A Malik

Plant and Soil **194**: 15–24, 1997.

Isolation of endophytic diazotrophic bacteria from wetland rice

W.L. Barraquio, L. Revilla and J.K. Ladha[1]
Soil Microbiology Laboratory, SWSD, International Rice Research Institute, Los Banos, Laguna, Philippines.
[1]*Corresponding author**

Key words: endophytic diazotrophs, rice endophytes, surface sterilization

Abstract

Endophytic nitrogen-fixing bacteria are believed to contribute substantial amounts of N to certain gramineous crops. We have been interested to find (a) a diazotroph(s) in rice which can aggressively and stably persist and fix nitrogen in interior tissues and (b) unique rice-diazotrophic endophyte combinations. To achieve these objectives, it has been essential to find an efficient method to surface sterilize rice tissues. The method described here consists of exposing tissues to 1% Chloramine T for 15 min followed by shaking with glass beads. It has proven very efficient since (a) surface bacterial populations on the root and culm were found to be reduced by more than 90%, (b) the number of the internal colonizers was found to be significantly higher than the number of surface bacteria, and (c) colonization of root but not subepidermal tissue by *gusA*-marked *Herbaspirillum seropedicae* Z67 bacteria was found to be virtually eliminated. Nitrogen-fixing putative endophytic populations (MPN g dry wt^{-1}) in the root (7.94×10^7) and culm (2.57×10^6) on field-grown IR72 plants grown in the absence of N fertilizer was found to be significantly higher near heading stage. The corresponding total putative endophyte populations in the tissues of 25 highly diverse genotypes of rice and their relatives was found to range from 10^5–10^8 and 10^4–10^9, in the roots and culms, respectively. Generally, the resident bacteria were found to be non-diazotrophic, although in isolated cases diazotrophs were found, for example in the roots and culm of IR72 rice plants, or the culm of *Zizaniopsis villanensis* plants. The size of populations of diazotrophic bacteria in different rice genotypes was found to be 10^3–10^7 for the roots and 10^4–10^6 for the culms, respectively. The rice genera-related plants *Potamophila pariffora* and *Rhynchoryza subulata* showed the highest levels.

Introduction

A variety of nitrogen-fixing bacteria have been found to colonize the root interior of axenically and field grown rice, maize and grass plants (Diem et al., 1978; Hurek et al., 1994; James et al., 1994; Magalhaes et al., 1979; Patriquin and Dobereiner, 1978; Umali et al., 1980; You and Zhou, 1989). Endophytic diazotrophs have been proposed to be responsible for the supply of biologically fixed N to their host plant (Boddey et al., 1995). It is not known presently whether a predominant endophytic diazotroph resides in rice which can aggressively and persistently establish itself and fix nitrogen in interior tissues and whether a particular rice genotype exists that is inhabited naturally by a specific diazotrophic endophyte. It does appear that endophytic diazotrophs, such as *Acetobacter diazotrophi-*

cus (Cavalcante and Dobereiner, 1988), *Azoarcus* spp. (Reinhold et al., 1986) and *Herbaspirillum seropedicae* (Baldani et al., 1986) are prevalent in the roots and stems of certain gramineous crops.

One of the central problems with the identification of endophytic bacteria is the elimination of surface bacteria. Sterilizing by chemical means has been the usual strategy used to recover endophytic bacteria that persist intercellularly or intracellularly in plant tissues. However, it is clear that techniques to surface sterilize root tissues are subject to wide variations due to different growing conditions, the age of the plants, and structures of the roots which demand experimentally determined surface sterilization methods. Thus, different kinds of surface sterilization methods have been employed using various types of plants (Bell et al., 1995; Fisher et al., 1992; Gagne et al., 1987; Gardner et al., 1982; Jacobs et al., 1985; McClung et al., 1983;

* FAX No: +6328911292. E-mail: J.K.LADHA@CGNET.COM

McInroy and Kloepper, 1995 a, b; Philipson and Blair, 1957; Reinhold et al., 1986; Shishido et al., 1995). We found that concentration of 1:1000 mercuric bichloride, which was used to isolate endophytes from clover roots (Philipson and Blair,1957) and 5-6% NaOCl, as used for kallar grass (Reinhold et al., 1986) and cherry tree (Cameron, 1970) endophytes was too harsh (very few or no survivors) for wetland rice roots. Surface sterilization of sugarcane roots with 1% Chloramine T has been used to isolate endophytic and diazotrophic *A. diazotrophicus* and *H. seropedicae* bacteria (Baldani et al., 1986; Cavalcante and Dobereiner, 1988): Chloramine T has also been used to identify microbes inside of rice root (Diem et al., 1978).

This study is a part of the Project on Assessing Opportunities of Nitrogen Fixation in Rice (Ladha and Reddy, 1995) and is designed to identify (a) a diazotroph(s) in rice tissues which can aggressively and stably persist and fix nitrogen in interior rice tissues and (b) a unique rice genotype-diazotrophic endophyte combination. To realize these objectives, we developed a surface sterilization protocol that is suitable for rice tissues. The efficiency of the method we developed was evaluated on naturally and artificially grown rice plants by MPN (most probable number) estimation of putative endophytes and by following the fate of *gusA*-marked *Herbaspirillum seropedicae* Z67 bacteria in surface sterilized and non-sterilized tissues. The method was also used to examine the population dynamics of diazotrophic putative endophytes at different growth stages of rice and to enumerate and isolate these organisms from highly diverse rice genotypes and their relatives.

Materials and methods

Bacteria

Herbaspirillum seropedicae Z67 was a gift from J Dobereiner. It was used as the recipient strain in mating experiments with *E. coli* strain S17-1 λpir (mTn5SS*gusA*21), carrying the constitutively expressed *gusA* gene in a transposon (Wilson et al., 1995). This strain was kindly provided to us by Dr K Wilson (formerly of CAMBIA, Australia). Conjugation experiments were carried out according to the method of Wilson (1996). A derivative carrying the transposon with GUS activity inserted into the genome was isolated. The growth behavior (specific growth rate and generation time) and nitrogen-fixing activity

(acetylene reduction) of the tagged strain were found to be identical to those of the wild type strain.

Culture media

Semisolid media used in MPN counting of bacteria based on their acetylene reduction activity (ARA) were N-deficient glucose with 50 mg L^{-1} yeast extract (Watanabe et al., 1979), JNfb (Olivares et al., 1996), Nfb (Dobereiner, 1980), LGIP (Reis et al., 1994), arabinose-N (basal medium of Nfb but with arabinose instead of malate), and culm extract + water agar. The culm extract was prepared by blending 100 g culm in 100 mL phosphate buffered saline (PBS) followed by mixing with agar (4 g L^{-1}) on a 1:1 basis. Tryptic soy broth (TSB, 0.1%) was used for MPN-based enumeration of total counts of aerobic heterotrophic bacteria.

Rice genotypes

Highly diverse genotypes were included in the study consisting of cultivated rice varieties (long duration, traditional and improved, wetland, upland, and deepwater genotypes), wild rice varieties, and genera related to *Oryza* (Table 1). Seeds of these varieties were obtained from the International Rice Germplasm Center at IRRI. The plants were grown in 15-L Wagner pots in the greenhouse and in the field, using unfertilized Maahas clay. When required plants were grown aseptically for about a month in test tubes (200 × 25 mm) with 30 mL semisolid, N-deficient, Fahreus medium (pH 6.5), kept in the growth room (14 h light/10 h dark). These gnotobiotic plants were derived from surface sterilized seeds using 70% ethanol for 4 min and then 0.1% $HgCl_2$ for 4 min. Most of the cultivated varieties in the field and pots were sampled at heading stage. The plants used to study the dynamics of endophyte population at different growth stages were raised in N-unfertilized field and sampled at 1–2 weeks interval starting from 14 days after transplanting of one-week old seedlings from the greenhouse to the field. The wild rice varieties and related plants were raised vegetatively from stem cuttings maintained in unfertilized pots in the greenhouse.

Surface sterilization and MPN counting

Soil was removed from the roots under running demineralized water until the washings were very clear. Roots and culms from 5 plants each were pooled and five-gram samples were excised and surface sterilized.

Table 1. Rice genotypes and related genera used in the experiment

Genotype	Cultural character[a]	Growth duration(days)/ Life cycle[b]
Cultivated rice (*Oryza sativa*)		
IR72	W, I	112
IR42	W, I	120-135
Azucena	W, T	100
Peta	W, T	120–140
IR47686-1-4-8	U, I	89
IR63372-02	U, I	118
Salumpikit	U, T	81
Vandana	U, T	91
IR42436-266-3-2-3	D, I	125–135
IR11141-6-1-4	D, I	130–140
Jalmagna	D, T	130–140
Madhukar	D, T	130
Wild rice		
O. alta	W	P
O. australiensis	W (r)	B-P
O. eichingeri	W	P
O. grandiglumis	W	P
O. latifolia	W	P
O. longiglumis	W	P
O. officinalis	W (r)	P
O. punctata	W	A-P
O. rhizomatis	W (r)	P
O. rufipogon	W	P
Genera related to *Oryza*		
Chikusichloa aquatica	W	P
Rhynchoryza subulata	W	P
Zizaniopsis villanensis	W	P
Potamophila pariffora	W	P

[a] W, I – wetland, improved; W, T – wetland, traditional; U, I – upland, improved; U, T – upland, traditional; D, I – deepwater, improved; D, T – deepwater, traditional; r – rhizomatous. Some information on wild rices and genera related to *Oryza* were obtained from Vaughan (1994).
[b] A – annual, B – biennial, P – perennial.

Surface sterilization was carried out using 1% Chloramine T for 15 min, followed by washes with sterile PBS. Subsequently, the samples were shaken with sterile glass beads (20 g) in PBS for 20 min, washed 4 times with PBS, and blended in PBS. Rhizoplane bacteria (RB) were obtained from the suspension after shaking the roots with glass beads in PBS. Serial 5-fold dilutions of the rhizoplane bacterial suspension and the root and culm homogenates were prepared and each of the six appropriate dilutions was introduced into four replicate tubes (Woomer, 1994), containing 3 mL semisolid N-deficient medium and 0.1% tryptic soy broth (TSB). Efficiency of surface sterilization was based on % of the rhizoplane bacterial population removed after exposure of roots to 1% Chloramine T and was calculated as follows:

% Reduction in RB population
after exposure to Chlor T

$$\frac{\textit{No. of RB from unexposed roots} - \textit{No of RB from exposed roots}}{\text{No. of RB from unexposed roots}} \times 100 =$$

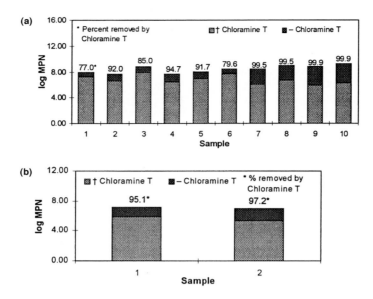

Figure 1. **(a)** Reduction in rhizoplane total bacterial population (log MPN per g dry tissue) after exposure of roots to 1% Chloramine T. Sample 1, PSBRC-18/field, fertilized, heading; 2, IR1552/field, fertilized, heading; 3, PSBRC-18/field, fertilized, 2 weeks before harvesting; 4, PSBRC-18/field, 3 weeks before harvesting; 5, IR72/field, past heading, unfertilized; 6, IR72/field, harvesting stage, unfertilized; 7, IR42/axenic, 20 days after inoculation with *Herbaspirillum seropedicae* Z67; 8, IR63372-02/axenic, 30 days after inoculation with *H. seropedicae* Z67; 9, Salumpikit/axenic 30 days after inoculation with *H. seropedicae* Z67; 10, Jalmagna/axenic 30 days after inoculation with *H. seropedicae* Z67. **(b)** Reduction in total surface bacterial population on stem (log MPN per gram dry tissue) after exposure to 1% Chloramine T. Sample 1, PSBRC-18/field,fertilized, 2 weeks before harvesting; 2, PSBRC-18/field, fertilized, heading stage.

Table 2. Reduction in counts of root surface bacteria after exposure to 1% Chloramine T

Sample	MPN ($\times 10^5$) g dry root^{-1}
Rhizoplane[a]	
(a) Washing after shaking Chloramine T-treated roots with glass beads	117*
(b) 1st washing of roots from sample (a)	19*
(c) 2nd washing of roots from sample (a)	4.8*
(d) 3rd washing of roots from sample (a)	1.6*
(e) 4th washing of roots from sample (a)	7.6*
Interior	
(f) Homogenate from sample (e)	5500

[a] Initial rhizoplane bacterial population from roots unexposed to Chloramine T was 10^8 MPN g dry root^{-1}.
* Significantly different from homogenate (sample f) at $p = 0.05$ based on confidence limits determined using confidence factor of 2.88 (Woomer, 1994).

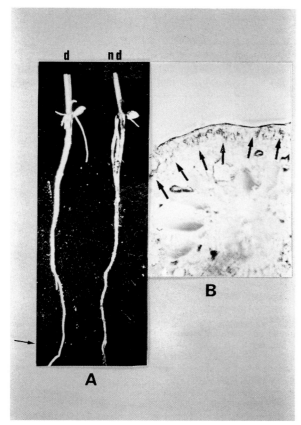

Figure 2. GUS staining of root of IR42 inoculated with *gusA*-labeled *Herbaspirillum seropedicae* Z67. Intense and very slight (arrow) GUS staining of non-disinfected (nd) and disinfected (d) roots, respectively, is shown (**A**). Transverse section of the stained part of the disinfected root showing subepidermal GUS activity (arrows) (**B**).

GUS assays and microscopic analyses

Two-day old germinated seeds in test tubes were inoculated with 1 mL of 10^9 stationary phase cells of a *gusA*-labeled *H. seropedicae* Z67 strain and incubated for a month in the growth room. Rice roots colonized by these bacteria were examined by histochemical staining, using X-glucA as the chromogenic substrate (Wilson, 1996). Staining was carried out by dipping the roots in 50 mM sodium phosphate buffer (pH 7.0) solution containing chloramphenicol (100 μg mL^{-1}), 0.5 M EDTA, and X-glucA (250 μg mL^{-1}) overnight at 30 °C. Transverse sections (20 μm thick; Microcut H1250; Energy Beam Sciences, Inc.) of stained unfixed roots were analyzed using an Olympus light microscope, under bright field optics.

Results and discussion

Efficiency of the surface sterilization method

In preliminary experiments, the use of 0.1% HgCl$_2$ and 5% NaOCl as surface sterilization agents was found to be too harsh (very few or no survivors) for rice roots. However, a modified surface sterilization method using 1% Chloramine T was found to remove more than 90% of the surface bacteria from the rice root and culm (Figures 1a, b). The recovered bacteria are treated here as putative endophytes until proven true endophytes by more rigid experimentation. The method enabled the recovery of a significantly larger number of putative endophytes (5.49×10^8 MPN g dry wt^{-1}) compared to that from the surface (7.6×10^5 MPN g dry wt^{-1}) (Table 2). Roots inoculated with the *gusA*-marked *H. seropedicae* strain, grown for a month and then surface sterilized are shown in Figure 2. The non-surface sterilized roots were intensely stained (blue color) due to external colonization by the bacteria. However, most of the surface colonizers was effectively removed by Chloramine T treatment, as shown by the pattern of GUS staining on the sterilized roots. Some portions of the sterilized root exhibited very slight random GUS staining. When these root segments were sectioned and examined under the microscope, GUS activity was observed in the subepidermal regions but not on the surface of the roots. To investigate the possibility that the subepidermal staining was an artifact due to diffusion of GUS stain in the freshly cut samples, thin sections(1 μm) were cut, fixed, stained with 1% toluidine blue, and examined under bright field optics. Bacteria were seen in the GUS-stained subepidermal regions (data not shown) indicating that the sterilization method resulted in eliminating the surface bacteria but not the putative endophytes. Earlier reports had also revealed that treatment of roots of rice and maize for 1–1.5 h and 6 h, respectively, with 1% Chloramine T did not affect the bacteria in the stele nor eliminate totally the bacteria in the cortex (Diem et al., 1978; Magalhaes et al., 1979; Patriquin and Dobereiner, 1978). It was surprising, however, that colonization was detected only in the outer, but not in the inner cortical region. Recently, we found similar extent of internal colonization of rice root by *Azorhizobium caulinodans* (Reddy et al., 1997). The primary mode of rhizobial invasion of rice root was found to be through cracks in the epidermis and fissures created during emergence of lateral roots.

Table 3. MPN ($\times 10^3$) g dry wt^{-1} of putative endophytes in the root and culm of IR72 grown unfertilized in the field

Tissue	Days after transplanting	Total endophytes (A)	Diazotrophic endophytes[a] (B)	% Incidence of diazotrophs (B/A × 100)
Root	14	$2.57 \times 10^3 *$	$9.12 \times 10^2 *$	35
	21	$1.91 \times 10^2 *$	$9.55 \times 10^1 *$	50
	28	$2.00 \times 10^3 *$	$8.32 \times 10^2 *$	42
	35	9.77×10^4	$2.82 \times 10^3 *$	3
	56	$3.39 \times 10^3 *$	$3.72 \times 10^3 *$	110
	71 (heading)	5.37×10^4 ns	7.94×10^4	148
	80	$2.82 \times 10^3 *$	$5.62 \times 10^2 *$	20
Culm	14	2.82×10^3 ns	$5.50 \times 10^1 *$	2
	21	1.70×10^4	6.76×10^2 ns	4
	28	$2.82 \times 10^2 *$	$1.58 \times 10^1 *$	6
	35	2.57×10^3 ns	1.91×10^3 ns	74
	56	2.29×10^3 ns	2.57×10^3	112
	71	$3.47 \times 10^2 *$	5.25×10^2 ns	151
	80	8.91×10^3 ns	$1.90 \times 10^2 *$	2

[a] The highest diazotrophic MPN estimate from the six semisolid media are presented.

* For the root, total putative endophyte number is significantly different from the highest number (9.77×10^7 g dry wt^{-1}) and diazotrophic putative endophyte number is significantly different from the highest number (7.94×10^7 g dry wt^{-1}). For the culm, total putative endophyte number is significantly different from the highest number (1.70×10^7 g dry wt^{-1}) and diazotrophic putative endophyte number is significantly different from the highest number (2.57×10^6 g dry wt^{-1}); ns = not significant. Significance ($p = 0.05$) of data was based on confidence limits established using confidence factor of 2.88 (Woomer, 1994).

Dynamics of and rice genotype differences in endophyte populations

The highest MPN estimates of diazotrophic putative endophytes obtained from the six different semisolid media used were compiled and are presented in Tables 3, 4a and b. Each of the six media interchangeably but not consistently gave the highest relative counts. This observation suggests the existence of different classes of diazotrophs that preferred different media. Different compounds produced by the genotypes may have enhanced enrichment of different bacteria. Therefore, the need to develop a general medium was obvious. The total putative endophyte number fluctuated between 10^5 and 10^8 MPN g dry wt^{-1} (Table 3). Nitrogen-fixing putative endophytic populations were detected in roots (7.94×10^7 MPN g dry wt^{-1}) or culms (2.57×10^6 MPN g dry wt^{-1}) early at 14 days after transplanting in field-grown IR72 plants that had not been fertilized and were significantly higher at or near heading. The number of diazotrophic putative endophytes was found to fluctuate between 10^5 and 10^6 before and after heading stage. In certain cases, a very high incidence (more than 100% of the total putative endophyte population) of N_2-fixing putative endophytes was observed, both in roots and culms, especially near or at the heading stage. The occurrence of high population of putative endophytes in rice root at heading stage may explain the earlier results of maximum acetylene-reducing activity in rice roots at this stage (Ladha et al., 1986, 1987). Perhaps, rice roots and culms in the field can more easily be internally colonized by bacteria through several natural entry points at specific stages of the growth cycle.

The roots and culms of 25 highly diverse rice genotypes and their wild relatives were found to have a putative endophyte population ranging from 10^5 to 10^8 and 10^4 to 10^9 MPN g dry wt^{-1}, respectively (Tables 4a and b). The fact that the plant can accommodate 10^9 putative endophytes suggests that total N_2 fixation may

Table 4a. MPN ($\times$ 10^3) g dry wt^{-1} of putative endophytes and percent incidence of nitrogen-fixers in roots of different rice genotypes and related genera

Variety[a]	Total putative endophytes (A)	Diazotrophic putative endophytes[b] (B)	% Incidence of diazotrophs (B/A $\times$ 100)
IR42	1.29×10^4*	1.41×10^3ns	11
Azucena	3.39×10^3*	1.29×10^3ns	38
Peta	3.47×10^4*	1.29×10^3ns	4
IR47686-1-4-8	1.51×10^3*	5.25×10^2*	35
IR63372-02	1.51×10^3*	8.32×10^1*	6
Salumpikit	1.51×10^3*	5.62×10^2*	37
Vandana	1.51×10^3*	4.36*	<1
IR42436-266-3-2-3	7.76×10^2*	1.58×10^2*	20
IR11141-6-1-4	2.29×10^3*	3.63×10^2*	16
Jalmagna	1.29×10^3*	6.76×10^2*	52
Madhukar	1.51×10^4*	1.20×10^3*	8
Oryza punctata	2.88×10^5ns	3.72×10^2*	<1
O. grandiglumis	3.80×10^4*	3.47×10^3ns	9
O. rhizomatis	1.41×10^5ns	8.71×10^1*	<1
O. eichingeri	7.24×10^4ns	2.09×10^2*	<1
O. longiglumis	7.76×10^2*	5.01×10^1*	6
O. rufipogon	5.25×10^4*	3.39×10^3ns	6
O. alta	1.29×10^3*	2.88×10^2*	22
O. latifolia	4.36×10^3*	5.25×10^1*	1
O. officinalis	2.19×10^3*	6.61×10^2*	30
O. australiensis	4.68×10^5	2.29×10^3ns	<1
Chikushichloa aquatica	7.41×10^3*	8.13×10^2*	11
Rhynchoryza subulata	4.36×10^2*	2.51×10^2*	58
Zizaniopsis villanensis	3.02×10^3*	1.33×10^3ns	44
Potamophila pariffora	1.58×10^4*	1.00×10^4	63

[a] Plants were grown in N-unfertilized Maahas clay pots and were near or at heading stage when used.

[b] The highest MPN count from the six semisolid media is presented.

* For total putative endophytes, significantly different from *O. australienses*; for diazotrophic putative endophytes, significantly different from *P. pariffora* at $p = 0.05$; ns = not significant. Significance of data was based on confidence limits established using 2.88 as confidence factor (Woomer, 1994).

be improved by ensuring all these endophytic bacteria are diazotrophs. In our studies, the putative endophyte populations were found to contain less than 10% diazotrophs. Only four out of total of 25 genotypes tested (Jalmagna, *Potamophila pariffora, Rhynchoryza subulata,* and *Zizaniopsis villanensis*) showed an incidence of diazotrophs of more than 50%. The range of diazotrophic putative endophyte populations in the roots was found to be between 10^3 and 10^7 MPN g dry wt^{-1} in *P. pariffora* (the highest) and in genotypes such as *Oryza australiensis, O. grandiglumis, O. rufipogon,* IR42, Azucena, and Peta similar levels were found. We noticed that not one of four upland genotypes yielded

significantly high numbers of diazotrophic bacteria. In the culms, populations ranged from 10^4-10^6 MPN g dry wt^{-1} in *R. subulata* (the highest) and in genotypes such as *Z. villanensis, O. rhizomatis, O. alta, O. latifolia, O. longiglumis, O. rufipogon,* and *O. australiensis* similar population numbers were found. The highest numbers of diazotrophic putative endophytes observed in this study are similar to those of *H. seropedicae* in sorghum, sugarcane, and forage grasses (about 10^5–10^7 g dry root^{-1}) (Baldani et al., 1992), *Azoarcus* in Kallar grass (7.3×10^7 g dry root^{-1}) (Reinhold et al., 1986), and *Acetobacter diazotrophicus* in sugarcane (10^6–10^7 g dry root^{-1}) (Li and MacRae, 1992). Plat-

Table 4b. MPN ($\times 10^3$) g dry wt^{-1} of putative endophytes and percent diazotrophs in the culms of different rice genotypes and related genera

Variety[a]	Total endophytes putative (A)	Diazotrophic putative endophytes[b] (B)	% Incidence of diazotrophs (B/A $\times$ 100)
PSBRC-18 (wetland, improved)[c]	1.51×10^3*	9.12×10^1*	6
Oryza alta	1.10×10^4*	8.13×10^2ns	7
O. latifolia	1.10×10^5*	6.45×10^2ns	<1
O. officinalis	2.19×10^6	1.10×10^2*	<1
O. eichingeri	5.13×10^3*	1.10×10^2*	2
O. longiglumis	2.52×10^3*	2.51×10^2ns	10
O. rufipogon	3.16×10^4*	3.09×10^2ns	1
O. australisensis	2.00×10^4*	3.31×10^2ns	2
O. grandiglumis	1.10×10^3*	6.45×10^1*	6
O. punctata	3.02×10^4*	2.19×10^2*	<1
O. rhizomatis	2.19×10^4*	1.41×10^3ns	6
Chikusichloa aquatica	7.08×10^3*	3.09×10^1*	<1
Potamophila pariffora	4.68×10^1*	3.47×10^1*	74
Rhynchoryza subulata	1.29×10^4*	2.00×10^3	16
Zizaniopsis villanensis	7.76×10^2*	1.20×10^3ns	155

[a] Plants were grown in N-unfertilized Maahas clay pots and were near or at heading stage when used.
[b] The highest MPN count from the six semisolid media is presented.
[c] Grown in the field, 3 weeks before harvesting.
* For total putative endophytes, significantly different from *O. officinalis*; for diazotrophic putative endophytes, significantly different from *R. subulata* at $p = 0.05$; ns = not significant. Significance of data was based on confidence limits determined using 2.88 as confidence factor (Woomer, 1994).

ing of highly diluted acetylene-reducing cultures from roots and culms yielded bacteria which were acid- and gas-producing, constituting perhaps members of Enterobacteriaceae, and *Herbaspirillum*-like organisms based on growth behavior and acetylene reduction activity in acidic JNfb medium. No *A. diazotrophicus*-like bacteria were isolated from ARA positive cultures from semisolid LGIP medium. The observation that each of six media interchangeably but not consistently yielded the highest number of nitrogen-fixing putative endophytes and that pure isolates from a particular medium seldom grew on any of the other media (data not shown) suggest that the diversity of diazotrophic putative endophytes in rice roots and culms is large, as has been reported by Ueda et al. (1995a, b) for diazotrophs in the rhizosphere of wetland rice. Genomic diversity of the putative endophytes isolated as part of this study has been confirmed by genomic fingerprinting methods (see Stoltzfus et al., 1997).

The percentage of nitrogen-fixing endophytes was sometimes found to be higher in the culms than in roots (Tables 3, 4b). Surface disinfected and non-disinfected culms have previously been shown to exhibit nitrogen-fixation activity and to harbor nitrogen-fixing bacteria

(Ito et al., 1979; Watanabe et al., 1981, unpublished). It is possible that the culm is a more suitable niche for nitrogen-fixing endophytes than the root, because it has a seemingly less crowded microbial environment and photosynthates being transported downward through the phloem may reach the culm before the root. Stems have previously been shown to be good sources of inocula for emerging roots (Hurek et al., 1991). Moreover, *A. diazotrophicus* was shown to be localized in sugarcane stem nodes, where sucrose is stored (James et al., 1994). *H. seropedicae* and *Azoarcus* spp. have also been found in the stems of sugarcane and Kallar grass (Hurek et al., 1991; James et al., 1994).

The results of this study show that (a) efficient removal of surface bacteria from root and culm of rice can be achieved by treatment with 1% Chloramine T and shaking with glass beads, (b) the number of diazotrophic putative endophytes in roots and culms fluctuates at different growth stages of rice but seems to peak near or at heading, and (c) certain rice genotypes, in particular wetland cultivated, wild rice varieties and related genera, harbor diazotrophic putative endophytic populations. No evidence for the existence of true endophytes that actually fix nitrogen within the rice tissues has been obtained yet, but we are currently examining pure cultures of diazotrophic and non-diazotrophic strains obtained from the different rice genotypes for their colonizing and growth-promoting activities in the laboratory or greenhouse and are examining their genetic diversity (see Stoltzfus et al., 1997).

Acknowledgement

This work was supported by DANIDA. We thank F J de Bruijn and B Rolfe for comments on the manuscript.

References

Bell C R, Dickie G A, Harvey W L G and Chan J W Y F 1995 Endophytic bacteria in grapevine. Can. J. Microbiol. 41, 46–53.

Baldani V L D, Baldani J I, Olivares F and Dobereiner J 1992 Identification and ecology of *Herbaspirillum seropedicae* and the closely related *Pseudomonas rubrisubalbicans*. Symbiosis 13, 65–73.

Baldani J I, Baldani V L D, Seldin L and Dobereiner J 1986 Characterization of *Herbaspirillum seropedicae* gen. nov., sp. nov., a root-associated nitrogen-fixing bacterium. Int. J. Syst. Bacteriol. 36, 86–93.

Boddey R M, de Oliveira O C, Urquiga S, Reis V M, de Olivares F L, Baldani V L D and Dobereiner J 1995 Biological nitrogen fixation associated with sugar cane and rice: contributions and prospects for improvement. Plant Soil 174, 195–209.

Cameron, H. R. 1970. *Pseudomonas* content of cherry trees. Phytopathology 60, 1343–1346.

Cavalcante V A and Dobereiner J 1988 A new acid-tolerant nitrogen-fixing bacterium associated with sugarcane. Plant Soil 108, 23–31.

Diem G, Rougier M, Hamad-Fares I, Balandreau J P and Dommergues Y R. 1978 Colonization of rice roots by diazotroph bacteria. *In* Environmental Role of Nitrogen-fixing Blue-green Algae and Asymbiotic Bacteria. Ed. U Granhall. Ecol. Bull. (Stockholm) 26, 305–311.

Dobereiner J 1980 Forage grasses and grain crops. *In* Methods for Evaluating Biological Nitrogen Fixation. Ed. F J Bergerson. pp 535–555. John Wiley and Sons, New York.

Fisher P J, Petrini O and Lappinscott H M 1992 The distribution of some fungal and bacterial endophytes in maize (*Zea mays* L.). New Phytol. 122, 299–305.

Gagne S, Richard C, Rousseau H and Antoun H 1987 Xylem-residing bacteria in alfalfa roots. Can. J. Microbiol. 33, 996–1000.

Gardner J M, Feldman A W and Zablotowicz R M 1982 Identity and behavior of xylem-residing bacteria in rough lemon roots of Florida citrus trees. Appl. Environ. Microbiol. 43, 1335–1342.

Hurek T, Reinhold-Hurek B, van Montagu M and Kellenberger E 1991 Infection of intact roots of Kallar grass and rice seedlings by *Azoarcus*. *In* Developments in Plant and Soil Sciences, Vol. 48. Nitrogen fixation. Eds. M Posinelli, R Materassi and M Vincenzini. pp 235–242. Kluwer Acad. Publ., Dordrecht.

Hurek T, Reinhold-Hurek B, van Montagu M and Kellenberger E 1994 Root colonization and systemic spreading of *Azoarcus* sp. strain BH72 in grasses. J. Bacteriol. 176, 1913–1923.

Ito O, Cabrera D A and Watanabe I 1980 Fixation of dinitrogen-15 associated with rice plants. Appl. Environ. Microbiol. 39, 554–558.

Jacobs M J, Bugbee W M and Gabrielson D A 1985 Enumeration, location, and characterization of endophytic bacteria within sugar beet roots. Can. J. Bot. 63, 1262–1265.

James E K, Reis V M, Olivares F L, Baldani J I and Dobereiner J 1994 Infection of sugarcane by the nitrogen-fixing bacterium *Acetobacter diazotrophicus*. J. Exp. Bot. 45, 757–766.

Ladha J K, Tirol-Padre A, Daroy M G, Punzalan G C, Ventura W and Watanabe I 1986 Plant-associative N_2 fixation (C_2H_2 reduction) by five rice varieties, and relationship with plant growth characters as affected by straw incorporation. Soil Sci. Plant Nutr. 32, 91–106.

Ladha J K, Tirol-Padre A, Punzalan and Watanabe I 1987 Nitrogen-fixing (C_2H_2-reducing) activity and plant growth characters of 16 wetland rice varieties. Soil Sci. Plant Nutr. 33, 187–200.

Ladha J K and Reddy P M 1995 Extension of nitrogen fixation to rice - necessity and possibilities. Geojournal 35, 363–372.

Li R P and MacRae I C 1992 Specific identification and enumeration of *Acetobacter diazotrophicus* in sugarcane. Soil Biol. Biochem. 24, 413–419.

Magalhaes F M M, Patriquin D and Dobereiner J 1979 Infection of field grown maize with *Azosprillum* spp. Rev. Brasil. Biol. 39, 587–596.

McClung C R, van Berkum P, Davis R E and Sloger C 1983 Enumeration and localization of N_2-fixing bacteria associated with roots of *Spartina alterniflora* Loisel. Appl. Environ. Microbiol. 45, 1914–1920.

McInroy J A and Kloepper J W 1995a Population dynamics of endophytic bacteria in field-grown sweet corn and cotton. Can. J. Microbiol. 41, 895–901.

McInroy J A and Kloepper J W 1995b Survey of indigenous bacterial endophytes from cotton and sweet corn. Plant Soil 173, 337–342.

Olivares F L, Baldani V L D, Reis V M, Baldani J I and Dobereiner J 1996 Occurrence of the endophytic diazotrophs *Herbaspirillum* spp. in roots, stems, and leaves, predominantly of Gramineae. Biol. Fert. Soils 21, 197–200.

Patriquin D G and Dobereiner J 1978 Light microscopy observation of tetrazolium reducing bacteria in endorhizosphere of maize and other grasses in Brazil. Can. J. Microbiol. 24, 734–742.

Philipson M N and Blair I D 1957 Bacteria in clover root tissue. Can. J. Microbiol. 3, 125–129.

Reddy P M, Ladha J K, So R, Hernandez R, Dazzo F B, Angeles O R, Ramos M C and de Bruijn F J 1997 Rhizobial communication with rice roots: induction of phenotypic changes, mode of invasion and extent of colonization. Plant Soil 194, 81–98.

Reinhold H B, Hurek T, Niemann E G and Fendrik I 1986 Close association of *Azospirillum* and diazotrophic rods with different root zones of Kallar grass. Appl. Environ. Microbiol. 52, 520–526.

Reis V M, Olivares F L and Dobereiner J 1994 Improved methodology for isolation of *Acetobacter diazotrophicus* and confirmation of its endophytic habitat. World J. Microbiol. Biotechnol. 10, 401–405.

Shishido M, Loeb B M and Chanway C P 1995 External and internal root colonization of lodgepole pine seedlings by two growth-promoting *Bacillus* strains originated from different root microsites. Can. J. Microbiol. 41, 707–713.

Stoltzfus J R, So R, Malarvizhi P P, Ladha J K and de Bruijn F J 1997 Isolation of endophytic bacteria from rice and assessment of their potential for supplying rice with biologically fixed nitrogen. Plant Soil 194, 25–36.

24

Ueda T, Suga Y, Yahiro N and Matsuguchi T 1995a Genetic diversity of N_2-fixing bacteria associated with rice roots by molecular evolutionary analysis of a *nifD* library. Can. J. Microbiol. 41, 235–240.

Ueda T, Suga Y, Yahiro N and Matsuguchi T 1995b Remarkable N_2-fixing bacterial diversity detected in rice roots by molecular evolutionary analysis of *nifH* gene sequences. J. Bacteriol. 177, 1414–1417.

Umali-Garcia M, Hubell D H, Gaskins M H and Dazzo F B 1980 Association of *Azospirillum* with grass roots. Appl. Environ. Microbiol. 39, 219–226.

Vaughan D A 1994 The Wild Relatives of Rice. IRRI, Manila.

Watanabe I, Barraquio W L, de Guzman M R and Cabrera D A 1979 Nitrogen-fixing (acetylene reduction activity) and population of aerobic heterotrophic nitrogen-fixing bacteria associated with wetland rice. Appl. Environ. Microbiol. 39, 813–819.

Watanabe I, Cabrera D A and Barraquio W L 1981 Contribution of basal portion of shoot to N_2 fixation associated with wetland rice. Plant Soil 59, 391–398.

Wilson K J, Sessitsch A, Corbo J C, Giller K E, Akkermans A D L and Jefferson R A 1995 β-glucuronidase activity in studies of rhizobia and other Gram-negative bacteria. Microbiology 141, 1691–1705.

Wilson K J 1996 *GusA* as a reporter gene to track microbes. *In* Molecular Microbial Ecology Manual. Eds. A D L Akkermans, F J de Bruijn and J D van Elsas. Kluwer Acad. Publishers, Dordrecht.

Woomer P L 1994 Most probable number counts. *In* Methods of Soil Analysis, part 2. Eds. R W Weaver, S Angle, P Bottomley, D Bezdicek, S Smith, A Tabatabai and A Wollum. pp 59–79. Soil Sci. Soc. Am., Inc., Wisconsin.

You C and Zhou F 1989 Non-nodular endorhizosphere nitrogen fixation in wetland rice. Can. J. Microbiol. 35, 403–408.

Guest editors: J K Ladha, F J de Bruijn and K A Malik

Plant and Soil **194**: 25–36, 1997.
© 1997 *Kluwer Academic Publishers. Printed in the Netherlands.*

Isolation of endophytic bacteria from rice and assessment of their potential for supplying rice with biologically fixed nitrogen

J.R. Stoltzfus[1], R. So[4], P.P. Malarvithi[4], J.K. Ladha[4] and F.J. de Bruijn[1,2,3,5]
[1] *MSU-DOE Plant Research Laboratory,* [2] *Department of Microbiology,* [3] *NSF Center for Microbial Ecology, Michigan State University, E. Lansing, MI 48824, U.S.A and* [4] *International Rice Research Institute, Los Banos, The Philippines.* [5] *Corresponding author**

Key words: endophytic microbes, endo-symbiosis, genomic fingerprinting, infection, marker genes, nitrogen fixation, PCR, rep-PCR, rice

Abstract

The extension of nitrogen-fixing symbioses to important crop plants such as the cereals has been a long-standing goal in the field of biological nitrogen fixation. One of the approaches that has been used to try to achieve this goal involves the isolation and characterization of stable endophytic bacteria from a variety of wild and cultivated rice species that either have a natural ability to fix nitrogen or can be engineered to do so. Here we present the results of our first screening effort for rice endophytes and their characterization using acetylene reduction assays (ARA), genomic fingerprinting with primers corresponding to naturally occurring repetitive DNA elements (rep-PCR), partial 16S rDNA sequence analysis and PCR mediated detection of nitrogen fixation (*nif*) genes with universal *nif* primers developed in our laboratory. We also describe our efforts to inoculate rice plants with the isolates obtained from the screening, in order to examine their invasiveness and persistence (stable endophytic maintenance). Lastly, we review our attempts to tag selected isolates with reporter genes/proteins, such as beta-glucuronidase (*gus*) or green fluorescent protein (*gfp*), in order to be able to track putative endophytes during colonization of rice tissues.

Introduction

Rice (*Oryza sativa* L.) is the staple in the diet of over 40% of the world's population making it the most important food crop currently produced (Hossain and Fischer, 1995). Much of this rice is grown in countries where rapidly growing populations, coupled to limited amounts of land and scarce resources, make high yields per hectare with reduced inputs essential to avoid food shortages. Crop productivity is based on numerous variables including weather, soil type, moisture, and nutrients. One of the most important factors in the generation of high yields from modern rice crops is nitrogen fertilizer. Without the addition of (fertilizer) nitrogen the yield of the present varieties is drastically limited. While biological nitrogen fixation in wetland rice fields contributes significantly to the long term fertility of these systems, it is not enough to produce maximum yields. Studies show that biological nitrogen fixation in flooded rice paddies can yield

up to 50 kg N ha^{-1} crop^{-1} (Roger and Ladha, 1992). This biologically fixed nitrogen has been sufficient to maintain traditional flooded rice systems for thousands of years. However, the productivity of these sustainable systems is low, with yields less than four tons per hectare. Modern agriculture has increased rice yields to five to eight tons per hectare, but requires the input of 60 to 100 kg ha^{-1} of fertilizer to supply additional nitrogen. Global population estimates predict the need for a 70% increase in rice production over the next thirty years. The use of current agronomic practices to generate this increase would require even larger inputs of fertilizer nitrogen.

The use of high levels of nitrogen fertilizers in crop production has several drawbacks. Most nitrogen fertilizer is produced via the Haber-Bosch process. This process requires large amounts of natural gas, coal, or petroleum, all non-renewable energy sources. In addition it produces CO_2, a gas implicated in the greenhouse effect. The chemical production of nitrogen fertilizer is also expensive, and in developing countries

---

* FAX No: + 15173539168. E-mail: debruijn@msu.edu

the additional costs often exceed the means of farmers, limiting the yield potential of their crops. Once chemical fertilizers are applied, additional problems can arise. Roughly one third of the nitrogen applied is used by the crop. The non-assimilated nitrogen from farming systems has been implicated in nitrate contamination of ground water supplies, a potential health hazard. In addition, excess nitrogen can also lead to production of nitrous oxide (N_2O), a potent "greenhouse" gas. Therefore, crop systems requiring large additions of fertilizer nitrogen are non-sustainable systems, since they require the use of non-renewable natural resources and can contribute to health hazards and environmental pollution.

Decreasing the amount of industrially produced fertilizer nitrogen needed in agricultural systems is an important goal of agricultural scientists in general (Bohlool et al., 1992). In the case of sustainable rice production, in particular, one important aim is to replace industrially fixed nitrogen with biologically fixed nitrogen. According to a National Research Council Report (1994), an estimated 100-175 million metric tons of biologically fixed nitrogen is produced annually. Most of this nitrogen is fixed by the legume/*Rhizobium* symbiosis. This is significantly more than the 10 million metric tons produced by lightning and 80 million tons produced in 1989 by industrial processes. Therefore, it is obvious that (symbiotic) biological nitrogen fixation has great potential for supplying nitrogen to crops.

However, the delivery of biologically fixed nitrogen to plants such as rice has been problematic. For example, supplying biologically fixed nitrogen through associative nitrogen fixing bacteria or green manuring has some of the same drawbacks that have been observed with industrially produced fertilizers. All three methods rely on plant uptake of nitrogen from the soil nitrogen pool. In order to meet the nitrogen requirements of the plant, a large excess nitrogen pool is needed. Green manure systems using *Sesbania rostrata, Aeschynomene afraspera,* or *Azolla* produce yields comparable to fertilized control plots (Diekmann et al., 1996; Kumarasinghe et al., 1986; Ladha et al., 1992). However, the percentage of the nitrogen from green manure incorporated by the rice plant is similar to that from fertilizer nitrogen (Clement et al., 1995; Kumarasinghe et al., 1986) and, therefore, nitrogen losses in green manure systems can be similar to those in fertilized systems (Harris et al., 1994). In legume crops this problem is avoided because the endo-symbiotic relationship with *Rhizobium* supplies

biologically fixed nitrogen directly to the plant, without requiring an excess pool of soil nitrogen, and the ammonia produced is rapidly and efficiently assimilated by the host plant.

Three approaches to achieve a more direct transfer of biologically fixed nitrogen to rice plants have been proposed: 1. Development of novel symbiotic interactions resulting in the formation of nitrogen-fixing nodules or nodule-like structures on rice roots; 2. Identification of stably maintained diazotrophic endophytic bacteria in rice tissues; and 3. Direct incorporation/expression of the required complement of nitrogen fixation genes in(to) rice (De Bruijn et al., 1995; Khush and Bennet, 1992; Ladha and Peoples, 1995; Ladha et al., 1997). While each of these methods has distinct merits, much of the basic knowledge needed to successfully develop any of these systems is still lacking. The development of nodules on legumes is a highly developed program involving specialized genes in both the plant and bacteria (Fisher and Long, 1992). While great progress has been made toward characterizing the genes involved in this process and the functions of the proteins they express, there are still large gaps in our knowledge, and extensive hurdles to be overcome (see De Bruijn et al., 1995). Exogenously applied hormones or cell wall degrading enzymes cause hypertrophies to form on the roots of several non-leguminous crops (Kennedy and Tchan, 1992). It has been reported that these structures can be invaded by diazotrophs, and that in some cases an increase in nitrogen uptake by the plant can be observed. However, these structures lack the complexity and specificity of nodules. More knowledge about the plant genes involved in nodulation and symbiotic nitrogen fixation and the function of their gene products is needed before the development of true nodulation of cereals can be seriously considered (see also De Bruijn et al., 1995; Reddy et al., 1997; Kennedy et al., 1997; Webster et al., 1997). The development of plants containing nitrogen fixing genes in their own genome may be even more complex (see De Bruijn et al., 1995; Dixon et al., 1997). While most of the genes necessary for nitrogen fixation in bacteria are well characterized, the transfer of the genes to the plant genome, along with the appropriate expression of all these genes is beyond our current technical ability. In addition, the creation of the proper environment, in terms of oxygen concentration/supply, energy provision to the bacteria, and efficient ammonia assimilation within plant cells may also present serious problems (see De Bruijn et al., 1995; Dixon et al., 1997).

Therefore, the employment of endophytic nitrogen fixing bacteria may involve the fewest technical challenges. Moreover, this approach has already been shown to be successful in the case of sugar cane (see Boddey et al., 1995; Kirchhof et al., 1997) and Kallar grass (see Reinhold-Hurek and Hurek, 1997). In addition, a beneficial natural endophytic association of rhizobia and rice grown in rotation with clover has recently been documented (Yanni et al., 1997). By using naturally occurring endophytes that colonize a niche in which conditions appropriate for nitrogen fixation exist, we would have 'allowed evolution to do some of the work for us'. Even if a stably endophytic microbe was to be identified that did not have the capacity to fix nitrogen, the process of introducing and expressing the *nif* gene complement in such an endophyte would be significantly easier than engineering the rice plant itself to fix nitrogen. However, it is clear that first some basic knowledge about the presence, predominance and stability of endophytic bacteria in different rice tissues must be obtained. Below we report some recent results of our ongoing study on rice-endophyte associations, that has the following objectives: a) isolating putative endophytic bacteria from diverse rice varieties grown in different soil types and assessing their diversity, b) developing molecular probes for the detection of putative N_2-fixing endophytes, and c) studying the internal colonization of rice tissue by putative endophytic bacteria.

Materials and methods

Isolation of endophytic bacteria

The isolation of bacteria from 8 varieties of rice grown in the greenhouse in pots containing one of five soil types was carried out at IRRI (Malarvithi, 1995; see Figure 1). The rice tissue was collected at the heading stage as follows. Plants were carefully removed from pots, washed to remove all soil and separated into stems and roots. The outer layer of the stems was removed, the stems were washed with tap water and deionized water, cut into sections 2–3 cm long, and dried on absorbent towels. Roots were cleaned thoroughly with tap water, rinsed with deionized water, and drained on absorbent towels. All tissues were surface sterilized in the following manner. Ten grams of tissue was shaken for thirty minutes in a 500 mL Erlenmeyer flask containing 250 mL sterile deionized water and 25 grams of glass beads. The tissue was transferred aseptically to a sterile beaker, washed two times with sterile distilled water, and sterilized using 0.2% $HgCl_2$ (30 seconds for roots, 60 seconds for stems). The tissue was washed 6 times with sterile distilled water, cut into small pieces and homogenized in a Warring blender containing 90 mL sterile distilled water. Serial dilutions were prepared and spread on plates containing bacterial growth media or tubes containing semisolid media described by Malarvithi (1995) and Barraquio et al. (1997).

The isolation of bacteria using an *in planta* selection step to reduce the number of non-endophytes (see Figure 2) was also carried out at IRRI. Initial isolation was carried out on field grown rice plants (variety IR72) as described above with the following modifications. The duration of surface sterilization of tissue was varied from 0 to 120 seconds. One hundred microliters of macerate from the 10^{-1} or 10^{-2} dilution were inoculated onto three day old aseptically grown rice seedlings generated as follows: Rice seeds (variety IR24, IR42, or Lemont) were gently dehulled, placed in 70% ethanol for five minutes, washed with sterile distilled water, sterilized by addition of 0.2% $HgCl_2$ for four minutes or a fresh solution of 30% Clorox (17.5 mL sterile distilled water, 7.5 mL Clorox Bleach, 30 mL tween 20) for 45 minutes, and washed six times with sterile distilled water. Seeds were placed on plates containing TY medium (Beringer, 1974) and incubated at 30 °C for two days to allow germination and check for contamination. Seeds showing no contamination were placed in 25 mm × 200 mm tubes containing either 25 mL of semisolid modified Fahraeus medium (Fahraeus, 1957) or sterile sand watered with Fahraeus medium. Seedlings grown in sand were watered with Fahraeus medium on a regular basis during the remainder of the growth period.

Bacteria were isolated from the inoculated seedlings as follows. Seedlings were grown from 23 to 25 days before re-isolation of the bacteria and carefully removed from the growth tubes to keep the roots intact. The roots and shoots were separated and the shoots were cut into small sections. The tissues were placed in 125 mL Erlenmeyer flasks containing 10 mL of sterile distilled water and 5 grams of glass beads, shaken for 30 minutes, and washed with 10 mL sterile distilled water. The tissue was sterilized by adding 10 mL of 0.2% $HgCl_2$ for 30 seconds or 30% Clorox solution for 15 minutes and washed six times with sterile distilled water. Tissues were macerated in one milliliter sterile distilled water and 20 μL of macerate was spread onto plates containing Tryptic Soy Agar (Difco), 0.1%

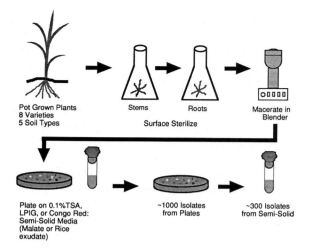

Figure 1. Isolation of putative rice endophytes. Scheme to isolate putative endophytic bacteria from rice tissues. Surface sterilization eliminates/reduces the presence of non-endophytic bacteria and maceration releases endophytic bacteria from the tissues. Over 1000 bacteria were isolated at IRRI using this procedure.

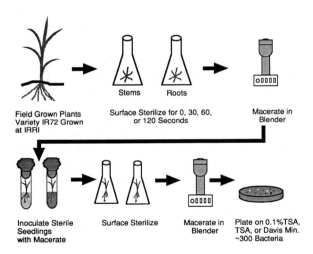

Figure 2. Isolation of endophytic bacteria using a re-infection step. Scheme adds an in planta selection step to eliminate non-endophytic bacteria. Approximately 300 bacteria were isolated at IRRI using this procedure.

Tryptic Soy Agar, TY medium (Beringer, 1974) or Davis and Mingolini minimal medium (Atlas, 1997).

Acetylene reduction assays

Acetylene reduction assays were performed at IRRI using malate semi-solid nitrogen-free media and standard protocols (see Barraquio et al., 1997).

PCR amplification of nifD-specific DNA fragments

Using sequence data from known diazotrophs (obtained from medline), the following universal *nifD* primers were derived: FdB261 (5'-TGGGGICCIRTIAARGAYATG-3') and FdB260 (5'-TCRTTIGCIATRTGRTGNCC-3'). These primers were synthesized at the MSU macromolecular facility and tested using a reference collection of known nitrogen-fixing and non-nitrogen fixing bacteria (see Results; J Stoltzfus and FJ De Bruijn, in preparation). A DNA fragment 390 bp in length was amplified from all diazotrophic bacteria tested, using the whole cell ERIC-PCR conditions previously described (Louws et al., 1996; Rademaker and de Bruijn, 1997). The DNA sequence of the amplified DNA band from *Rhizobium loti* NZP2235 was determined by automated fluorescent sequencing of the purified DNA at the MSU-DOE-PRL Plant Biochemistry Facility using the ABI Catalyst 800 for Taq cycle sequencing and the ABI 373A Sequencer and shown to be highly homologous to *nifD* sequences of a variety of diazotrophs. The identity of the PCR amplified fragments from the reference collection of diazotrophic microbes was confirmed by standard Southern hybridization (Maniatis et al., 1982), using the amplified *R. loti nifD* fragment as a probe.

rep-PCR genomic fingerprinting

Whole cell rep-PCR of bacteria and computer analysis of the resulting fingerprints was carried out as described by Louws et al. (1996) and Rademaker and De Bruijn (1997).

16S rRNA gene sequencing and phylogenetic studies

Two highly conserved eubacterial 16S rDNA primers (8F and 1492R) were used to amplify segments of the 16S rRNA genes of the members of the test collection (see Results and discussion), using rep-PCR conditions. The PCR amplification products were purified using ultra-Free columns (Millipore). Partial 16S rDNA sequences were obtained using primers 8F and 519RB using the ABI373A DNA sequenator (see above). The sequences obtained were submitted to the SSU RDP Database at the University of Illinois using Netscape Navigator and analyzed using the Similarity Rank method described by Maidak et al. (1994).

Introduction of marker transposons into putative endophytes and detection of GUS activity

Transposons containing the beta-glucuronidase (*uidA; gus*) marker gene were introduced into the genome of the rice isolates by conjugation with *Escherichia coli* strains harboring narrow host range plasmids carrying the respective transposons. Detection of Gus activity was carried out as described by Wilson et al. (1995).

Results and discussion

Isolation of rice endophytes

In order to isolate putative endophytes of rice, two distinct approaches were used. The first approach involved the use of root and stem tissues of diverse varieties grown in different soil types, as described in the Materials and methods (see Figure 1). One hundred thirty-three of these isolates were randomly selected for further study. In an attempt to enrich for invasive/endophytic bacteria a second approach incorporating an *in planta* selection step was utilized (see Materials and methods and Figure 2). This procedure led to the isolation of 300 microbial isolates, of which 175 were brought to MSU for further analysis. Two of the isolates from this collection [R061.S1.3. and R032.S2.3] were randomly chosen and included in the study presented below (Test Collection).

Diversity of putative endophytes in rice tissues

The diversity of a collection of one hundred thirty-three putative endophytic microbes isolated using the approach outlined in Figure 1 was assessed using rep-PCR genomic fingerprinting (Louws et al., 1996; Rademaker and De Bruijn, 1997; Versalovic et al., 1994). Both REP and ERIC primers were employed and the resulting fingerprints were combined and analyzed using the GelCompar software package, as described by Rademaker and de Bruijn (1997). The dendrogram derived from this analysis is shown in Figure 3. The results presented reveal a large degree of diversity in these putative endophytes, although some distinct clusters of closely related strains could be observed (see Figure 3). One large cluster of thirty-two similar fingerprints stands out in the dendrogram. The origin of these bacterial isolates is interesting from a standpoint of cosmopolitanism verses endemism. The thirty-two isolates in this cluster were isolated from

eleven of the rice variety/soil types tested. Bacteria were isolated from plants grown in all five soil types, and from six of the eight rice varieties used. The largest number of similar fingerprints in the cluster that originated from the same variety/soil combination was eight. The diversity of their origins suggests that the bacteria found in this cluster are cosmopolitan, being found in a wide distribution of geographically removed soil types and being able to colonize a number of different rice genotypes. Three strains from the large cluster [-R6a(126), R33(120) and R45 (42)] were selected for further analysis (Test Collection; see below).

Nitrogen fixation by and presence of nif *genes in the endophytic isolates.*

The nitrogen fixing ability of the one hundred thirty-three putative endophytic bacteria from the first collection (Figure 1) was examined using the acetylene reduction assay (ARA). In addition, highly conserved DNA primers for PCR mediated detection of *nifD* genes were developed and used to screen the collection for the presence of *nif* genes. In the case of diazotrophic bacteria, PCR amplification of genomic DNA using these primers generally leads to the generation of a single 390 bp fragment, which has been shown to be *nifD*-specific by examining a reference collection of thirty-three well characterized bacteria selected from our laboratory strain collection. This collection included twenty-four species of nitrogen-fixing bacteria and nine non-nitrogen-fixing bacteria. The identity of the amplified fragment was further confirmed by determining the DNA sequence of the amplified fragment from *Rhizobium loti* NZP2235 and by Southern blot analysis (J Stoltzfus and F J De Bruijn, in preparation). Seventeen of the one hundred thirty-three strains fingerprinted (Figure 3) displayed a characteristic amplification band when tested using the *nifD* primers. Of the seventeen strains shown to contain *nifD* using PCR, thirteen were found to reduce acetylene (Nif plus). Four additional isolates, which were scored as *nifD* minus in the PCR reactions, were nevertheless found to be ARA plus, suggesting that they contain highly divergent *nifD* genes, or alternative *nif* genes. Figure 4 shows fingerprints and *nifD* PCR results for ten bacteria which were both *nifD* plus and Nif plus, four bacteria that were *nifD* plus and Nif minus, and four bacteria that were *nifD* minus and Nif plus. The remaining one hundred twelve strains were scored as *nifD* minus and Nif minus. Three of the isolates that were found to be both *nifD* and Nif plus [T105, R90(8), R100(64)] and

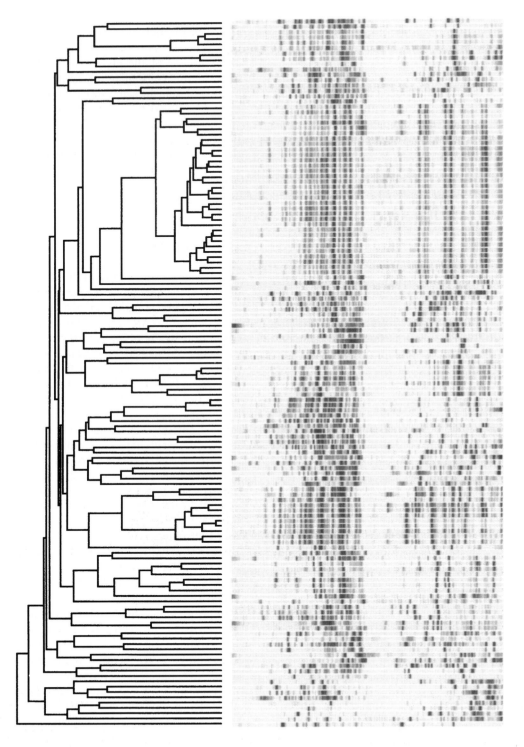

Figure 3. Gel profiles and dendrogram of rep-PCR generated genomic fingerprints of putative endophytic bacteria isolated from rice roots. The dendrogram of 133 putative endophytes from rice generated from the combined BOX- and ERIC fingerprints using the GelCompar 4.0 program is shown. The fingerprint patterns revealed a high degree of diversity, as well as several clusters of related bacteria. The analysis was performed using Pearson correlation applied to the densitometric curves formed by the fingerprints (see Rademaker and de Bruijn, 1997), followed by clustering analysis using the unweighted pair-grouping method based on arithmetic averages (UPGMA).

31

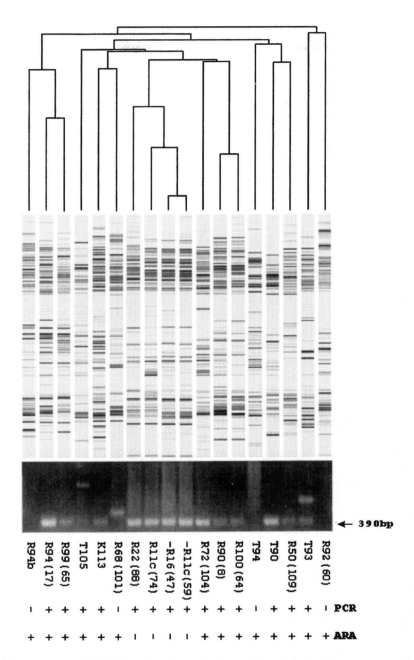

Figure 4. rep-PCR genomic fingerprints and *nifD*-PCR analysis of putative diazotrophic endophytes. The dendrogram of the rep-PCR fingerprints of 18 putative nitrogen fixing strains is shown. The dendrogram was generated using the combined BOX and ERIC and REP fingerprint patterns and the Gelcompar 4.0 software program. This dendrogram shows both the diversity of organisms, as well as clustering of related bacteria. A representative gel on which *nifD*-PCR fragments were electrophoretically separated is shown below the fingerprints. (+) indicates bacteria from which an amplified *nifD* fragment could be obtained. (-) indicates bacteria from which no amplified *nifD* PCR fragment could be obtained. Results from the acetylene reduction assays are also presented. (+) indicates bacteria capable of reducing acetylene in semi-solid malate media. (-) indicates bacteria which did not reduce acetylene in semi-solid malate media.

32

one isolate that was *nifD* plus but Nif minus [R22(88)] were chosen to be included in the Test Collection (see below).

Composition of a "test collection" and preliminary characterization of its members

A "Test Collection" of bacteria was composed for further studies (see Table 1). The nine isolates noted above [-R6a(126), R33(120), R45(42), T105, R22(88), R90(8), R100(64), R032.S2.3 and R061.S1.3] were included, as well as five previously characterized "control" bacteria. *Escherichia coli* DH5α was chosen as a negative control because it is unlikely that this bacterium would colonize rice roots endophytically. *Azorhizobium caulinodans* ORS571 was chosen because it has been reported to invade the roots of wheat and rice (Cocking et al., 1995; see also Reddy et al., 1997; Webster et al., 1997). *Rhizobium meliloti* 1021 was included to compare a *Rhizobium* with *Azorhizobium caulinodans* ORS571 in its ability to colonize rice roots. *Azoarcus indigens* LMG9092 and *Herbaspirillum seropedicae* Z67 were included since these two bacteria have previously been reported to invade rice tissues (see Reinhold-Hurek and Hurek, 1997).

In order to further characterize the endophytic isolates from rice, a partial DNA sequence of their 16s rDNA was determined. The DNA sequences obtained were submitted to the SSU RDP database at University of Illinois using the Similarity Rank function (Maidak et al., 1994). The results of this analysis are shown in Table 1. The most similar 16s rDNA sequence in the database for each of the "control" bacteria was found to be "itself", except for the case of *Herbaspirillum seropedicae* Z67, where 16S rDNA sequences are not yet available for comparison.

Re-colonization of rice tissue by putative endophytes

To study the endophytic nature of the bacteria in the Test Collection, experiments were performed to verify that the same bacteria inoculated to sterile rice seedlings could be re-isolated from these seedlings (to fulfill Koch's postulate), and to examine their endophytic competence (infection and persistence characteristics). Gnotobiotically grown rice seedlings were inoculated with each of the bacteria from the test collection and bacteria were re-isolated as described in materials and methods (see Figure 5). The identity of the colonies was verified by comparing the ERIC-PCR fingerprints of these bacteria with the ERIC-PCR fin-

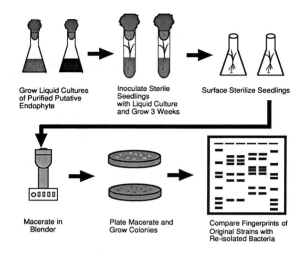

Figure 5. Re-infection potential of putative endophytes. The scheme for re-infecting rice plants with and re-isolation of putative endophytes is shown. This experiment was carried out to fulfill Koch's postulate that true endophytes should be re-isolatable after infection of gnotobiotically grown rice seedlings. After re-isolation the identity of the bacteria was confirmed using rep-PCR fingerprinting.

gerprint of the original inocula. In some cases, no bacteria could be re-isolated from the inoculated seedlings. In other cases a large number of bacteria (>25,000), could be re-isolated from a single seedling.

As expected, no bacteria were re-isolated from seedlings inoculated with *Escherichia coli*. Surprisingly, no bacteria could be re-isolated from seedlings inoculated with either *Herbaspirillum seropedicea* or *Azoarcus indigens*. In the case of the rice strains and other control strains large differences in numbers of re-isolatable bacteria were noted. However, large differences in the number of bacteria re-isolated from different seedlings inoculated with the same bacteria were also observed. In spite of these variations, certain endophytic isolates [R22(88), R33(120), R45(42), and R061.S1.3] appeared to be considerably more aggressive colonizers than others. The most aggressive colonizer [R22(88)] was found to cause stunting of seedling growth and therefore may be perceived by the plant as potential pathogen. Presently, the experimental parameters of these tests are being further standardized and the experiments are being repeated using sufficient replicas to ensure statistically significant results.

The analysis of bacteria re-isolated from seedlings inoculated with *Azorhizobium caulinodans* ORS571, R22(88), R100(64), -R6a(126), R33(120), R45(42), R032.S2.3, and R061.S1.3 using rep-PCR genomic fingerprinting (see Figure 5 for the experimental

Table 1. Bacteria listed in the first colunm were used to test various methods for evaluation of colonization. Partial SSU rDNA sequences from these bacteria were submitted to the SSU RDP database (version 5.0) at University of Illinois using Similarity Rank (version 2.5). The first and second most similar SSU rDNA sequences are shown. The similarity rank function of the SSU RDP finds the database sequences that are most similar to a sequence submitted by the user. The program counts the number of unique 7 base oligomers in the submitted sequence that are common to each sequence in the data base. The Sab value is the result expressed as the number of shared 7 base oligomers divided by either the number of unique oligomers in the submitted sequence or the database sequence, whichever is lower. This approach cannot be used to carry out a phylogenetic analysis *per se*, but allows assesment of the similarities between the sequences

Bacteria	Most similar 16S rRNA	Sab	2nd Most similar 16S rRNA	Sab
Escherichia coli DH5α	*Escherichia coli*	0.945	*Escherichia coli* subsp. K-12 str. MG 1665	0.945
Rhizobium meliloti 1021	*Sinorhizobium xinjiangensis*	0.949	*Rhizobium meliloti*	0.949
Azorhizobium caulinodans ORS571	*Azorhizobium caulinodans*	0.972	*Aquabacter spiritensis* str. SPL-1	0.791
Azoarcus indigens I M G 9092	*Azoarcus indigens* str. VB32	0.929	*Pseudomonas* sp. str. K172 clone K1	0.780
Herbaspirillum seropedicae Z67	*Zoogloea ramigera*	0.762	*Telluria chitinolytica*	0.688
T105-IRRI collection	*Azorhizobium caulinodans*	0.852	*Aquabacter spiritensis* str. SPL - 1	0/809
R22(88)-IRRI collection	*Pseudomonas cepacia* str. G4	0.952	*Burkholderia gladioli* subsp. *pathovar gladiolis*	0.899
R90(8)-IRRI collection	*Agrobacterium vitis*	0.896	*Agrobacterium vitis*	0.878
R100(64)-IRRI collection	*Agrobacterium vitis*	0.878	*Agrobacterium vitis*	0.878
-R6a(126)-IRRI collection	*Azospirillum* sp*	0.913	*Pseudomonas Putida* str. mt-2	0.826
R33(120)-IRRI collection	*Azospirillum* sp*	0.919	*Pseudomonas Putida* str. mt-2	0.801
R45(42)-IRRI collection	*Azospirillum* sp*	0.913	*Pseudomonas Putida* str. mt-2	0.826
R032.S2.3-JRS selection	*Bacillus subtilis* str 168	0.839	*Bacillus subtilis* str 168	0.835
R061.S1.3-JRS selection	*Bacillus megaterium*	0.880	*Bacillus megaterium*	0.822

*This bacteria is given the name *Azospirillum* in the RDP database. However, it is not closely related to other *Azospirillum* strains based on phylogenetic trees made using SSU rDNA sequences.

approach) revealed that Koch's postulate was fulfilled, in that bacteria with the same fingerprint as the original inoculum strain could be re-isolated (data not shown).

Use of marker genes to track endophytic bacteria in rice tissues

To track bacteria during colonization of rice roots and subsequent inter- or intracellular persistence, specific marker gene constructs were selected (see Figure 6). A number of different marker genes are available for the tracking of genetically modified organisms (see Jansson, 1995). Initially, we focused on the use of the beta-glucuronidase(*uidA; gus*) gene, since Gus activity can be easily detected using a colorometric Gus enzyme assays or *in situ* staining methods (Wilson et al., 1995). In order to obtain stably tagged strains, we used the Tn5*gus* transposons pCAM111 or pCAM121, kindly provided by Kate Wilson (Wilson et al., 1995). One or the other of these transposons was introduced into 8 out of the 14 test collection strains, and Gus activity of the tagged bacteria was verified. The remaining six strains exhibit multiple intrinsic antibiotic resistances, and modified transposons/suicide vectors are being developed to tag these strains.

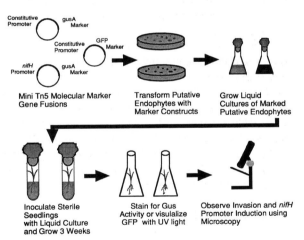

Figure 6. In planta visualization of endophyte colonization in rice. The scheme for the utilization of the molecular markers beta-glucuronidase (GUS) and green fluorescent protein (GFP) to characterize endophyte invasion of rice tissues is shown. A constitutive promoter fused to the *gus* or GFP reporter gene will be used to visualize colonization. A *nifH* promoter fused to the *gus* reporter gene will be used to monitor *nif* gene expression during colonization.

Initial results with a limited number of tagged strains indicate that global colonization patterns of *gus* tagged microbes in rice roots can indeed be monitored by in situ Gus staining (J Stoltzfus and FJ De Bruijn, unpublished results). However, the work required for sectioning tissues in order to localize individual bacteria using Gus activity may be prohibitive to the screening of large sample numbers. Therefore, in collaboration with the laboratory of J. Jansson at Stockholm University, we have developed Tn5 derived transposons carrying the gene encoding the green fluorescent protein (*gfp*), originally derived from the jellyfish *Aequorea* (Chalfie et al., 1994). Using the constitutively expressed Tn5*gfp* marker transposon, we have shown that GFP is stably and highly expressed in single cells of *Pseudomonas fluorescence*, and that single *gfp* tagged cells can be detected on the root hairs of plants using laser confocal microscopy (Tombolini et al., 1996; Unge et al., 1997). The use of fluorescent laser scanning confocal microscopy allows optical sectioning reducing the work required to screen large sample numbers. We are presently attempting to tag the rice endophytes and control strains with this marker gene and will use the resulting isolates to study rice plant colonization.

Do the proper physiological conditions exist in rice tissues to derepress nif *genes harbored by endophytic bacteria?*

Even when stably maintained diazotrophic endophytes (endosymbionts) of rice are identified, it still remains to be shown that the proper physiological environment exists at the site(s) of colonization to allow expression of the *nif* genes and functioning of nitrogenase (see de Bruijn et al., 1995; Dixon et al., 1997). As a first step in this analysis we have made use of a plasmid carrying a translational fusion of the 5' end of the *Azospirillum brasilense nifH* gene (including its promoter region) to *gus* (Vande Broek et al., 1992; 1996). Preliminary results show that selected endophytic strains from the test collection harboring the *nifH-gus* fusion do express the reporter gene in the free living state under low oxygen tensions, as expected (see Vande Broek et al., 1996). These strains will be used to infect rice seedlings, and in situ staining of infected tissues for Gus activity will be performed in order to determine if and where the proper physiological conditions exist for *nif* gene derepression, and therefore, by inference, for nitrogen fixation per se, although the latter correlation has not always been found to hold true (Vande Broek et al., 1996).

Conclusions

In this paper we have described preliminary results from our first screens for and characterization of naturally occurring diazotrophic endophytes of rice. The approach we have used here differs from strategies to assess the potential for nitrogen fixation in rice we have reported previously (De Bruijn et al., 1995) and in this volume (Yanni et al., 1997). The latter strategies are based on the assessment of the potential of known rhizobia to induce (nitrogen-fixing) nodule-like structures or root hypertrophies on rice roots, or the ability of specific naturally-occurring endophytic rhizobia to stimulate rice growth under different nitrogen regimes and display acetylene reduction activity, respectively. As a basis for the screen described here, we reasoned (hypothesized) that rather than, or in addition to, attempting to induce or modify known rhizobia to engage in symbiotic or endophytic interactions with rice plants resulting in biological nitrogen fixation for plant growth enhancement, we would try to isolate and characterize naturally occurring, stably endophytic microbes from a variety of different modern and more primitive varieties of rice, and subsequently determine their nitrogen fixation potential or genetically modify them to fix nitrogen. A major difference between this study and that of Yanni et al. (1997) was the strategy used to isolate diazotrophs from the rice endophytic community. Whereas Yanni et al. used an appropriate legume trap host to specifically enumerate and isolate endophytic rhizobia from rice roots, this study used various plating media to isolate a more diverse collection of microbes belonging to the bacterial community that colonizes rice roots. In fact, these studies are complementary and resulted in the isolation of a large collection of rhizobial and non-rhizobial native microbes that are capable of developing endophytic relationships with rice.

The results presented here, and those described by Barraquio et al. (1997) indicate that a large diversity of apparently diazotrophic ($\sim$10%) and non-diazotrophic endophytic bacteria can be isolated from rice tissues, some of which are capable of re-colonizing their host when re-inoculated onto sterile rice seedlings. Past isolations of nitrogen fixing bacteria from rice roots have already revealed a broad diversity of diazotrophs inhabiting the rice rhizosphere (Bally et al., 1983; Ladha, 1986; Lakshmi Kumari et al., 1976; Oyaizu-Masuchi et al., 1988; Watanabe and Barraquio, 1979; see also Yanni et al., 1997). In addition, recent studies based on the molecular phylogeny of the DNA sequences generated by PCR amplification of nitrogen fixing genes found in the rice rhizosphere also suggest the existence of a broad range of different rhizosphere diazotrophs (Ueda et al., 1995a, 1995b). However, the ability of these bacteria to colonize rice tissue endophytically and de-repress genes needed for nitrogen fixation remains virtually unstudied.

Having developed the tools to mark and track some of the promising stable endophytes and monitor the physiological environment in their colonization niches with regard to expressing *nif* genes, we are now able to further test our hypothesis that natural or genetically modified rice endophytes may be useful to achieve the long term goal of developing rice-microbe interactions capable of supplying biologically fixed nitrogen for higher yields, in the absence of additional chemical fertilizer. In addition, information gained during these studies may yield interesting new insights into basic mechanisms of plant-microbe interactions and microbial colonization of plant tissues, and be helpful in designing strategies for the discovery and adaptation of nitrogen fixing bacteria that could provide nitrogen to other major cereal crops such as maize and wheat.

Acknowledgements

We would like to thank Dr Kate Wilson for her gift of several GusA transposons for use as molecular markers, Jan Rademaker for his help with the Gelcompar 4.0 program, Dr Tom Schmidt for his help with the analysis of the SSU rDNA sequence data and Mary Ellen Davey for her contributions to the GFP portion of this project. This project has been supported by grants from the Danish International Development Agency (DANIDA), and the US Department of Energy (DE-FG02-91 ER20021).

References

Atlas R M 1997 Handbook of Microbial Media. 2nd ed. Ed. L C Parks. CRC Press, Inc. Boca Raton, FL. 401 p.

Bally R, Thomas-Bauzon D, Heulin T and Balandreau J 1983 Determination of the most frequent N₂-fixing bacteria in a rice rhizosphere. Can. J. Microbiol. 29, 881–887.

Barraquio W L, Revilla L and Ladha J K 1997 Isolation of endophytic diazotrophic bacteria from wetland rice. Plant Soil 194, 15–24.

Beringer J E 1974 R factor transfer in *Rhizobium leguminosarum*. J. Gen. Microbiol. 84, 188–189.

Boddey R M, de Oliveira O C, Urquiaga S, Reis V M, de Olivares F L, Baldani V L D and Dobereiner J 1995 Biological nitrogen fixation associated with sugar cane and rice: Contributions and prospects for improvement. Plant Soil 174, 195–209.

Bohlool B B, Ladha J K, Garrity D P and George T 1992 Biological nitrogen fixation for sustainable agriculture: A perspective. Plant Soil 141, 1–11.

Chalfie M, Tu Y, Euskirchen G, Ward W W and Prasher D C 1994 Green fluorescent protein as a marker for gene expression. Science 263, 802–805.

Clement A, Ladha J K and Chalifour F P 1995 Crop residue effects on nitrogen mineralization, microbial biomass, and rice yield in submerged soils. Soil Sci. Soc. Am. J. 59, 1595–1603.

Cocking E C, Kothari S I, Batchelor C A, Jain S, Webster G, Jones J, Jotham J and Davey M R 1995 Interaction of rhizobia with non-legume crops for symbiotic nitrogen fixation nodulation. *In* Azospirillum VI and related Microorganisms: Genetics, Physiology, Ecology. Eds. I Fendrik, M del Gallo, J Vanderleyden and M de Zamaroczy. pp 197–205. Springer-Verlag, New York.

De Bruijn F J, Jing Y and Dazzo F B 1995 Potential and pitfalls of trying to extend symbiotic interactions of nitrogen-fixing organisms to presently non-nodulated plants, such as rice. Plant Soil 174, 225–240.

Diekmann K H, Ottow J C G and De Datta S K 1996 Yield and nitrogen response of lowland rice (*Oryza sativa* L.) to *Sesbania rostrata* and *Aeschynomene afraspera* green manure in different marginally productive soils in the Philippines. Biol. Fertil. Soils 21, 103–108.

Dixon R. Cheng Q, Shen G-F, Day A and Dowson-Day M 1997 Nif gene transfer and expression in chloroplasts: prospects and problems. Plant Soil 194, 193–203.

Fahraeus A 1957 The infection of clover root hairs by nodule bacteria studied by a simple glass slide technique. J. Gen. Microbiol. 16, 374–381.

Fisher R F and Long S R 1992 Rhizobium-plant signal exchange. Nature 357, 655–660.

Harris G H, Hesterman O B, Paul E A, Peters S E and Janke R R 1994 Fate of legume and fertilizer nitrogen-15 in a long-term cropping systems experiment. Agron. J. 86, 910–915.

Hossain M and Fischer K S 1995 Rice research for food security and sustainable agricultural development in Asia: Achievements and future challenges. GeoJournal 35, 286–295.

Jansson J K 1995. Tracking genetically engineered microorganisms in nature. Curr. Opin. Biotechnol. 6, 275–283.

Kennedy I R and Tchan Y 1992 Biological nitrogen fixation in non-leguminous field crops: Recent advances. Plant Soil 141, 93–118.

Kennedy I R, Pereg-Gerk L, Wood C, Deaker R, Gilchrest K and Katupitiya S 1997 Biological nitrogen fixation in non-leguminous field crops: facilitating the evolution of an effective association between Azospirillum and wheat. Plant Soil 194, 65–79.

Kirchhof G, Reis V M, Baldani J I, Eckert B, Döbereiner J and Hartmann A 1997 Occurence, physiological and molecular analysis of endophytic diazotrophic bacteria in gramineous energy plants. Plant Soil 194, 45–55.

Kumarasinghe K S, Zapata F, Kovacs G, Eskew D L and Danso S K A 1986 Evaluation of the availability of Azolla N and urea N to rice using ¹⁵N. Plant Soil 90, 293–299.

Kush G S and Bennet J (eds) 1992 Nodulation and nitrogen fixation in rice: Potential and prospects. IRRI, Manila.

Ladha J K 1986 Studies on nitrogen fixation by free-living and rice-plant associated bacteria in wetland rice field. Bionature 6, 47–58.

36

Ladha J K, Pareek R and Becker M 1992 Stem-nodulating legume-*Rhizobium* symbiosis and its agronomic use in lowland rice. Adv. Soil Sci. 20, 147–192.

Ladha J K and Peoples M B 1995 Management of biological nitrogen fixation for the development of more productive and sustainable agricultural systems. Plant Soil 174, 1–286.

Ladha J K, De Bruijn F J and Malik K A 1997 Introduction: assessing opportunities for nitrogen fixation in rice - a frontier project. Plant Soil 194, 1–10.

Lakshmi Kumari M, Kavimandan S K and Subba Rao N S 1976 Occurrence of nitrogen-fixing *Spirillum* in roots of rice, sorghum, maize, and other plants. Indian J. Exp. Biol. 14, 638–639.

Louws F J, Schneider M and De Bruijn F J 1996 Assessing genetic diversity of microbes using repetitive-sequence-based PCR (rep-PCR). *In* Nucleic Acid Amplification Methods for the Analysis of Environmental Samples. Ed. G Toranzos. pp 63–93. Technomic Publishing Co., Lancaster, PA.

Maidak B L, Larsen N, McCaughey M J, Overbeek R, Olsen G J, Fogel K, Blandy J and Woese C R 1994 The ribosomal database project. Nucleic Acids Res. 22, 3485–3487.

Malarvithi P P 1995 Association of endophytic diazotrophs in rice genotypes from various soil environments and estimation of their BNF potential using ^{15}N dilution methods. PhD thesis submitted to the University of Philippines. pp 195.

Maniatis T, Fritsch E F and Sambrook J 1982 Molecular Cloning: a Laboratory Manual. Cold Spring Harbor Laboratory Press, Cold Spring Harbor, NY.

National Research Council 1994 Biological Nitrogen Fixation: Research Challenges. National Academy, Press, Washington DC.

Oyaizu-Masuchi Y and Komagata K 1988 Isolation of free-living nitrogen-fixing bacteria from the rhizosphere of rice. J. Gen. Appl. Microbiol. 34, 127–164.

Rademaker J L W and De Bruijn F J 1997 Characterization and classification of microbes by rep-PCR genomic fingerprinting and computer assisted pattern analysis. *In* DNA markers: Protocols, Applications, and Overviews. Eds. G Caetano-Anolles and P M Gresshoff. John Wiley and Sons, Inc. New York. (*In press*).

Reddy P M, Ladha J K, So R, Hernandez R, Ramos M C, Angeles O R, Dazzo F B and De Bruijn F J 1997 Rhizobial communication with rice roots: induction of phenotypic changes, mode of invasion and extent of colonization. Plant Soil 194, 81–98.

Reinhold-Hurek B and Hurek T 1997 *Azoarcus* spp. and their interactions with grass roots. Plant Soil 194, 57–64.

Roger P A and Ladha J K 1992 Biological N_2 fixation in wetland rice fields: Estimation and contribution to nitrogen balance. Plant Soil 141, 41–55.

Tombolini R, Unge A, Davey M E, De Bruijn F J and Jansson J K 1997 Flow cytometric and microscopic analysis of GFP-tagged *Pseudomonas fluorescens* bacteria. FEMS Microbiol. Ecol. 22, 17–28.

Ueda T, Suga Y, Yahiro N and Matsuguchi T 1995a. Remarkable N_2-fixing bacterial diversity detected in rice roots by molecular evolutionary analysis of *nifH* gene sequences. J. Bacteriol. 177, 1414–1417.

Ueda T, Suga Y, Yahiro N and Matsuguchi T 1995b Genetic diversity of N_2-fixing bacteria associated with rice roots by molecular evolutionary analysis of *nifD* library. Can. J. Microbiol. 41, 235–240.

Unge A, Tombolini R, Davey M E, De Bruijn F J and Jansson J K 1997 GFP as a marker gene. *In* Molecular Microbial Ecology Manual. 2nd ed. Eds. A D L Akkermans, J D van Elsas and F J De Bruijn. Kluwer Academic Publishers, Dordrecht. (*In press*).

Vande Broek A, Michiels J, de Faria S M, Milcamps A and Vanderleyden J 1992 Transcription of the *Azospirillum brasilense nifH* gene is positively regulated by *nifA* and *ntrA* and is negatively controlled by the cellular nitrogen status. Mol. Gen. Genet. 232, 279–283.

Vande Broek A, Keijers V and Vanderleyden J 1996 Effect of oxygen on the free-living nitrogen fixation activity and expression of the *Azospirillum brasilense nifH* gene in various plant-associated diazotrophs. Symbiosis 21, 25–40.

Versalovic J, Schneider M, De Bruijn F J and Lupski J R 1994 Genomic fingerprinting of bacteria using repetitive sequence-based polymerase chain reaction. Methods Mol. Cell Biol. 5, 25–40.

Watanabe I and Barraquio W 1979 Low levels of fixed nitrogen required for isolation of free-living N_2-fixing organisms from rice roots. Nature 277, 565–566.

Webster G, Gough C, Vasse J, Batchelor C A, O'Callaghan K J, Kothari S L, Davey M R, Dinarii J and Cocking E C 1997 Interactions of rhizobia with rice and wheat. Plant Soil 194, 115–122.

Wilson K J, Sessitsch A, Corbo J C, Giller K E, Akkermans A D L and Jefferson R A 1995 β-Glucuronidase (GUS) transposons for ecological and genetic studies of rhizobia and other Gram-negative bacteria. Microbiology 141, 1691–1705.

Yanni Y G, Rizk R Y, Corich V, Squartini A, Ninke K, Philip-Hollingsworth S, Orgambide G, De Bruijn F, Stoltzfus J, Buckley D, Schmidt T M, Mateos P F, Ladha J K and Dazzo F B 1997 Natural endophytic association between *Rhizobium leguminosarum* bv. *trifolii* and rice roots and assessment of its potential to promote rice growth. Plant Soil 194, 99–114.

Guest editors: J K Ladha, F J De Bruijn and K A Malik

Plant and Soil **194**: 37–44, 1997.

Association of nitrogen-fixing, plant-growth-promoting rhizobacteria (PGPR) with kallar grass and rice

K.A. Malik, Rakhshanda Bilal[1], Samina Mehnaz, G. Rasul, M.S. Mirza and S. Ali
National Institute for Biotechnology and Genetic Engineering, P.O. Box 577, Faisalabad, Pakistan. * [1]*Present address: PINSTECH, Rawalpindi, Pakistan*

Key words: acetylene reduction, *Azospirillum*, *Azoarcus*, grasses, ^{15}N isotope dilution, root-associated bacterial population

Abstract

Leptochloa fusca (L.) Kunth (kallar grass) has previously been found to exhibit high rates of nitrogen fixation. A series of experiments to determine the level of biological nitrogen fixation using ^{15}N isotopic dilution were carried out in nutrient solution and saline soil. These studies indicated an agronomically significant amount of nitrogen being fixed in soil. Kallar grass has a similar growth habitat to rice. Therefore similar studies were carried out with rice after isolating various diazotrophs from the roots which were also screened for their ability to produce auxin (IAA). Five such strains namely *Azospirillum lipoferum* N-4, *Azospirillum brasilense* Wb-3, *Azoarcus* K-1, *Pseudomonas* 96-51, *Zoogloea* Ky-1 were selected for inoculating two rice varieties i.e. NIAB-6 and BAS-370 under aseptic laboratory conditions. The nitrogen fixed was quantified using the ^{15}N isotopic dilution method. Variety BAS-370 had nearly 70% nitrogen derived from atmosphere (Ndfa) when inoculated with *Azospirillum* N-4. Similar studies with the mixed inoculum using ^{15}N fertilizer in the micro plots indicated that nearly 29% of plant nitrogen was derived from the atmosphere.

Introduction

Nitrogen fixation associated with roots of grasses has been recognized as an important component of the nitrogen cycle in a range of ecosystems including several extreme environments (Chalk, 1991). Salinity represents an extreme environment and is characterized by low organic matter and very low nitrogen contents in the soil. However, under these conditions certain plants especially kallar grass (*Leptochloa fusca*) have been found to grow well (Malik et al., 1986). Since the advent of acetylene reduction methodology (ARA), many plants have been shown to harbour diazotrophs in and around their roots (Boddey et al., 1996; Malik et al., 1991). Though certain limitations exist in the application of ARA methodology as reported by Van Berkum (1980) and Witty (1979), its usefulness in screening plants and microorganisms for presence of nitrogenase activity is beyond doubt. Among the plants growing in saline environment kallar grass has been

extensively studied with regard to associative nitrogen fixation.

Kallar grass, a highly salt-tolerant plant species, has been recommended as a primary colonizer of salt-affected soils and grows luxuriantly without addition of nitrogenous fertilizers under waterlogged conditions (Sandhu and Malik, 1975). This observation led to the investigation of its rhizosphere for the presence of root-associated nitrogen fixation. Nitrogenase activity associated with the roots in this grass was first reported by Malik et al. (1980, 1982) and a number of diazotrophic bacteria were isolated and characterized from various root fractions (Bilal and Malik, 1987; Bilal et al., 1990; Zafar et al., 1986, 1987). The growth habitat of kallar grass is similar to that of lowland rice. Moreover, rice is generally recommended to be grown on the saline soils where kallar grass had been cultivated. Thus basic information and knowledge obtained from diazotroph-kallar grass association could be directly applied to rice which also grows under similar waterlogged conditions.

---
* FAX No: +924116571472.
E-mail:kamalik@nibge.fsb.erum.com.pk

Auxins are growth regulators which are essential for plant growth. Some bacteria are also reported to produce these growth hormones affecting plant growth (Bric et al., 1991; Lifshitz et al., 1987; Tien et al., 1979; Zimmer et al., 1988). The bacteria are known to change the root morphology and increase their biomass, thus enabling them to exploit more of the soil volume and take up soil nutrients. Thus, such bacteria can complement the beneficial effects of nitrogen fixation. Studies have therefore been conducted on the beneficial effects of these bacteria on rice productivity.

Nitrogenase activity and nitrogen fixing bacteria associated with kallar grass

Kallar grass is a perennial grass that has shown high salt-tolerance and can be cultivated with brackish water. Its high biomass yield led us to study its rhizosphere for possible inputs due to biological nitrogen fixation. Our results showed that excised roots of kallar grass exhibited high nitrogenase activity as estimated by ARA (Malik et al., 1982). High acetylene-reducing activities associated with soil and roots (unwashed, washed and surface sterilized) of kallar grass were found (Zafar et al., 1986). Enumeration of diazotrophic bacteria present in the rhizosphere was also carried out (Table 1) using four different culture media (Bilal, 1988). In these studies rhizosphere was separated into various fractions like (1) soil away from roots (NRS), (2) soil adhering to roots (RS), (3) surface of roots (rhizoplane, RP) and (4) interior of roots (histoplane, HP). Nutrient agar medium was used for the enumeration of the total heterotrophic population. Microaerophilic conditions were provided using 0.7% soft agar plating and semisolid media were prepared by using 0.2% agar. Enumerations were based on spread plate technique for aerobes, soft agar overlay for micro aerobic bacteria and most probable number (Cochran, 1950) of ARA positive vials for diazotrophs.

The combined carbon medium (CCM) of Rennie (1981) invariably gave the highest counts of N_2-fixers in all fractions, followed by the malate medium (NFM). Populations differed significantly in rhizosphere soil (RS) and rhizoplane (RP) fractions on CCM medium. The most probable number (MPN) values based on ARA were lower than those obtained by the dilution plate count. The distribution of bacteria was variable in these different fractions. Highest counts were always observed in the rhizoplane fraction on all four media tested (Table 1 and 2). Various diazotrophic bacteria

were isolated and characterized (Bilal et al., 1990; Malik et al., 1991). These belong to the genera of *Azospirillum*, *Azoarcus*, *Enterobacter*, *Klebsiella* and *Zoogloea*. It was found that azospirillia were more numerous in the root interior whereas there was preponderance of the genera belonging to Enterobacteriaceae on the root surface. Survival and colonization of the inoculated bacteria in the root rhizosphere was also studied using immunofluorescent techniques (Malik and Bilal, 1989). Some of the common genera of agronomic significance are described below.

Table 1. Enumeration of diazotrophic bacterial populations in various rhizosphere fractions of kallar grass using different media

Medium*	Fraction**	Log no. of bacteria/g dry root or soil			
		MPN*** (ARA-based)	Aerobic	Micro-aerobic	Mean
			CFU****		
CCM	NRS	7.00	6.03	8.77	6.81 ab
	RS	6.82	8.23	8.17	7.61 a
	RP	9.61	9.0	8.88	9.44 c
	HP	4.17	6.07	5.95	5.33 a
NFM	NRS	5.05	6.95	7.16	6.32 b
	RS	6.60	7.55	7.88	7.32 bc
	RP	7.77	8.51	9.00	8.47 c
	HP	1.00	5.34	5.30	3.88 a
NFDM	NRS	3.10	8.69	9.14	5.25 a
	RS	5.39	7.88	6.47	6.45 a
	RP	7.47	6.60	8.04	7.51 a
	HP	0.00	5.68	5.57	4.07 a
SSM	NRS	3.00	4.58	7.60	3.80 a
	RS	6.60	7.93	7.92	7.13 bc
	RP	7.92	8.98	9.38	8.76 c
	HP	0.00	5.43	5.64	4.10 a

*= CCM, Rennie, 1981; NFM, Döbereiner et al., 1976; NFDM, Cannon et al., 1974; SSM, Reinhold et al., 1986.
**NRS=soil away from roots; RS=rhizosphere soil; RP=surface of roots; HP=interior of roots.
***=ARA-based most probable number.
****=CFU, colony forming units.

Table 2. N_2 fixers in various fractions of kallar grass, estimated by ARA-based MPN method

Medium**	% N_2 fixers*			
	NRS	RS	RP	HP
CCM	1.5	0.014	100	1.36
NFM	0.07	0.080	19	0.00
NFDM	0.01	0.001	4	0.00
SSM	0.01	0.009	10	0.00

*Percent of the total bacterial population; NRS=soil away from roots; RS=rhizosphere soil; RP=surface of roots; HP=interior of roots.
**=CCM, Rennie, 1981; NFM, Döbereiner et al., 1976; NFDM, Cannon et al., 1974; SSM, Reinhold et al., 1986.

Azospirillum

The availability of selective media for isolating diazotrophs belonging to the genus *Azospirillum* and the ease of its detection by its characteristic sub-surface white pellicle in semi-solid agar medium has helped in isolations from the rhizosphere and root surface (Baldani et al., 1986; Döbereiner et al., 1976; Hegazi et al., 1979; Ladha et al., 1987; Sundaram et al., 1988). Several isolates were obtained on semi-solid nitrogen-free media (NFM, CCM) from roots of kallar grass (Bilal and Malik, 1987; Bilal et al., 1990; Malik et al., 1991; Zafar et al., 1987). These isolates formed a fine sub-surface white pellicle in nitrogen-free malate medium within 24 hours, which gradually moved to the surface. The growth in NFM was always accompanied by alkali production and high rates of nitrogenase activity (more than 100 nmol C_2H_4/hour/culture vial). Phase-contrast microscopic examination of wet mounts of the actively growing cultures showed helicle cells resembling *Azospirillum*. The cells were actively motile and exhibited typical spinning motility. Morphologically, *Azospirillum* can be differentiated from all the other bacteria because of its helicle shape from which it derives its name. A high percentage of the isolates accumulated poly-β-hydroxy butyrate (PHB) which was also observed in *Azospirillum* strains by other workers (Reinhold et al., 1986). The majority of the isolates formed light pink colonies when grown on nutrient agar medium, but the pigment was more characteristic and dark pink when nutrient broth-grown cells were pelleted by centrifugation.

None of the isolates used glucose, sucrose and mannitol as carbon sources but grew on lactate, malate and fructose. The biochemical characteristics of the isolates showed that they closely resembled *Azospirillum brasilense* which differs from *Azospirillum lipoferum* in its biotin requirement and ability to utilize glucose as a carbon source (Tarrand et al., 1978). The isolates were compared with various *Azospirillum* species by using one dimensional SDS-PAGE of total bacterial proteins. Comparison with *A. lipoferum*, *A. brasilense* and *A. amazonense* showed that all isolates closely resembled *A. brasilense* strain CdJA (Reinhold et al., 1987).

In a study carried out by Reinhold et al. (1986), nitrogen fixing bacteria were found to form root-zone specific associations with kallar grass, with different populations colonizing the surface and the interior of roots. Two *Azospirillum* species were dominant on the rhizoplane and one of these organisms has been described as a new salt-tolerant species, *Azospirillum halopraferens* (Reinhold et al., 1987a).

Azoarcus

In most of the attempts to isolate diazotrophs from kallar grass, gram negative motile rods were isolated which did not belong to any of the known diazotrophic genera on the basis of the commonly used morphological and physiological tests (Bilal and Malik, 1987; Bilal et al., 1990; Malik et al., 1991; Reinhold et al., 1986). These bacteria have a strictly aerobic type of metabolism, fix nitrogen micro aerobically, and grow well on salts of organic acids but not on carbohydrates. As a result of a polyphasic study carried out by Reinhold-Hurek et al. (1993), a new genus *Azoarcus* was generated to accomodate these isolates from roots of kallar grass collected from Pakistan. One isolate (K-1) from kallar grass was found to belong to this genus by using 16S rRNA oligonucleotide probes (Hassan et al., 1996). Indirect evidence for the colonization of the root interior by these organisms was confirmed by fluorescent antibodies (Reinhold et al., 1987) and immunogold electron microscopy (Hurek et al., 1991). Based on the results of a polyphasic study in which morphological, nutritional and biochemical features and several molecular biological techniques were employed, a new genus *Azoarcus* was proposed with two species *A. communis* and *A. indigenous* (Reinhold-Hurek et al., 1993).

Zoogloea

This nitrogen fixing organism was isolated from the histoplane fraction of kallar grass on CCM medium (Bilal and Malik, 1987). The organism designated as Ky-1 is not one of those commonly occurring on the kallar grass histoplane. Identification of this organism as *Zoogloea* (family Pseudomonadaceae) was primarily based on the presence of a characteristic extracellular capsule or zoogloeal matrix, and also because of its peculiar growth in shaken broth culture, where it aggregates to form macroscopic star-like flocs similar to those of *Zoogloea ramigera* (Krieg and Holt, 1984). The extra-cellular slime produced by Ky-1 is water-insoluble as the culture does not show any turbidity during growth and floc formation. Under the microscope the flocs showed finger-like dendritic out growths, each originating from a cluster of cells. The cells are embedded in a matrix and are static, but some exhibit rhythmic movements along the margins, whereas free cells in the suspension are actively motile.

Klebsiella, Beijerinckia *and other diazotrophs*

Three diazotrophic bacterial strains were isolated on CCM and named as NIAB-1, C-2 and Iso-2 (Zafar et al., 1987). These bacteria were characterized on the basis of physiological tests and the determination of mole % G+C values of the DNAs. Isolates NIAB-1, C-2 and Iso-2 have mol % G+C contents of 57, 64 and 53, respectively, which were lower than that of the genus *Azospirillum* (68–70%). The characteristics of three isolates were compared with some known nitrogen fixers *Azotobacter, Azomonas, Beijerinckia* and *Klebsiella*. NIAB-1 shared maximum characteristics with the genus *Klebsiella*. Because of the differences within *Klebsiella pneumoniae* in physiological behavior, protein pattern and other molecular characteristics (e.g. 60–70% DNA homology, presence of plasmids), it was proposed that NIAB-1 is a new species of *Klebsiella* (Qureshi et al., 1988) and Iso-2 was identified as *Beijerinckia* .

Quantification of nitrogen fixation in kallar grass by ^{15}N isotope dilution

Quantification of biologically fixed nitrogen in legumes and in grasses has been the most important factor in determining the overall benefit of this process to the cropping system. Nitrogen balance studies have been used most widely as an indication of the extent of nitrogen fixation (Day et al., 1975; Rennie and Larson, 1979). However, the use of ^{15}N$_2$ has given values of direct incorporation of fixed N into plants (Eskew et al., 1981; Rennie et al., 1983). This technique can only be used for a short period, during which plants are grown under an enclosed atmosphere and cannot be used in the field. These limitations have made this technique inappropriate for routine use. Techniques based on ^{15}N dilution are however, more versatile and can be adapted to various experimental conditions (Boddey et al., 1996; Roger and Ladha, 1992). ^{15}N isotope dilution techniques have earlier been used for quantifying associative N$_2$-fixation in wheat (Lethbridge and Davidson, 1983; Rennie et al., 1983), maize (Rennie, 1980), rice (Shrestha and Ladha, 1996; Ventura and Watanabe, 1983), sugarcane (Boddey et al., 1996; Ruschel et al., 1975) and other grasses (Boddey et al., 1983; Boddey

and Victoria, 1986; Malik and Zafar, 1985; Malik et al., 1987, 1988).

Several experiments to quantify the amount of nitrogen fixed (Table 3) were carried out both under aseptic conditions and in the field, using the ^{15}N isotopic dilution technique (Malik and Zafar, 1985; Malik et al., 1987, 1988). It has been shown that under sterile conditions 60-80% of the N in aerial parts of kallar grass was derived from fixation by the inoculated bacteria (Malik et al., 1987). In these experiments no additional carbon source was added indicating that root exudation was able to sustain the proliferation of the inoculated bacteria. In the case of a field experiment, the indigenous nitrogen-fixing ability of soil was inhibited by applying a high dose of ^{15}N fertilizer which was taken as the reference treatment. The estimation based on 'A' values (Fried and Broeshart, 1975) indicated that an amount of 32 kg N/ha was derived from fixation resulting in% Ndfa of 26 (Malik et al., 1988).

All the studies reported on kallar grass have indicated the presence of nitrogenase activity associated with its roots and substantial uptake of biologically fixed nitrogen by the plant as demonstrated by ^{15}N dilution technique. Studies on the survival, colonization and attachment of the bacteria to the kallar grass roots have also been reported (Bilal et al., 1993; Malik and Bilal 1989) which indicated that the bacteria colonized and proliferated on the root surface and were also present inside the root hair cells.The bacteria were abundant in the mucigel and around the points of emergence of lateral roots.

N$_2$-fixation in association with rice

In addition to the kallar grass, diazotroph isolations from rice roots and its rhizosphere were made (Malik et al., 1993). All these isolates were characterized and were found to belong to the genera of *Azospirillum, Azotobacter, Flavobacterium, Pseudomonas, Xanthomonas,* and *Zoogloea*. Five bacterial strains isolated earlier from roots of kallar grass and rice were selected on the basis of their nitrogenase activity and ability to produce IAA for studying their effect on plant growth both individually and as mixed inoculum in pot and field experiments (Table 4). Among the selected strains *A. lipoferum* (N-4) had maximum nitrogenase activity as indicated by ARA values. *Pseudomonas* 96-51 did not show any nitrogenase activity but had maximum IAA production (35 μg/mL). Rice seedlings of variety NIAB-6 and BAS-370 were grown in pots

Table 3. Quantification of N$_2$-fixation by mixed inoculum of 3 diazotrophic strains in association with roots of kallar grass based on ^{15}N isotopic dilution

Treatment	Control	%Ndfa	Reference	Remark
Lab experiment under sterile conditions	Heat killed E. coli	65–80%	Malik et al., 1987	Plants grown in vermiculite in tubes
Pot experiment	Uninoculated fumigated	Roots 32% shoot 6%	Malik et al., 1987	Saline sodic soil
Field experiment	60 kg N/ha	26%	Malik et al., 1988	Saline sodic soil

Table 4. Characterization of bacterial strains included in the mixed inoculum

Bacterial strain	ARA (nmole C$_2$H$_4$/h/mg protein)	IAA production (μg/mL)
Azospirillum lipoferum N-4	686	6.3
Azospirillum brasilense Wb-3	215	16.1
Azoarcus K-1	290	10.5
Pseudomonas 96-51	0	35.7
Zoogloea Ky-1	544	5.1

containing vermiculite and N-free Hoagland nutrient solution. ^{15}N labelled ammonium sulphate of 10 atom % excess (34 mg/pot) was added as a tracer to quantify nitrogen fixation. The plants were harvested after six weeks and plant samples were analyzed for ^{15}N excess on a double inlet Mass Spectrometer (MAT GD 150).

Among the two rice varieties tested, the beneficial effect of bacterial inoculation was more prominent in variety BAS-370 as compared to NIAB-6 (Figure 1). In association with the rice variety BAS-370, *Azospirillum* strain N-4 fixed considerably higher nitrogen as compared to other strains, contributing about 66% of the total N in plants. For rice variety NIAB-6, *Azoarcus* strain K-1 proved to be more efficient, and maximum plant biomass was also recorded with the same strain K-1. Two bacterial strains *Azoarcus* K-1 and *Zoogloea* Ky-1 used in the present study were isolated from kallar grass growing in saline/waterlogged soils of Pakistan (Bilal and Malik, 1987).

Another experiment was performed in cemented micro plots of 1.5 m × 1.5 m size. Mixed inoculum comprising the five strains listed in Table 4 was used. The inoculum was applied to rice by dipping the roots of one-month old seedlings for 30 minutes before transplanting. Rice variety NIAB-6 was used. The field experiment was carried out in a randomized complete block design with four replicates for each

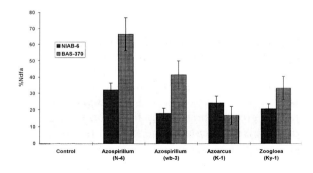

Figure 1. Quantification of N$_2$ fixation (%Ndfa) by inoculated bacteria in rice.

treatment. Inoculated treatments received 5 atom % ^{15}N fertilizer in the form of ammonium sulfate at 30 kg N/ha. The treatment designated as the non-fixing control received ^{15}N labelled fertilizer at 60 kg N/ha so that the indigenous nitrogen fixing bacteria could be inhibited (Malik et al., 1988). The results of biomass yield, total nitrogen and nitrogen fixed calculated on the basis of ^{15}N isotopic dilution are summarized in Table 5. In the inoculated treatment, relatively higher biomass and nitrogen yield as compared to uninoculated treatment were observed. There was also no significant increase in rice yield in case of treatment where 60 kg N/ha was applied (data not presented).

Table 5. Effect of PGPR* inoculation on rice biomass and nitrogen fixation, calculated on the basis of ^{15}N dilution data (microplot experiment)

Treatment	Straw+grain (kg/ha)	Total N (kg/ha)	Ndfa (%)	N fixed (kg/ha)
30kg^{15}N/ha Uninoculated	15541	151	2.3	3.5
30kg^{15}N/ha Inoculated	16202	157	28.9	45.5
LSD (0.05%)	775	12	20	4

*PGPR= A mixed inoculum of *Azospirillum lipoferum* strain N-4, *Azospirillum brasilense* strain Wb-3, *Azoarcus* strain K-1, *Zoogloea* strain Ky-1 and *Pseudomonas* strain 96-51.

Such a response may be due to the available N (NH$_4$ 11.5 mg/kg; NO$_3$ 13.4 mg/kg) present in the microplot soil. However, statistically significant increase in the amount of nitrogen fixed was observed in the case of the inoculated treatment. The estimates were based on the calculation of "A" values of fixing and non-fixing control (treatment having 60 kg N/ha application). In case of uninoculated treatment with 30 kg N/ha application, only 3.5 kg N/ha was biologically fixed which is due to the natural rhizospheric microflora. From this data, we can infer that in case of inoculated treatment, plants are deriving nearly 30% of its nitrogen requirement from the atmospheric N, thus resulting in conservation of soil fertility by reducing the depletion of soil nitrogen. At the time of harvest, an increase in the root biomass (data not presented) in the inoculated treatments was also observed indicating the possible role of growth hormone producing bacteria included in the mixed inoculum. The survival of the inoculated bacteria was also studied by using selective media containing antibiotics, morphological characters and reaction with fluorescent antibody stain (Mehnaz et al., 1996). Maximum nitrogenase activity as estimated by acetylene reduction assays of the rice roots in the inoculated treatments was observed after one month of transplantation. Although it decreased to one third before harvesting, it was significantly higher than the uninoculated treatment (Mehnaz et al., 1996).

Conclusions

The presence of associative nitrogen fixation in grasses is a random process. There has been no evidence of the presence of any specificity of the endosymbiont for the host. However, it can be safely said that there is a group of microorganisms which colonize the root preferentially and these may thus be called "rhizotrophs".

Although much work related to the inoculation of the diazotrophs has been done on the grass rhizosphere, our knowledge about root colonization is still not complete, as little is known about the mechanism of initial attachment of bacteria to the root surface. There are probably specific sites on the root surface where colonization occurs more frequently. Further studies are required to elucidate the colonization process.

References

Baldani J I, Baldani V L D, Seldin L and Döbereiner J 1986 Characterization of *Herbaspirillum seropedicae* gen. nov., sp. nov., a root associated N$_2$-fixing bacteria. Int. J. Syst. Bacteriol. 36, 86–93.

Bilal R 1988 Associative nitrogen fixation in plants growing in saline environments. Ph.D. Thesis, Punjab University Lahore, Pakistan. 248 p.

Bilal R and Malik K A 1987 Isolation and identification of a N$_2$-fixing Zoogloea-forming bacterium from kallar grass histoplane. J. Appl. Bacteriol. 62, 289–294.

Bilal R, Rasul G, Arshad M and Malik K A 1993 Attachment, colonization and proliferation of *Azospirillum brasilense* and *Enterobacter* spp. on root surface of grasses. World J. Microbiol. Biotechnol. 9, 63–69.

Bilal R, Rasul G, Qureshi J A and Malik K A 1990 Characterization of *Azospirillum* and related diazotrophs associated with roots of plants growing in saline soils. World J. Microbiol. Biotechnol. 6, 46–52.

Boddey R M and Victoria R L 1986 Estimation of biological nitrogen fixation associated with *Brachiaria* and *Paspalum notatum* using ^{15}N labelled organic matter and fertilizer. Plant Soil 90, 265–292.

Boddey R M, Alves B J R and Urquiaga S 1996 Evaluation of biological nitrogen fixation associated with non-legumes. *In* Nitrogen Fixation with Non-Legumes. Eds. K A Malik, M S Mirza and J K Ladha, Proceedings of 7th International Symposium on BNF with non-legumes. Oct. 16–21, 1996. Faisalabad, Pakistan. *(In press).*

Boddey R M, Chalk P M, Victoria R L and Matsui E 1983 The ^{15}N isotope dilution technique applied to the estimation of biological nitrogen fixation associated with *Paspalum notatum* cv. batata in the field. Soil Biol. Biochem. 15, 25–32.

Bric J M, Bostock R M and Silverstone S E 1991 Rapid in-situ assay for indoleacetic acid production by bacteria immobilized on a nitrocellulose membrane. Appl. Environ. Microbiol. 57, 535–538.

Cannon F C, Dixon R A and Postgate J R 1974 Chromosomal integration of *Klebsiella* nitrogen fixation gene in *Escherichia coli*. J. Gen. Microbiol. 80, 227–239.

Chalk P M 1991 The contribution of associative and symbiotic nitrogen fixation to the nitrogen nutrition of non-legumes. Plant Soil 132, 29–39.

Cochran W G 1950 Estimation of bacterial densities by means of the "Most Probable Number". Biometrics 6, 105–116.

Day J M, Harris D, Dart P J and Van Berkum P 1975 The Broadbalk experiment. An investigation of nitrogen gains from non-symbiotic fixation. *In* Nitrogen fixation by free-living Microorganisms. Ed. W D P Stewart Vol. 6, pp 71–84. International

Biological programme Series, Cambridge Univ. Press, Cambridge.

Döbereiner J, Marriel E and Nery M 1976 Ecological distribution of *Spirillum lipoferum* Beijerinck. Can. J. Microbiol. 22, 1464–1473.

Eskew D L, Eaglesham A R J and App A A 1981 Heterotrophic ^{15}N-fixation and distribution of newly fixed nitrogen in a rice-flooded soil system. Plant Physiol. 68, 48–52.

Fried M and Broeshart H 1975 An independent measurement of the amount of the nitrogen fixed by a legume crop. Plant Soil 43, 707–711.

Hassan U, Mirza M S, Mehnaz S, Rasul G and Malik K A 1996 Isolation and identification of diazotrophic bacteria from rice, wheat and kallar grass. *In* Nitrogen Fixation with Non-Legumes. Eds. K A Malik, M S Mirza and J K Ladha, Proceedings of 7th International Symposium on BNF with non-legumes. Oct. 16–21, 1996. Faisalabad, Pakistan. *(In press)*.

Hegazi N A, Amer H A and Monib M 1979 Enumeration of N_2-fixing Spirilla. Soil Biol. Biochem. 11, 437–438.

Hurek T, Reinhold-Hurek B, van Montagu M and Kellenberger E 1991 Infection of intact roots of kallar grass and rice seedlings by "*Azoarcus*". *In* Nitrogen Fixation. Proceedings of the 5th International Symposium on Nitrogen Fixation with Non-Legumes. Eds. M Polsinelli, R Materassi and M Vincenzeni. Kluwer Academic Publishers, Dordrecht.

Kreig N R and Holt J G (Eds.) 1984 The genus *Zoogloea*. *In* Shorter Bergey's Manual of Determinative Bacteriology, 8th Edition, pp 83–84. Williams & Wilkins Co., Baltimore, USA.

Ladha J K, So R B and Watanabe I 1987 Composition of *Azospirillum* species associated with wetland rice plant grown in different soils. Plant Soil 102, 127–129.

Lethbridge G and Davidson M S 1983 Root-associated nitrogen-fixing bacteria and their role in the nitrogen nutrition of wheat estimated by ^{15}N isotope dilution. Soil Biol. Biochem. 15, 365–374.

Lifshitz R, Kloepper J W, Kozlowski M, Simonson C, Carlson J, Tipping E M and Zaleska I 1987 Growth promotion of canola (rapeseed) seedlings by a strain of *Pseudomonas putida* under gnotobiotic conditions. Can. J. Microbiol. 33, 390–395.

Malik K A, Aslam Z and Naqvi S H M 1986 Kallar grass: a plant for saline land. Nuclear Institute for Agriculture and Biology (NIAB), Faisalabad, Pakistan.

Malik K A and Bilal R 1989 Survival and colonization of inoculated bacteria in kallar grass rhizosphere and quantification of N_2-fixation. *In* Nitrogen Fixation with Non-Legumes. Eds. F A Skinner, R M Boddey and I Fendrik. pp 301–310. Kluwer Academic Publishers, Dordrecht.

Malik K A, Bilal R, Azam F and Sajjad M I 1988 Quantification of N_2-fixation and survival of inoculated diazotrophs associated with roots of kallar grass. Plant Soil 108, 43–51.

Malik K A, Bilal R, Rasul G, Mahmood K and Sajjad M I 1991 N_2-fixation in plants growing in saline sodic soils and its relative quantification based on ^{15}N natural abundance. Plant Soil 137, 67–74.

Malik K A, Rasul G, Hassan U, Mehnaz S and Ashraf M 1993 Role of N_2-fixing and growth hormones producing bacteria in improving growth of wheat and rice. *In* Nitrogen Fixation with non-legumes. Eds. N A Hegazi, M Fayez and M Monib. pp 409–422. The Am. Univ. in Cairo Press, Cairo.

Malik K A and Zafar Y 1985 Quantification of root associated nitrogen fixation in kallar grass as estimated by ^{15}N isotope dilution. *In* Nitrogen and the Environment. Eds. K A Malik, S H M Naqvi, M I H Aleem. pp 161–171. NIAB, Faisalabad, Pakistan.

Malik K A, Zafar Y and Hussain A 1980 Nitrogenase activity in the rhizosphere of kallar grass (*Diplachne fusca* Linn Beauv). Biologia 26, 107–112.

Malik K A, Zafar Y and Hussain A 1982 Associative dinitrogen fixation in *Diplachne fusca* (kallar grass). *In* Biological Nitrogen Fixation Technology for Tropical Agriculture. Eds. P H Graham and S C Harris. pp 503–507. CIAT, Cali, Columbia.

Malik K A, Zafar Y, Bilal R and Azam F 1987 Use of ^{15}N isotope dilution for quantification of N_2-fixation associated with roots of kallar grass (*Leptochloa fusca* (L.) Biol. Fertil. Soils 4, 103–108.

Mehnaz S, Mirza M S, Hassan U and Malik K A 1996 Detection of inoculated plant growth promoting rhizobacteria in the rhizosphere of rice. *In* Nitrogen Fixation with Non-Legumes. Eds. K A Malik, M S Mirza and J K Ladha, Proceedings of 7th International Symposium on BNF with non-legumes. Oct. 16–21, 1996. Faisalabad, Pakistan *(In press)*.

Qureshi J A, Zafar Y and Malik K A 1988 *Klebsiella* sp. NIAB-1: A new diazotroph, associated with the roots of kallar grass from saline sodic soils. Plant Soil 110, 219–224.

Reinhold B, Hurek T and Fendrik I 1987 Cross-reaction of predominant nitrogen-fixing bacteria with enveloped, round bodies in the root interior of kallar grass. Appl. Environ. Microbiol. 53, 889–891.

Reinhold B, Hurek T, Fendrik I, Pot B, Gillis M, Kersters K, Thielemans S and De Ley J 1987a *Azospirillum halopraeferens* sp. Nov., a nitrogen-fixing organism associated with the roots of kallar grass (*Leptochloa fusca* (L.) Kunth). Int. J. Syst. Bacteriol. 37, 43–51.

Reinhold B, Hurek T, Niemann E G and Fendrik I 1986 Close association of *Azospirillum* and diazotrophic rods with different root zones of kallar grass. Appl. Environ. Microbiol. 52, 520–526.

Reinhold-Hurek B, Hurek T, Gillis M, Hoste B, Vancanneyt M, Kersters K and De Ley J 1993 *Azoarcus* gen. nov., nitrogen fixing proteobacteria associated with roots of kallar grass (*Leptochloa fusca* (L.) Kunth), and description of two species, *Azoarcus indigens* sp. nov., and *Azoarcus communis* sp. nov. Int. J. Syst. Bacteriol. 43, 574–584.

Rennie R J 1980 Isotope dilution as a measure of dinitrogen fixation by *Azospirillum brasilense* associated with maize. Can. J. Bot. 58, 21–24.

Rennie R J 1981 A single medium for the isolation of acetylene reducing (dinitrogen fixing) bacteria from soil. Can. J. Microbiol. 27, 8–14.

Rennie R J and Larson R I 1979 Dinitrogen fixation with disomic chromosome substitution lines of spring wheat. Can. J. Bot. 57, 2771–2775.

Rennie R J, de Freitas J R, Ruschel A P and Vose P B 1983 ^{15}N dilution to quantify dinitrogen (N_2)-fixation associated with Canadian and Brazilian wheat. Can. J. Bot. 61, 1667–1671.

Roger P A and Ladha J K 1992 Biological N_2 fixation in wetland rice fields: estimation and contribution to nitrogen balance. Plant Soil 141, 41–55.

Ruschel A, Henis Y and Salati E 1975 Nitrogen ^{15}N tracing of N_2-fixation with soil grown sugarcane seedlings. Soil Biol. Biochem. 5, 83–89.

Sandhu G R and Malik K A 1975 Plant succession – A key to utilization of saline soils. Nucleus 12, 35–38.

Shrestha R K and Ladha J K 1996 Genotypic variation in promotion of rice dinitrogen fixation as determined by nitrogen-15 dilution. Soil Sci. Soc. Am. J. 60, 1815–1821.

Sundaram S, Arunkumari A and Klucas R V 1988 Characterization of azospirilla isolated from seeds and roots of turf grass. Can. J. Microbiol. 34, 212–217.

44

Tarrand J J, Krieg N R and Döbereiner J 1978 A taxonomic study of *Spirillum lipoferum* group with descriptions of a new genus, *Azospirillum* gen. nov. and two species, *Azospirillum lipoferum* (Beijerinck) comb. nov. and *Azospirillum brasilense* sp. nov. Can. J. Microbiol. 24, 967–980.

Tien T M, Gaskins M H and Hubbell D H 1979 Plant growth substances produced by *Azospirillum brasilense* and their effect on the growth of pearl millet (*Pennisetum americanum* L.) Appl. Environ. Microbiol. 37, 1016–1024.

Van Berkum P 1980 Evaluation of acetylene reduction by excised roots for the determination of nitrogen fixation in grasses. Soil Biol. Biochem. 12, 141–145.

Ventura W and Watanabe I 1983 ^{15}N dilution of accessing the contribution of nitrogen fixation to rice plant. Soil Sci. Plant Nutr. 29, 123–131.

Witty J F 1979 Acetylene reduction assay can overestimate nitrogen fixation in soil. Soil Biol. Biochem. 11, 209–210.

Zafar Y, Ashraf M and Malik K A 1986 Nitrogen fixation associated with the roots of kallar grass (*Leptochloa fusca* (L.) Kunth). Plant Soil 90, 93–106.

Zafar Y, Malik K A and Niemann E G 1987 Studies on N_2-fixing bacteria associated with salt tolerant grass *Leptochloa fusca* (L.) Kunth. Mircen J. Appl. Microbiol. 3, 45–56.

Zimmer W, Roeben K and Boothe H (1988) An alternative explanation for plant growth promotion by bacteria of the genus *Azospirillum*. Planta 176, 333–342.

Guest editors: J K Ladha, F J de Bruijn and K A Malik

Plant and Soil **194**: 45–55, 1997.
© 1997 *Kluwer Academic Publishers. Printed in the Netherlands.*

Occurrence, physiological and molecular analysis of endophytic diazotrophic bacteria in gramineous energy plants

G. Kirchhof[1], V.M. Reis[2], J.I. Baldani[2], B. Eckert[1], J. Döbereiner[2] and A. Hartmann[1,3]
[1]*GSF-National Research Center for Environment and Health, Institute of Soil Ecology, Ingolstädter Landstraße 1, D-85764 Neuherberg, Germany and* [2]*EMBRAPA-Centro National de Pesquisa de Agrobiologia (CNPAB-EMBRAPA), Seropédica CEP 23851-970 Rio de Janeiro, Brazil.* [3]*Corresponding author**

Key words: Azospirillum, diversity, endophytic diazotrophic bacteria, *Herbaspirillum*, phenotypic and molecular taxonomy, renewable energy resources

Abstract

Endophytic diazotrophic bacteria could be isolated from the energy plants *Pennisetum purpureum, Miscanthus sinensis, Miscanthus sacchariflorus* and *Spartina pectinata* using semisolid nitrogen free media. Higher levels of diazotrophic bacteria were found if no nitrogen fertilizer was applied. The bacteria were characterized on the basis of typical morphology, physiological tests, and the use of phylogenetic oligonucleotide probes. They belong partially to the species *Azospirillum lipoferum* and *Herbaspirillum seropedicae* while others supposedly represent a new species of *Herbaspirillum*. Using PCR-fingerprinting techniques a limited genetic diversity of these isolates was found which may indicate an adaptation to the specific conditions of the interior of these plants.

Abbreviations: BNF – Biological Nitrogen Fixation; PCR – Polymerase Chain Reaction

Introduction

Many diazotrophic bacteria were isolated over the years from rhizosphere soil or the rhizoplane of a big variety of non leguminous plants (Döbereiner, 1992). However, the presence of a diazotrophic bacterium in any habitate does not necessarily mean that the bacteria are efficiently fixing nitrogen in association with plants and that the plants can finally obtain significant contributions from biologically fixed nitrogen. On the root surface, N_2-fixing bacteria are distant from the main source of assimilates of the plants and are in heavy competition with other microorganisms for root exudates as nutritional source. Bacteria which manage to enter and to colonize the inside of roots or even are able to spread to aerial parts of the plants are not suffering from these disadvantages. It is increasingly recognized that the internal colonization of plants by various endophytic bacteria is not necessarily pathogenic, but in contrast may be beneficial to the plant (Kloepper and Beauchamp, 1992). There are different ways of positive bacteria/plant

interactions. Endophytic plant-growth-promoting bacteria, which interact with pathogens or stimulate systemic resistance of the plants were reported (McInroy and Kloepper, 1994). In addition, a variety of diazotrophic bacteria have been found and quantified in different plants (Döbereiner et al., 1995) which may contribute substantially to the nitrogen nutrition of the plant. In several sugarcane cultivars, more than 60% of the nitrogen uptake (about 150 kg N ha^{-1} year^{-1}) were obtained from biological nitrogen fixation (BNF), as was demonstrated with ^{15}N isotope and N balance studies (Urquiaga et al., 1992).

A number of different species of diazotrophic bacteria were demonstrated to inhabit the interior of sugarcane. A newly discovered diazotrophic bacterium, named *Acetobacter diazotrophicus,* occurs in large numbers (up to 10^6 per gram fresh weight) in the roots and stems of sugarcane (Gillis et al., 1989). Electron microscopic studies of *A. diazotrophicus* after infection of sugarcane tissues clearly provided detailed information about the localization of *A. diazotrophicus* cells within sugarcane tissues of micropropagated plantlets (James et al., 1994). *Herbaspirillum sero-*

---

* FAX No: +498931873376. E-mail: hartmann@gsf.de

pedicae, a member of the β-Proteobacteria, was discovered in high numbers in leaves, stems and roots of sugarcane (Olivares et al., 1996). This bacterium seems to survive poorly in bulk soil (Baldani et al., 1992). Also *Herbaspirillum rubrisubalbicans*, formerly named *Pseudomonas rubrisubalbicans*, a close relative to *H. seropedicae* (Baldani et al., 1996) colonizes sugarcane endophytically and causes mottled stripe disease in some sugarcane varieties. *H. rubrisubalbicans*, too, is able to fix nitrogen (Döbereiner et al., 1995). Another new group of diazotrophic bacteria, provisionally named *Burkholderia brasilensis*, the formerly termed isolate E-bacteria, was frequently found in sugarcane tissue (Baldani, 1996; Hartmann et al., 1995). In addition, *Herbaspirillum seropedicae* and *Burkholderia* sp. were detected in maize plants (Olivares et al., 1993; Pereira, 1995; Kirchhof, unpublished).

Azospirillum spp. were repeatedly isolated from the inside of roots and shoots of many plants (Döbereiner et al., 1995). Using strain-specific monoclonal antibodies against *Azospirillum brasilense* strains Sp7 and Sp245 respectively, it was clearly demonstrated that the isolate Sp245 could efficiently enter and colonize roots and the vascular system after inoculation of soil-grown wheat plants (Schloter et al., 1994). This confirms a series of earlier studies, which claim the colonization of the root interior by certain *Azospirillum* strains and some other diazotrophs (Patriquin and Döbereiner, 1978). Specificities of wheat-*Azospirillum brasilense* interactions and a different degree of attachment and infections and their effects on plant growth and nitrate reduction with certain strains were reported by Jain and Patriquin (1984) and Ferreira et al. (1987). Recently, evidence was provided that specific components of the exopolysaccharide composition of *A. brasilense* Sp7 determine the degree of entry into roots (Katupipiya et al., 1995).

Azoarcus spp. were originally demonstrated as endophytic diazotrophs in Kallar grass (Reinhold-Hurek et al., 1993a). Azoarcus strain BH72 is able to invade rice roots and to colonize the cortex cells of roots as well as the stem bases and the shoot (Hurek et al., 1994). The systemic spreading of *Azoarcus* in the plant is facilitated by cellulolytic activities (Reinhold-Hurek et al., 1993b). The involvement of other diazotrophic bacteria, like *Herbaspirillum* sp. and *Burkholderia* spp. in rice colonization was recently summarized by Boddey et al. (1995).

The classification and identification of bacterial isolates recently became strongly supported from the application of molecular biological techniques. Much progress has been made in specific identification of bacteria with rRNA-targeted oligonucleotide probes. Such probes can be directed to highly variable sequence stretches of either the 23S-rDNA or the 16S-rDNA. A variety of phylogenetic oligonucleotide probes was developed for *Azospirillum* spp. (Kirchhof and Hartmann, 1992), *Herbaspirillum* spp. (Baldani et al., 1996), *Azoarcus* (Hurek et al., 1993), *Acetobacter diazotrophicus* and *Burkholderia* sp., as was summarized recently by Kirchhof et al. (1996). Probes identifying higher taxa, e.g. subgroups of the Proteobacteria (Manz et al., 1992), can be of great help to preliminary classify new isolates for which no specific probes are available yet. Some of these oligonucleotide probes were used for in situ staining of single bacterial cells after fluorescence labelling using a confocal laser scanning microscope (Aßmus et al., 1995; Aßmus et al., 1996). The application of monoclonal or monospecific antibodies can further focus the identification down to strain level (Schloter et al., 1995).

The identification of bacteria at the strain level is of major importance, since for an efficient, N_2-fixing bacterial-plant association a certain specificity needs to be involved both from the bacterial and the plant side. Several PCR-fingerprinting techniques have been developed to identify bacteria at the strain level. These methods are applied with cultured bacterial cells avoiding DNA-extraction and combining convenient analysis with universal applicability and the potential of automation (Versalovich et al., 1994). In our study of diazotrophic isolates, recently developed oligonucleotide primers derived from eukaryotic consensus LINE-sequences (LINEs = Long INterspersed Elements) were applied for PCR-fingerprinting (Smida et al., 1996). The electrophoretically separated band patterns were highly reproducible and strain-specific.

In the case of potential energy crop plants such as *Miscanthus sinensis*, *Miscanthus sacchariflorus*, *Spartina pectinata*, and *Pennisetum purpureum*, the energy balance is of high interest. Different agricultural practices, e.g. application of chemical nitrogen fertilizer, raise the costs to produce the crop plants and reduce the energy balance of growth which must be clearly positive to make energy crop plants economically and energetically viable. Therefore, it is most desirable to investigate, to which extent diazotrophic bacteria are living endophytically associated with energy crop plants. In this communication, first results of the isolation and taxonomic characterization of diazotrophic bacteria from gramineous energy crops with physiological and molecular tools are presented.

Materials and methods

Methods of isolation

Isolations were performed in nitrogen-free semisolid media with different carbon sources and pH values as described in Döbereiner (1995). Semisolid nitrogen free media offer the possibility for diazotrophic bacteria to find the right niche within an oxygen gradient exhibiting optimal conditions for nitrogen fixation. There they form a growth pellicle. Different media favour the enrichment of various bacterial species.

Spartina pectinata and *Miscanthus sinensis* cv. "Giganteus", "Goliath", and *Miscanthus sacchariflorus*, were grown with 0 or 150 kg ha^{-1} dosages of nitrogen fertilizer at LBP Freising, Germany. *Pennisetum purpureum* cv. "Cameroon", "Caipim Cana D'Africa", "Gramafante", "Guaçu", "Merker", "Merker x 239", "Mineiro", "Mott", "Piracicaba", "Roxo", "s/pelo", and "Taiwan" were cultivated at EMBRAPA/CNPGL, Minas Gerais, Brazil. Specimen of washed roots, stems, and leaves were macerated, and 0.1 mL of serial dilutions in 4% (commercial) sugar solution up to 10^{-6} were inoculated into vials containing NFb, JNFb or LGI semisolid media. The composition of **NFb**-medium per liter is: malic acid (5.0 g); K_2HPO_4 (0.5 g); $MgSO_4 \cdot 7H_2O$ (0.2 g); NaCl (0.1 g); $CaCl_2$ (0.02 g); bromthymol blue 0.5% in KOH 0.2 N (2 mL); vitamin solution (1 mL); micronutrient solution (2 mL); 1.64% FeEDTA solution (4 mL); KOH (4.5 g). The vitamin solution contained in 100 mL: biotin (10 mg), pyridoxol-HCl (20 mg) and the micronutrient solution consisted of (quantities per liter): $CuSO_4$ (0.4 g); $ZnSO_4 \cdot 7H_2O$ (0.12 g); H_2BO_3 (1.4 g); $Na_2MoO_4 \cdot 2H_2O$ (1.0 g) $MnSO_4 \cdot H_2O$ (1.5 g). The pH was adjusted to 6.8 and 1.9 g L^{-1} of agar was added. The medium called **JNFb** consisted of the same components as NFb-medium except following different ingredients and amounts per liter: K_2HPO_4 (0.6 g); KH_2PO_4 (1.8 g) and the pH was adjusted to 5.8. A third medium, called **LGI**, was composed of (quantities per liter): sucrose (5.0 g); K_2HPO_4 (0.2 g); KH_2PO_4 (0.6 g); $mgSO_4 \cdot 7H_2O$ (0.2 g); $CaCl_2$ (0.02 g); $Na_2MoO_4 \cdot 2H_2O$ (0.002 g); $FeCl_3$ (0.01 g); bromthymol blue 0.5% in KOH 0.2 N (5 mL). The pH was adjusted to 6.0 and 1.9 g L^{-1} of agar was added. After four to six days of incubation at 30 °C the population size was estimated by the MPN technique and pellicle forming bacteria were subjected to further purification steps by streaking on agar plates with additional 20 mg L^{-1} of yeast extract and again colonies transferred to cultivation in semisolid N-free media.

Phenotypic tests

Morphological characteristics were investigated microscopically. Physiological properties were screened with the API20 NE galleries (bioMerieux SA, Lyon, France) according to the manufacturers instructions. The pellicle forming capability (microaerophilic, in occasion dinitrogen fixation dependent growth) with different carbon sources was screened with selected strains. For this purpose, malate in semisolid JNFb medium was replaced by N-acetylglucosamine, arabinose, benzoate, fructose, glucose, inositol, lactate, manitol, L-rhamnose, and tartrate respectively. Growth with *m*-erythritol as carbon source and addition of NH_4Cl served as physiological differentiation marker between *H. seropedicae* (-) and *H. rubrisubalbicans* (+) (Baldani et al., 1996). Three replicates of the media were inoculated with 10 μL of an overnight culture in rich medium and incubated at 30°C over a period of 3-4 days.

Phylogenetic identification

For phylogenetic identification the extracted, immobilized bulk nucleic acids of the isolates were subjected to hybridization experiments with ^{32}P-labeled oligonucleotide probes specific for the α- and β-subgroups of Proteobacteria (Manz et al., 1992) as well as species-specific probes for diazotrophic bacteria (Kirchhof et al., 1996) following the methodology described in Kirchhof and Hartmann (1992). Nucleic acids of isolates, which did not hybridize with one of the species-specific probes, were subjected to further 23S rDNA sequence analysis of stretches within domain III known to be highly variable in base composition (Ludwig et al., 1994). Direct sequencing of amplified 23S rDNA fragments was performed. The resulting sequences were aligned and analysed with homologous sequences from other bacteria using the EMBL/DKFZ data base entries or the ARB sequence analysis package (Strunk and Ludwig, 1997). Referring to the obtained sequence data, a new oligonucleotide probe was designed and applied in dot-blot hybridization experiments.

PCR fingerprinting

The genomic diversity of the isolates was examined by PCR fingerprinting using recently developed primers derived from repetitive mammalian sequences (LINEs) (Smida et al., 1996). One of the primers was slightly modified (GRK: 5'-GAG TTT GGC AAA GAC CC) and used in an arbitrary PCR, producing reproducible, probably strain-specific patterns. 1 mL of an overnight culture of the diazotrophic isolates was harvested by centrifugation, resuspended in 1 mL of H_2O, boiled for 10 minutes and centrifuged again (10 min, 13.000 rpm). 2 and 10 μL of the supernatant were used in the PCR reaction as template. Each 50 μL PCR reaction contained 2 μm of the oligonucleotide primer, 1.25 mm of each of four dNTPs (dATP, dCTP, dGTP, dTTP) and 1.5 mm $mgCl_2$. After an initial denaturation step of 95°C, 10 min, 2 units of Taq-polymerase were added and the following temperature profile was carried out 5 times: 95°C/1 min, 37°C/1 min, 72°C/1.5 min, followed by another 25 cycles with an increased annealing temperature of 52°C and a final extension step of 10 min at 72°C. Amplified products were separated by electrophoresis on 5% non-denaturing polyacrylamide gels at constant temperature of 20°C and visualized by silver staining.

Results

Occurrence of diazotrophic bacteria

The isolation experiments of diazotrophic bacteria from *Pennisetum purpureum* cultivars were carried out in Brazil during April/May. Population sizes up to 10^5-10^6 per gram fresh weight were found. Different *Pennisetum purpureum* cultivars exhibited 10^5-10^6 dinitrogen fixing bacteria per gram fresh weight in the roots and 10^4-10^5 cells in aerial parts (Silva et al., 1995). The first isolation attempts of diazotrophic bacteria from *Miscanthus sinensis, Miscanthus sacchariflorus* and *Spartina pectinata* grown in Germany yielded rather low bacteria numbers. For all plant tissues, the number was about 10^2-10^3 per gram fresh weight. The numbers of diazotrophic bacteria from the plants grown with nitrogen fertilizer dosages were always below 10^2 per gram fresh weight or not detectable (Table 1).

An entire number of 34 bacterial isolates from different *Pennisetum purpureum* genotypes, 17 isolates from *Spartina pectinata* and 19 isolates from *Miscan-*

thus sinensis and *Miscanthus sacchariflorus,* also different genotypes, was studied further. Table 2 summarizes the origin and morphology of the isolates.

Phenotypic tests

Applying the API 20 galleries, coherent patterns of the use of carbon sources were achieved with the *Herbaspirillum*-like isolates from *Miscanthus sinensis, Miscanthus sacchariflorus* and *Pennisetum purpureum* and the type strain *H. seropedicae* Z67. All strains were able to assimilate N-actylglucosamine, arabinose, caprate, citrate, glucose, gluconate, malate, mannose, mannitol, and phenylacetate, whereas they were unable to use maltose and adipate. All screened isolates as well as the type strain were able to reduce nitrate and possess a cytochrome oxidase. Part of the *Herbaspirillum*-like isolates showed arginine dihydrolase, urease and β-glucosidase activity. The test for pellicle forming capacity on different carbon sources resulted in a unique utility pattern (Table 3): all tested isolates were able to form a pellicle with arabinose, N-acetylglucosamine, fructose, glucose, malate, manitol, and tartrate (except strain 84B) and lacked this property if inositol or L-rhamnose was used as carbon source. In this test, differences compared with the type strains of *H. seropedicae* and *H. rubrisubalbicans* could be observed. In contrast to the fibre plant isolates and *H. rubrisubalbicans* type strain (M4), strain Z67 exhibited pellicular growth with inositol and L-rhamnose. Strain M4 lacked this property if N-acetylglucosamine was offered as carbon source (see Table 3). Growth with *m*-erythritol and N addition was found only with *H. rubrisubalbicans*.

Sequence analysis

For the purpose of probe design 23S rDNA sequence analysis of stretches within domain III known to be highly variable in base composition (Ludwig et al., 1994) were performed with several *Herbaspirillum*-like isolates from *Pennisetum* and *Miscanthus*. The resulting sequences were aligned with homologous sequence parts of other bacteria. In Figure 1 the aligned partial 23S rRNA sequences of the already known *Herbaspirillum* spp. and the *Herbaspirillum*-like isolates from energy crops are shown. A suitable site for an oligonucleotide probe is marked in bold and underlined letters. Isolates from one plant species did not show any sequence differences. The sequence stretches of *Herbaspirillum*-like isolates from *Pennisetum* and

Table 1. MPN-counts per g fresh weight washed plant material of diazotrophic bacteria from *Miscanthus sinensis, Miscanthus sacchariflorus, Spartina pectinata,* and *Pennisetum purpureum*

	Pennisetum purpureum	*Miscanthus sinensis, M. sacchariflorus, Spartina pectinata*
no N-fertilizer application	10^5-10^6	10^2-10^3
150 kg ha^{-1} N-fertilizer application	not determined	$< 10^2$

Table 2. Origin and morphology of the isolates

Isolate	Plant	Tissue	Cell morphology
GSF 1, 2, 4, 5, 6, 7, 8, 11, 12, 13, 14, 15, 16, 17	*Spartina pectinata*	washed roots	*Azospirillum lipoferum*-like
GSF 9, 10	*S. pectinata*	washed roots	*Azospirillum*-like
GSF 3	*S. pectinata*	washed roots	*Herbaspirillum*-like
GSF 36	*S. pectinata*	washed stems	*Herbaspirillum*-like
GSF 29	*Miscanthus sinensis* cv. Giganteus	washed stems	*Azospirillum lipoferum*-like
GSF 18, 31	*M. sinensis* cv. Giganteus	washed stems	*Azospirillum*-like
GSF 20	*M. sinensis* cv. Giganteus	washed stems	*Herbaspirillum*-like
GSF 26	*M. sacchariflorus*	washed roots	*Azospirillum lipoferum*-like
GSF 21	*M. sacchariflorus*	washed roots	*Azospirillum*-like
GSF 28	*M. sacchariflorus*	washed roots	*Herbaspirillum*-like
GSF 22, 24, 27, 35	*M. sacchariflorus*	washed stems	*Herbaspirillum*-like
GSF 19, 32, 33	*M. sacchariflorus*	washed leaves	*Azospirillum lipoferum*-like
GSF 23, 25, 30, 34	*M. sacchariflorus*	washed leaves	*Herbaspirillum*-like
80 B	*Pennisetum purpureum* cv. Cameroon	root	*Herbaspirillum*-like
70 B, 74B, 96B	*P. purpureum* cv. Cana Africa	root	*Herbaspirillum*-like
37	*P. purpureum* cv.Gramafante	aerial part (stem mostly)	*Herbaspirillum*-like
61 B	*P. purpureum* cv. Guaçu	aerial part (stem mostly)	*Herbaspirillum*-like
114 B	*P. purpureum* cv. Guaçu	root	*Herbaspirillum*-like
72 B	*P. purpureum* cv. Merker	aerial part (stem mostly)	*Herbaspirillum*-like
16 B	*P. purpureum* cv. Merker	root	*Herbaspirillum*-like
65 H, 73B, 90B, 101H, 121B	*P. purpureum* cv. Merker x 239	root	*Herbaspirillum*-like
11 B	*P. purpureum* cv. Mineiro	aerial part (stem mostly)	*Herbaspirillum*-like
75 B, 118B	*P. purpureum* cv. Minero	root	*Herbaspirillum*-like
112 B, 115B	*P. purpureum* cv. Mott	aerial part (stem mostly)	*Herbaspirillum*-like
106 B	*P. purpureum* cv. Mott	root	*Herbaspirillum*-like
60 B, 69B, 122B	*P. purpureum* cv. Piracicaba	aerial part (stem mostly)	*Herbaspirillum*- like
78 B, 91B	*P. purpureum* cv. Piracicaba	root	*Herbaspirillum*-like
36 B, 41B	*P. purpureum* cv. Roxo	aerial part (stem mostly)	*Herbaspirillum*-like
81 B	*P. purpureum* cv. Roxo	root	*Herbaspirillum*-like
48 H	*P. purpureum* cv. s/pelo	aerial part (stem mostly)	*Herbaspirillum*-like
95 B	*P. purpureum* cv. s/pelo	root	*Herbaspirillum*-like
84 B, 103H, 113B	*P. purpureum* cv. Taiwan	aerial part (stem mostly)	*Herbaspirillum*- like
67 B, 117B	*P. purpureum* cv. Taiwan	root	*Herbaspirillum*-like

Miscanthus differ only at one single base position (see Figure; 1 base no. 124)

Dot blot hybridization

Miscanthus sinensis, Miscanthus sacchariflorus and *Spartina pectinata*: 24 isolates showed positive hybridization signals with an oligonucleotide probe

Table 3. Pellicle formation in semi-solid media by various *Herbaspirillum* strains with different carbon sources with and without nitrogen source added

Carbon source	M4[a]	Z67[b]	GSF3	GSF30	75B	84B
malate	+++[c]	+++[c]	+++[c]	+++[c]	+++[c]	+++[c]
glucose	+++	+++	++	++	++	++
N-acetyglucosamine	-	+++[c]	+++[c]	+++[c]	+++[c]	+++[c]
arabinose	+++	+++	+++	+++	+++	+++
mannitol	+++	+++	+++	+++	+++	+++
fructose	+	+	+	+	+	+
fructose + N	+++	+++	+++	+++	+++	+++
inositol	-	++	-	-	-	-
tartrate	+	+	++[c]	++[c]	+	-
L-rhamnose	-	+++	-	-	-	-
m-erythritol + N	+++	-	-	-	-	-

[a] Type strain of *Herbaspirillum rubrisubalbicans*
[b] Type strain of *Herbaspirillum seropedicae*
+++ = thick pellicle coming to the medium surface
++ = pellicle formation some mm under the medium surface;   alkalization of the medium
+ = very fine pellicle
- = no growth; + N = addition of NH$_4$Cl (1g/l)

specific for the genus *Azospirillum* (not shown). 21 of these strains could be assigned to the species *Azospirillum lipoferum* (see Figure 2A), the other strains did not deliver a signal with any *Azospirillum* species-specific probe (not shown). The remaining 12 isolates belong to the β-subgroup of Proteobacteria as was shown using the probe specific for the β-subgroup of Proteobacteria (not shown). With the newly developed 23S rRNA-targeting oligonucleotide probe (named **beta 20**: 5'-GAT ACA AGA ACC GGG AC-3') all of them showed positive signals (see Figure 2B), while the bulk nucleic acids did not hybridize with one of the already developed probes specific for *H. seropedicae* or *H. rubrisubalbicans* (Baldani et al., 1996). The idiosyncrasy of the beta 20 probe sequence was checked with the data base of accessible rDNA sequences (GenBank/DKFZ, Heidelberg, F.R.G.). Overmore, the specificity of the designed oligonucleotide was tested further in hybridization experiments with membrane bound bulk nucleic acid of 112 different ecologically relevant bacteria and found to be able to differentiate the new *Herbaspirillum*-like strains from members of the genus *Herbaspirillum* and other diazotrophic and nondiazotrophic plant-associated bacteria (data not shown).

Pennisetum purpureum: The immobilized nucleic acid of all analyzed isolates hybridized with the probe specific for the β-subgroup of the Proteobacteria (not shown). Four isolates showed positive hybridization signals with the oligonucleotide probe specific for *Herbaspirillum seropedicae* (Figure 2C). All others hybridized with the newly developed probe "beta 20" obtained from *Miscanthus* isolates mentioned above (Figure 2D).

PCR fingerprinting

The isolates were further investigated by PCR fingerprinting using oligonucleotides directed to sequences which are derived from LINEs (Long Interspersed Elements) and are supposedly conserved in all cells. Figure 3 shows the pattern of the PCR products separated by polyacrylamide gel electrophoresis of a representative selection of the isolates. It can be observed that all isolates exhibit similar banding pattern with only few bands varying in their molecular weight, even if they are derived from different plant genotypes. This can be found for *Pennisetum purpureum* isolates (lanes 1-3: *H. seropedicae* isolates; lanes: 4-19: *Herbaspirillum*-like isolates) as well as for isolates of *Miscanthus sinensis* and *Miscanthus sacchariflorus* (lanes 21-28: *Herbaspirillum*-like isolates). Equivalent observations could be gained with the *Azospirillum lipoferum* isolates of *Miscanthus* and *Spartina* (data not shown).

```
            10        20        30        40        50        60
            .         .         .         .         .         .
H.ser       GTCGTTAGATGCGATGGGGGGACGGATCGCGGAAGGTTGTCCGGGTGTTGGAAGTCCCGG
H.rub       GTCGTTAGATGCGATGGGGGGACGGATCGCGGAAGGTTGTCCGACTGTTGGAATAGTCGG
iso. Miscanthus  GTCGTTAGATGCGATGGGGGGACGGATCGCGGAAGGTTGTCCGGGTGTTGGAAGTCCCGG
iso. Pennisetum  GTCGTTAGATGCGATGGGGGGACGGATCGCGGAAGGTTGTCCGGGTGTTGGAAGTCCCGG
            *************************************************  *******   ***

            70        80        90        100       120       130
            .         .         .         .         .         .
H.ser       TTTTTGCATCGAAGAAGGCTGTTAGGCAAATCCGGCAGCGTAATTCAAGGGTGTGAGACG
H.rub       TTTTTGCATCGAAGAAGGCTGTTAGGCAAATCCGGCAGCGTAATTCAAGGGTGTGAGACG
iso. Miscanthus  TTCTTGTATCGAAGAAGGCTGTTAGGCAAATCCGGCAGCGTAATTCAAGGGTATGAGACG
iso. Pennisetum  TTCTTGTATCGAAGAAGGCTGTTAGGCAAATCCGGCAGCGTAATTCAAGGGTACGAGACG
            **  ***  ************************************************  ******

            140       150       160       170       180       190
            .         .         .         .         .         .
H.ser       AGCGAACTTGTTCGCGAAGCAATCGGAAGTGGTTCCAAGAAAAGCCTCTAAGCTTCAGTC
H.rub       AGCGAACTTGTTCGCGAAGCAATCGGAAGTGGTTCCAAGAAAAGCCTCTAAGCTTCAGTC
iso. Miscanthus  AGCGAACTTGTTCGCGAAGCAATCGGAAGTGGTTCCAAGAAAAGCCTCTAAGCTTCAGTC
iso. Pennisetum  AGCGAACTTGTTCGCGAAGCAATCGGAAGTGGTTCCAAGAAAAGCCTCTAAGCTTCAGTC
            ************************************************************


H.ser       TAAC
H.rub       TAAC
iso. Miscanthus  TAAC
iso. Pennisetum  TAAC
            ****
```

Abbreviations: H.ser: *Herbaspirillum seropedicae* Z67; H.rub: *Herbaspirillum rubrisubalbicans* M4; iso Miscanthus: *Herbaspirillum*-like isolate GSF 20 and GSF 30 of *Miscanthus sinensis*; iso Pennisetum: *Herbaspirillum*-like isolate 60B, 75B, and 84B of *Pennisetum purpureum*;

Figure 1. Multiple alignment of 23S rDNA partial sequences of *Herbaspirillum* spp. and *Herbaspirillum*-like isolates from energy plants.

Discussion

The scope of this communication is to demonstrate the discovery of diazotrophic bacteria from *Miscanthus sinensis, Miscanthus sacchariflorus, Spartina pectinata* and *Pennisetum purpureum*. From all these potential energy crop plants diazotrophic bacteria could be isolated, especially if the plants had been grown without chemical nitrogen fertilizer. Therefore, these C4-plants harbour bacterial populations which have the potential to support the nitrogen nutrition of the plants as already described for other gramineous crops (e.g. sugarcane; Döbereiner et al., 1995). However, the population size in *Miscanthus sinensis, Miscanthus sacchariflorus* and *Spartina pectinata* was much too low for substantial nutritive contributions. This may partially be explained by the fact, that the experiments

with *Miscanthus sinensis, Miscanthus sacchariflorus* and *Spartina pectinata* were carried out in late summer and autumn (August/October) when the vegetative activity of the plant was low in Germany. During the shooting stage size and structure of the bacterial community may be different. Furthermore, it remains to be examined, whether the associations between these plants and diazotrophic bacteria are limited when nitrogen fertilizer is supplemented in excess (Fuentes-Ramirez et al., 1993), as it has been shown in the case of sugarcane (120 kg ha^{-1}).

The nitrogen fixing bacterial isolates from *Pennisetum, Spartina* and *Miscanthus* could be assigned to the genera *Herbaspirillum* and *Azospirillum*. Most of the immobilized nucleic acid of *Azospirillum* produced a positive signal with the oligonucleotide probe specific for *Azospirillum lipoferum*.

2A: probe "*Azospirillum lipoferum*"

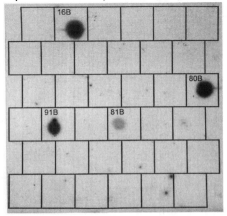

2B: probe "beta 20"

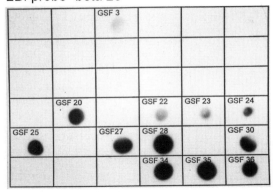

2C: probe "*Herbaspirillum seropedicae*"

2D: probe "beta 20"

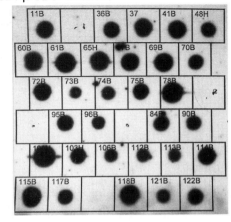

Figure 2. Dot blot hybridization of bulk nucleic acids of diazotrophic bacteria isolated from energy crops. **A**: Hybridization signals of *Miscanthus sinensis, Miscanthus sacchariflorus* and *Spartina pectinata* isolates with the oligonucleotide probe AL specific for *Azospirillum lipoferum*. **B**: Hybridization result of the identical (2A) membrane with the oligonucleotide probe "beta 20". **C**: hybridization signals of *Pennisetum purpureum* isolates with the oligonucleotide probe HS specific for *Herbaspirillum seropedicae*. **D**: Hybridization result of a similar membrane with the oligonucleotide probe "beta 20".

Several isolates could be grouped to the genus *Herbaspirillum* but differ from the already known species *Herbaspirillum seropedicae* or *H. rubrisubalbicans*. The auxanographic analyses of these isolates indicated a strong relationship to the known *Herbaspirillum* species but the combination of N-acetylglucosamine assimilation and the lack of inositol and L-rhamnose assimilation capacity has not been found with anyone of the known strains of *H. seropedicae* or *H. rubrisubalbicans* (Baldani et al., 1996) so far. In addition, the oligonucleotide probes specific for the two known herbaspirilla (Hartmann et al., 1995) fail to produce positive hybridization results. Furthermore, the sequence of a highly variable 23S rDNA stretch within domain III (helix 52b-54b; nummeration

according to Ludwig et al., 1994) showed close similarity to the known *Herbaspirillum* spp. but exhibited significant variations in base composition (see Figure1). Exploiting these data, an oligonucleotide probe specific for the newly discovered *Herbaspirillum*-like organism could be designed and its successful use to identify diazotrophic bacterial isolates was demonstrated (see Figure 2B and 2D). These observations indicate that the diazotrophic bacterial isolates associated with gramineous energy crops could probably represent a new species within the genus *Herbaspirillum*. Further studies, especially DNA:DNA hybridization experiments and total 16S rDNA sequencing, will be necessary to confirm this hypothesis.

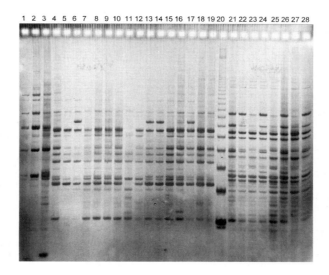

Figure 3. Inter-LINE PCR fingerprinting of diazotrophic bacteria isolated from energy crops. Lanes 1-3: *Herbaspirillum seropedicae* isolates from *Pennisetum purpureum*; lanes 4-19: *Herbaspirillum*-like isolates from *Pennisetum purpureum;* lane 20: 123 bp molecular weight standard; lanes 21-28: *Herbaspirillum*-like isolates from *Miscanthus sinensis/M. sacchariflorus.*

The diversity within certain species of nitrogen fixing soil bacteria has been already considered including *Azospirillum* (Gündisch et al., 1993, Fani et al., 1993), *Bacillus* (Mavingui et al., 1992) and *Rhizobium* (Piñero et al., 1988) using genomic fingerprinting approches like RFLP (restriction fragment length polymorphism) or PCR with primers directed to arbitrary or repetitive sequences. The diversity of the recently discovered endophytic bacterial populations is studied to a lesser extent. With multilocus enzyme electrophoresis (MLE), RFLP and plasmid pattern the variability of Brazilian and Mexican isolates of *Acetobacter diazotrophicus* was investigated and a limited genetic diversity was found (Caballero-Mellado and Martinez-Romero, 1994). For the nitrogen fixing bacterial isolates analyzed in the present communication similar findings could be obtained by Inter-LINE PCR fingerprinting (see Figure 3). The levels of genetic diversity in species of bacteria may be related to their habitat (McArthur et al., 1988). It could be a general characteristic for plant- associated bacteria or plant endophytes that the genetic structure in these bacterial populations is limited and highly conserved, probably due to the selective conditions inside the plant. C4 fibre plants are usually reproduced via rhizomes and the endophytic bacterial population may be distributed by this procedure. This could also contribute to the striking genomic homogeny of diazotrophic plant associated microorganisms.

In situ localization and activity studies of the endophytic, diazotrophic bacterial populations during a vegetation period will help to further elucidate the population dynamics and the mechanism of plant-bacteria interaction. Resulting conclusions may faciliate the finding of beneficial and ecologically competent bacterial strains and the possibility of pretreating axenic plants with optimal bacterial inocula should be considered.

Acknowledgements

This work was partly supported by the Brazilian-German scientific and technological research cooperation (project BRA ENV 34) and CNPq/PADCT (process no. 620512/94-6)

References

Aßmus B, Hutzler P, Kirchhof G, Amann R, Lawrence J R and Hartmann A 1995 In situ localization of *Azospirillum brasilense* in the rhizosphere of wheat with fluorescently labeled, rRNA-targeted oligonucleotide probes and scanning confocal laser microscopy. Appl. Environ. Microbiol. 61, 1013–1019.

Aßmus B, Schloter M, Kirchhof G, Hutzler P and Hartmann A 1997 Improved in situ tracking of rhizosphere bacteria using dual staining with fluorescence-labeled antibodies and rRNA-targeted oligonucleotide probes. Microb. Ecol. 33, 32–40.

Baldani J I, Pot B, Kirchhof G, Falsen E, Baldani V L D, Olivares F L, Hoste B, Kersters K, Hartmann A, Gillis M and Döbereiner J 1996 Emended description of *Herbaspirillum*; inclusion of *[Pseodomonas] rubrisubalbicans*, a mild plant pathogen, as *Herbaspirillum rubrisubalbicans* comb. nov., and classification of a group of clinical isolates (EF group 1) as *Herbaspirillum* species 3. Int. J. Syst. Bacteriol. 46, 802-810.

Baldani V L D 1996 Caracterização parcial de uma nova bacteria diazotrofica e endofitica do genero *Burkholderia*. PhD thesis. Universidade Federal Rual no Rio de Janeiro, Brazil.

Baldani V L D, Baldani J I, Olivares F L. and Döbereiner J 1992 Identification and ecology of *Herbaspirillum seropedicae* and the closely related *Pseudomonas rubrisubalbicans*. Symbiosis 13, 65–73.

Boddey R M, de Oliveira O C, Urquiaga S, Reis V M, Olivares F L, Baldani, V L D and Döbereiner J 1995 Biological nitrogen fixation associated with sugar cane and rice: Contributions and prospects of improvement. Plant Soil 174, 195–209.

Caballero-Medallo J and Martinez-Romero E 1994 Limited genetic diversity in the endophytic sugarcane bacterium *Acetobacter diazotrophicus*. Appl. Environ. Microbiol. 60, 1532–1537.

Döbereiner J 1992 History and new perspectives of diazotrophs in association with non-leguminous plants. Symbiosis 13, 1–13.

Döbereiner J 1995 Isolation and identification of aerobic nitrogen-fixing bacteria from soil and plants. *In* Methods in Applied Soil Microbiology and Biochemistry Eds. K Alef and P Nannipieri. pp 134–141. Academic Press, London, San Diego, New York, Boston, Sydney, Tokyo, Toronto.

Döbereiner J, Baldani V L D and Reis V M 1995 Endophytic occurrence of diazotrophic bacteria in non-leguminous crops. *In* Azospirillum VI and related microorganisms. Eds. I Fendrik, M del Gallo, J Vanderleyden and M de Zamarocy. pp 3–14. Springer Verlag, Berlin, Germany.

Fani R, Bandi C, Bardin M G, Comincini S, Damiani G, Grifoni A and Bazzicalupo M 1993 RAPD fingerprinting is useful for identification of *Azospirillum* strains. Microb. Releases 1, 217–221.

Ferreira M C B, Fernandes M S and Döbereiner J 1987 Role of *Azospirillum brasilense* nitrate reductase in nitrate assimilation by wheat plants. Biol. Fert. Soils 4, 47–53.

Fuentez-Ramirez L E, Jimenez-Salgado T, Abarca-Ocampo I R and Caballero-Medallo J 1993 *Acetobacter diazotrophicus*, an indolacetic acid producing bacterium isolated from sugarcane cultivars of Mexico. Plant Soil 154, 145–150.

Gillis M, Kersters K, Hoste B, Janssen D, Kroppenstedt R M, Stephan M P, Teixera K R S, Döbereiner J and DeLey J 1989 *Acetobacter diazotrophicus* sp. nov., a nitrogen-fixing acetic acid bacterium associated with sugarcane. Int. J. Syst. Bacteriol. 39, 361–364.

Gündisch C, Kirchhof G, Baur M, Bode W and Hartmann A 1993 Identification of *Azospirillum* species by RFLP and pulsed-field gel electrophoresis. Microb. Releases 2, 41–45.

Hartmann A, Baldani J I, Kirchhof G, Aßmus B, Hutzler P, Springer N, Ludwig W, Baldani V L D and Döbereiner J 1995 Taxonomic and ecologic studies of diazotrophic rhizosphere bacteria using phylogenetic probes. *In* Azospirillum VI and related microorganisms. Eds. I Fendrik, M del Gallo, J Vanderleyden and M de Zamarocy. pp 415–427. Springer-Verlag, Berlin.

Hurek T, Reinhold-Hurek B, van Montagu M and Kellenberger E 1994 Root colonization and systemic spreading of *Azoarcus* sp. Strain BH72 in grasses. J. Bacteriol. 176, 1913–1923.

Jain D and Patriquin D 1984 Root hair deformation, bacterial attachment, and plant growth in weat-*Azospirillum* interaction. Appl. Environ. Microbiol. 48, 1208–1213.

James E K, Reis V M, Olivares F L, Baldani J I and Döbereiner J 1994 Infection of sugar cane by the nitrogen-fixing bacterium *Acetobacter diazotrophicus*. J. Exp. Bot. 45, 757–766.

Katupipiya S, Millet M, Vesk M, Viccars L, Zeman A, Lidong Z, Elmerich C and Kennedy I R 1995 A mutant of *Azospirillum brasilense* Sp7 impaired in flocculation with a modified colonization pattern and superior nitrogen fixation in association with wheat. Appl. Environ. Microbiol. 61, 1987–1995.

Kirchhof G and Hartmann A 1992 Development of gene-probes for *Azospirillum* based on 23S-rRNA sequences. Symbiosis 13, 27–35.

Kirchhof G, Schloter M, Aßmus B and Hartmann A 1997 Molecular microbial ecology approaches applied to diazotrophs associated with non-legumes. Soil Biol. Biochem. (*In press*).

Kloepper J W and Beauchamp C J 1992 A review of issues related to measuring colonization of plant roots by bacteria. Can. J. Microbiol. 38, 1219–1232.

Ludwig W, Dorn S, Springer N, Kirchhof G and Schleifer K H 1994 PCR-based preparation of 23S rRNA-targeted group-specific polynucleotide probes. Appl. Environ. Microbiol. 60, 3236–3244.

Manz W, Amann R, Ludwig W, Wagner M and Schleifer K H 1992 Phylogenetic oligodeoxynucleotide probes for the major subclasses of Proteobacteria: problems and solutions. Syst. Appl. Microbiol. 15, 593–600.

Mavingui P, Laguerre G, Berge O and Heulin T 1992 Genetic and phenotypic diversity of *Bacillus polymyxa* in soil and wheat rhizosphere. Appl. Environ. Microbiol. 58, 1894–1903.

McArthur J V, Kovacic D A and Smith M H 1988 Genetic diversity in natural populations of a soil bacterium across a landscape gradient. Proc. Nat. Acad. Sci. 85, 9621–9624.

McInroy J A and Kloepper J W 1994 Novel bacterial taxa inhabiting internal tissues of sweet corn and cotton. *In* Improving plant productivity with rhizosphere bacteria. Eds. M H Ryder, P M Stephens and G D Bowe. p 190. CSIRO Division of Soils, Adelaide, Australia.

Olivares F L, Baldani V L D, Reis V M, Baldani J I and Döbereiner J 1993 Ecology of the endophytic diazotrophs *Herbaspirillum* spp.. Fitopat. Bras. 18, 313.

Olivares F L, Baldani V L D, Reis V M, Baldani J I and Döbereiner J 1996 Occurrence of the endophytic diazotrophs *Herbaspirillum* spp. in roots, stems and leaves predominantly of gramineae. Biol. Fertil. Soils 21, 197–200.

Patriquin D G and Döbereiner J 1978 Light microscopy observations of tetrazolium reducing bacteria in the endorhizosphere of maize and other grasses in Brazil. Can. J. Microbiol. 24, 734–742.

Pereira J A R 1995 Bacterias diazotroficas e fungos micorrizicos arbusculares em diferentes genotipos de milho (*Zea mays* L.). PhD thesis. Universidade Federal de Lavras, Brazil.

Pinero D, Martinez E and Selander K 1988 Genetic diversity and relationships among isolates of *Rhizobium leguminusarum* biovar *phaseoli*. Appl. Environ. Microbiol. 54, 2825–2832.

Reinhold-Hurek B, Hurek T, Gillis M, Hoste B, Vancanneyt M, Kersters K and DeLey J 1993a Azoarcus, gen.nov., nitrogen-fixing Proteobacteria associated with roots of Kallar grass (*Leptochloa fusca* (L.) Kunth), and description of two species, *Azoarcus indigens* sp. nov. and *Azoarcus communis* sp.nov.. Int. J. Syst. Bacteriol. 43, 574–584.

Reinhold-Hurek B, Hurek T, Claeyssens M and van Montagu M 1993b Cloning, expression in *Escherichia coli*, and characterization of cellulolytic enzymes of *Azoarcus* sp., a root-invading diazotroph. J. Bacteriol. 175, 7056-7065.

Schloter M, Aßmus B and Hartmann A 1995 The use of immunological methods to detect and identify bacteria in the environment. Biotech. Adv. 13, 75–90.

Schloter M, Kirchhof G, Heinzmann U, Döbereiner J and Hartmann A 1994 Immunological studies of the weat-root colonization by the *Azospirillum brasilense* strains Sp7 and Sp 245 using strain specific monoclonal antibodies. *In* Nitrogen Fixation with nonlegumes. Eds. N Hegazi, M Fayez and M Monib. pp 290–295, The American University in Cairo Press, Cairo, Egypt.

Silva M M P, Reis V M, Urquiaga S, Boddey R M, Xavier D F and Döbereiner J 1995 Screening *Pennisetum* ecotypes (*Pennisetum purpureum*, Schum.) for Biological Nitrogen Fixation. *In* Int. Symp. of Sustainable Agriculure for the Tropics-The Role of Nitrogen Fixation (Book of abstracts). Eds. R M Boddey and A S de Resende. pp 236–237. Angra dos Reis, Brazil.

Smida J, Leibhard S, Nickel A M, Eckardt-Schupp F and Hieber L 1996 Application of repetitive sequence-based PCR (Inter-LINE PCR) for the analysis of genomic rearrangements and for the genome characterization on different taxonomic levels. Gen. Anal. (Biomolecular Engineering) 13, 95–98.

Strunk O and Ludwig W 1997 ARB: A software environment for sequence data. Nucleic Acid Res. (*In press*)

Urquiaga S, Cruz K H S and Boddey R M 1992 Contribution of nitrogen fixation to sugarcane: Nitrogen-15 and nitrogen balance estimates. Soil Sci. Soc. Am. Proc. 56, 105–114.

Versalovich J, Schneider M, deBruijn F and Lupski J R 1994 Genomic fingerprinting of bacteria using repetitive sequence-based polymerase chain reaction. Meth. Mol. Cellul. Biol. 5, 25–40.

Guest editors: J K Ladha, F J de Bruijn and K A Malik

Plant and Soil **194**: 57–64, 1997.

Azoarcus spp. and their interactions with grass roots

Barbara Reinhold-Hurek* and Thomas Hurek
Max-Planck-Institut für terrestrische Mikrobiologie, Arbeitsgruppe Symbioseforschung, Karl-von-Frisch-Straße, D-35043 Marburg, Germany

Key words: Azoarcus, endorhizosphere, *nif*-genes, nitrogen fixation, 16S ribosomal RNA

Abstract

The current knowledge on the divergence within the genus *Azoarcus* and about interactions with grasses is summarized. Grass-associated members of this genus of diazotrophs have only been isolated from a salt- and flood-tolerant pioneer plant in Pakistan, Kallar grass (*Leptochloa fusca* (L.) Kunth). Members of these bacteria belong to the beta subclass of the *Proteobacteria*, most closely related to purple bacteria such as *Rhodocyclus purpureus*. The isolates from one single host plant showed a surprising divergence, consisting of five groups of *Azoarcus* distinct at the species level. Molecular diagnostic tests, which are based on 16S ribosomal RNA sequences, allowed preliminary assignment of isolates to *Azoarcus* by PCR amplification and sequencing of PCR products. Moreover, the moleculer tests enabled us to detect an unculturable strain in Kallar grass roots, stressing that classical cultivation techniques at times fail to detect some groups of the microbial population. Using similar techniques, sequences rooting in the *Azoarcus* clade were also detected in field-grown rice, indicating that the natural host range might extend to rice. In gnotobiotic laboratory cultures, a member of *Azoarcus* is able to colonize rice roots endophytically: bacteria invade the roots in the zone of elongation and differentiation, colonize the cortex intra- and inter-cellularly, and penetrate deeply into the vascular system, entering xylem vessels, allowing systemic spreading into the rice shoot. Recently, we detected expression of nitrogenase of *Azoarcus* cells inside roots of rice seedlings, a result encouraging us to analyze interactions with rice in detail.

Abbreviations: nifHDK – structural genes of nitrogenase; SSU rRNA – small subunit ribosomal RNA

Introduction

Nitrogen fixation and plant growth promotion by rhizosphere bacteria might be important factors for achieving a sustainable agriculture in the future. Gramineous plants such as rice, maize and wheat are major crops for food production. Since various diazotrophic bacteria have been found in association with gramineous plants, they are possible candidates for beneficial interactions with agriculturally important crops. Diazotrophic bacteria which have gained a lot of attention are *Azospirillum* spp., which occur in the rhizosphere of a large range of tropical, subtropical and temperate plants (Döbereiner and Pedrosa, 1987). *Herbaspirillum sero-*

pedicae and *Acetobacter diazotrophicus* were reported to colonize preferrably the interior of plants (e.g. sugar cane) as endophytes (Döbereiner et al., 1993). In this review, we summarize the current knowledge about another genus of diazotrophic *Proteobacteria*, *Azoarcus* spp., which shows endophytic colonization of Kallar grass and rice.

The natural host plant of Azoarcus *spp.*

As in many arid regions, soils in the Punjab of Pakistan are often salt-affected. To reclaim these soils, Kallar grass (*Leptochloa fusca* (L.) Kunth), a C_4 plant which has a high tolerance of waterlogged conditions, soil salinity and alkalinity (Khan, 1966), was introduced as a pioneer plant in the Punjab region (Sandhu

* Fax: +496421178109.
E-mail: reinhold@mailer.uni-marburg.de

and Malik, 1975). Its distribution ranges from Asia to Africa and Australia, where it occurrs as weed in agricultural lands or in patches at waterlogged conditions. It grows luxuriantly without any added nitrogeneous fertilizer, producing 20-40 t of green fodder per ha per year (Sandhu et al., 1981). This observation together with the acetylene reducing activity in its rhizosphere (Malik et al., 1980) indicated a possible contribution of nitrogen fixation to the nitrogen nutrition of these plants. Moreover, Kallar grass was never used for breeding selection for high responsiveness to nitrogen fertilizers, and is thus a promising system to study diazotrophs occurring naturally in its rhizosphere.

Diazotrophs in the rhizosphere of Kallar grass

Often several species of diazotrophs can be isolated from one plant (Patriquin et al., 1983). For Kallar grass, several diazotrophs have been described to occur in its rhizosphere as well, such as *Klebsiella* and *Beijerinckia* (Zafar et al., 1987), *Zoogloea* (Bilal and Malik, 1987), and *Azospirillum brasilense* (Bors et al., 1982). Analysis of the microbial population in different zones of the rhizosphere showed that the population of diazotrophic bacteria on the rhizoplane was found to be 35 fold higher than in non-rhizospheric soil, where they made up only 0.2% of the total aerobic microflora (Reinhold et al., 1986).

However, diazotrophic populations of the rhizoplane and the endorhizosphere were entirely different from each other. On the rhizoplane we found high numbers of diazotrophs (Table 1), which consisted in equal numbers of *Azospirillum lipoferum* and *Azospirillum halopraeferens*, a new species which has only been isolated from Kallar grass as yet (Reinhold et al., 1987). Since surface sterilzation effectively reduced the rhizoplane population, diazotrophs from homogenized roots which occurred in high numbers were regarded as endorhizospheric population (Table 1). Surprisingly, *Azospirillum* spp. could not be detected inside the roots, but unidentified diazotrophic rods predominated, which were later assigned to the new genus *Azoarcus* spp. (Reinhold-Hurek et al., 1993b). These bacteria seem to be highly adapted to growth in or on roots, since we were not able to isolate them from root-free soil.

Such a root-zone-specific association with diazotrophs raises the question, why certain bacteria can preferentially colonize the root interior. Since Kallar grass is a flood-tolerant plant with large aerenchymat-

ic air spaces in mature roots, one can assume that the root is a source of oxygen in flooded soils, as it has been described for rice (Tadano and Yoshida, 1978). Studies on sensitivity of nitrogen fixation to oxygen revealed that *Azospirillum* spp. from the rhizoplane were less resistant than diazotrophic rods dominating in the endorhizosphere (Hurek et al., 1987, see Table 2). If the efficient colonization of roots requires nitrogen-fixing growth, bacteria with a higher oxygen tolerance might have a competitive advantage in colonization of the root interior containing aerenchymatic tissue. (Hurek et al., 1987). In cortical spaces of wetland plants, oxygen concentrations of at least 2% (approx. 20–25 μM O_2) have been suggested (Armstrong and Gaynard, 1976), an oxygen concentration which would not allow nitrogen fixation of the rhizoplane isolates (Table 2 and Hurek, 1987). Moreover, enzymes degrading components of the plant cell walls might be involved in the infection process and thus be a reason for differential colonization of root zones (see below). Also plant root exudates might confer a selective advantage to those diazotrophs which show a strong chemotactic response to certain attractants of their host plant (Reinhold et al., 1985).

Divergence in the genus Azoarcus

Bacteria of the morphology and physiology similar to the unidentified diazotrophic rods from the root interior of Kallar grass were isolated repeatedly from this grass. Most isolates obtained during several years belong to a new genus, *Azoarcus* spp. (Reinhold-Hurek et al., 1993b). According to DNA-rRNA hybridization studies, eleven isolates constitute a separate rRNA branch in the beta subdivision of the *Proteobacteria*. These strains show a surprising diversity; they formed five groups distinct at the species level according to DNA-DNA hybridization studies. However, they are very similar in morphological, nutritional and biochemical features, making it difficult to differentiate them by classical tests. Therefore we proposed instead of five only two named species, *A. indigens* and *A. communis*, which could be easily distinguished by vitamin requirement or a slightly larger cell width, respectively (see Table 3 and Reinhold-Hurek et al., 1993b). The three species strain BH72, *A. indigens* and *A. communis* were more highly related to each other than to the other species represented by strains S5b2 and 6a3, as determined by $T_{m(e)}$ values (thermal stability of DNA-rRNA hybrids) and 16S rRNA sequence homology (Table 3). Despite the low rRNA sequence

Table 1. Recovery of nitrogen-fixing bacteria from different root zones of Kallar grass [a]

Fraction	Bacterial number (per g [dry wt.] of root)[b]	
	Azospirillum spp.	*Azoarcus* spp.
Rhizoplane	$(2.0 \pm 1,3) \times 10^7$	n.d.
Rhizoplane after surface sterilization	$(2.2 \pm 1,8) \times 10^4$	n.d.
Endorhizosphere	n.d.	$(7.3 \pm 7.0) \times 10^7$
Soil	$(1,2 \pm 0.4) \times 10^4$	n.d.

[a] Table from Reinhold et al. (1986), modified.
[b] Bacterial number from MPN (most probable number) counts given with standard deviation; n.d., not detected in high MPN dilutions.

Table 2. Oxygen tolerance of nitrogen fixation of diazotrophs isolated from different root zones of Kallar grass[a]

	Rhizoplane isolates	Endorhizosphere isolates		
	Azospirillum lipoferum Rp5	*Azospirillum halopraeferens* Au4	*Azoarcus* sp. BH72	strain H6a2
Threshold concentrations of d.o.t. for steady-state nitrogen fixation[b]	6	5	27	30
Nitrogen fixation at 20 μm d.o.t.	–	–	+	+

[a] Data from Hurek et al. (1987), from dissolved-oxygen (d.o.t.)-controlled continuous cultures under malate limitation.
[b] Growth rate $\mu = 0.083$ h^{-1} on potassium malate at 37°C.

homologies between species represented by strains S5b2 and 6a3 (Table 3), they were included into the same genus *Azoarcus* due to the lack of sufficient phenotypic differences. Occurrence of the five species of plant-associated *Azoarcus* is not restricted to Pakistan, since an isolate from polluted soil in France could be assigned to *A. communis* (Reinhold-Hurek et al., 1993b). At least one strain of *Azoarcus*, BH72, has a genetic element which is very unusual for *Bacteria*, a self-splicing group I intron (Reinhold and Shub, 1992). The role and distribution of this element within this genus is still not clear.

Recently, two new species of *Azoarcus* have been described by other groups: *A. tolulyticus* (Zhou et al., 1995) and *A. evansii* (Anders et al., 1995). They are not associated with plants, but originate from various non-contaminated or polluted soils or sediments from the US or South America (Anders et al., 1995; Fries et al., 1994). In contrast to the plant-associated species, they are capable of degrading aromatic hydrocarbons anaerobically with nitrate as an electron acceptor. Moreover, they grow on sugars such as fructose or glucose as a sole carbon source, while none out of 49 carbohydrates tested can be used as a carbon source by the other five species (Table 3). Whether the soil isolates colonize plants is not clear. Comparative studies,

however, might reveal specific mechanisms of plant-microbe interactions.

The genotypic diversity of plant-associated *Azoarcus* strains seems surprising, since five species were present in and on plants of the same species in one field (Reinhold-Hurek et al., 1993b). However, it is consistent with the observation that often the diversity of microorganisms appears to reflect obligate or facultative associations with higher organisms, and to be determined by spatial-temporal diversity of their hosts or associates (Whithcomb and Hackett, 1989). Moreover, the presence of several slightly different genotypes with similar functions might give a buffering capacity to the ecosystem against the loss of species (Perry et al., 1989). Accordingly, in many studies on ecosystems which use phylogenetic analyses of PCR-amplified genes instead of classical cultivation techniques, clades of genotypically similar organisms are detected which show some sequence variation (Giovannoni et al., 1990; Ueda et al., 1995).

Detection of Azoarcus *spp. in the environment by molecular diagnostic tests*

Since *Azoarcus* spp. do not grow on carbohydrates and show negative reactions also in many other classical

Table 3. Characteristics of *Azoarcus* species [a]

Characteristic	*Azoarcus indigens* VB32[T], VW35a, VW34c	*Azoarcus communis* SWub3[T], S2	*Azoarcus* sp. BH72	*Azoarcus* sp. S5b2, S5b1, SSa3	*Azoarcus* sp. 6a3, 6a2, 5c1	*Azoarcus tolulyticus*	*Azoarcus evansii*
Cell width (μm)	0.5–0.7	0.8–1.0	0.6–0.8	0.6–0.8	0.4–0.6	n.d.	0.6–0.8
Requirement for p-aminobenzoic acid	+	–	–	–	–	–	–
Growth on L-malate, acetate, and DL-lactate	+	+	+	+	+	+	+
Growth on glucose	–	–	–	–	–	+	n.d.
Growth on fructose	–	–	–	–	–	n.d.	+
rRNA relatedness to strain BH72							
$T_{m(e)}$ -values [°C]	80.4	79.4	81.5	71.4	71.6	n.d.	n.d.
rRNA sequence homology	97.1%	96.2%	100%	93.7%	92.3%	95.3%	95.2%

[a] Data from Reinhold-Hurek et al. (1993b), Hurek et al. (1993), Anders et al. (1995), Zhou et al. (1995), Reinhold-Hurek and Hurek (1995), and unpublished sequence data.

tests such as ß-glucosidase or other enzymatic activities (Reinhold-Hurek et al., 1993b), new isolates are difficult to assign to the known taxa. New methods for classification of bacteria are often DNA-based; e.g. sequence analysis of genes for 16S ribosomal RNA is applicable to all taxa. Moreover, molecular phylogeny is a useful tool to analyze microbial populations without prior cultivation of bacteria (Giovannoni et al., 1990; Amann et al., 1991) and may give new insights into a structure of a microbial community. Therefore, we developed molecular diagnostic tests for *Azoarcus* spp. using sequences of 16 ribosomal RNA (SSU rRNA) (Hurek et al., 1993; Hurek and Reinhold-Hurek, 1995). Based on total SSU rRNA sequences of *Azoarcus* spp., genus-specific molecular diagnostic tests were developed, which allow specific detection of four plant-associated species by PCR amplification using primer combination TH3/TH5. A somewhat less specific PCR assay (TH14/TH2) can detect all five plant-associated species including strain 6a3, however PCR products are also obtained with some other genera (Hurek et al., 1993).

These PCR assays may be used to detect *Azoarcus* in environmental DNA samples without prior cultivation. Direct sequencing of PCR products may even allow identification of the uncultivated or cultivated bacterial strain by phylogenetic analysis of partial SSU rDNA sequences (Hurek and Reinhold-Hurek, 1995). With this approach, a strain of *Azoarcus* was detected in Kallar grass roots when parallel attempts to culti-

vate these bacteria did fail (Hurek and Reinhold-Hurek, 1995). This underlines that in a given habitat such as roots, diazotrophs might be present which cannot be cultivated. Therefore, it is possible that viable counts of diazotrophs colonizing roots may not only underestimate their abundance, but also fail to detect the structure of their community.

An alternative approach for the assessment of a diazotrophic population is the phylogenetic analysis of the structural genes for nitrogenase (*nifHDK*) present in a given habitat. When primers targeted to conserved regions of *nifH* genes are used for PCR reactions, amplification products from e.g. plant roots can be analyzed by cloning, sequencing, and phylogenetic analysis (Ueda et al., 1995). Since phylogeny of the protein sequences deduced from *nifH* genes follows largely the organismal phylogeny based on 16S rDNA sequences (Young, 1992; Ueda et al., 1995), the identity of the organism harboring the *nifH* gene can often be deduced. By this approach, evidence was obtained for the first time that the natural host range of *Azoarcus* spp. might extend to plants other than Kallar grass. From rice roots grown in Japan, NifH protein sequences were obtained by Ueda et al. (1995), which could not be assigned to any known bacterial species or genus. When we sequenced *nifH* gene fragments of six species of *Azoarcus* as a reference, phylogenetic sequence analysis showed that one of the unassigned sequences from Japan rooted in the *Azoarcus* clade (Hurek, Egener, and Reinhold-Hurek, submitted for publication). This indicates that *Azoarcus* sp. might naturally colonize roots of field-grown rice, raising the question if they fix nitrogen in the rice rhizosphere.

Colonization of plants by Azoarcus *spp.*

Up to now, *Azoarcus* spp. have not been isolated from any plant other than Kallar grass. Most of the Kallar grass isolates were obtained from the endorhizosphere or from the culm after surface sterilization (Reinhold-Hurek et al., 1993b), indicating that they can colonize the interior of this plant. How do they invade a grass? In gnotobiotic experiments, *Azoarcus* sp. BH72 is indeed capable of infecting grass roots (Hurek et al., 1994b). Bacteria colonize their original host Kallar grass and and a new host, rice (*Oryza sativa* IR36), in a similar manner. By using a transposon mutant expressing ß-glucuronidase constitutively as a reporter gene, it was demonstrated that the apical region of roots behind the meristem is the region most intensively colonized. The bacteria penetrate the rhizoplane preferentially in the zones of elongation and differentiation. Colonization of the cortex region is inter- and intracellularly (Figure 1a), however bacteria appear not to occur within a live plant cytoplast, thus can not be regarded as true endosymbionts. In older parts of the roots, they also occur in aerenchymatic air spaces. *Azoarcus* sp. is capable of invading even the stele of rice and xylem vessels (Figure 1b, c). Transport in the transpiration stream might facilitate systemic spreading into the shoot, where they were detected in rice plants three weeks after inoculation (Hurek et al., 1994b). No damage of plants or growth inhibition was observed upon systemic infection, also a common hypersensitivity test on tobacco leaves provided no evidence of damage to the plant. Moreover, when no external carbon sources are added, growth of rice seedlings improved significantly (Table 4). Therefore, *Azoarcus* sp. BH72 apparently shows no sign of phytopathogenicity, but can be regarded as an endophytic bacterium (Hurek et al., 1994b).

In field-grown Kallar grass plants from Pakistan, immunofluorescence microscopic studies using antibodies generated against two endorhizospheric isolates including *Azoarcus* sp. BH72, located cross-reacting round bodies in the aerenchymatic spaces of roots which were of the size of plant cells (Reinhold et al., 1987a). These may contain microbial aggregates, in which respiring bacteria would face a steep oxygen gradient. In additional studies on resin-embedded roots, bacteria cross-reacting with *Azoarcus*-specific

or nitrogenase-specific antibodies were detected in the root aerenchyma, in the stelar tissue in pericycle cells, and in the culm next to vascular tissue (Hurek et al., 1994b). This demonstrates that endophytic *Azoarcus* spp. may colonize similar sites in situ as found in inoculation experiments in laboratory culture.

Penetration of roots requires degradation of plant cells walls, which might be an active process involving bacterial polymer-degrading enzymes. Indeed *Azoarcus* sp. BH72 expresses two cellulolytic enzymes, an endoglucanase and an exoglycanase (Reinhold-Hurek et al., 1993a). The exoglycanase has ß-glucosidase and cellobiohydrolase activity, but also degrades xylane, whereas the endoglucanase releases larger oligomers when attacking cellulose fibers internally. Since *Azoarcus* sp. cannot utilize glucose, cellobiose or cellulose for growth, it is remarkable that cellulolytic enzymes are expressed.

Several unusual features indicate that these cellulases might be used for infection of plants instead of metabolic processes. Most cellulases of cellulolytic bacteria are repressed or inactivated upon addition of the end product, glucose, or induced by the substrate cellulose, as in *Clostridium thermocellum* (Mishra et al., 1991). In contrast, both cellulases of *Azoarcus* sp. BH72 do not appear to be subject to strong induction by carboxymethylcellulose or repression by glucose. However, activities are significantly enhanced by ethanol and under microaerobic conditions (Reinhold-Hurek et al., 1993a). Both conditions might prevail in roots particularly when the root system is submerged. A rise in ethanol concentration in roots with flooding has been reported (Crawford and Baines, 1977). In addition, ethanol has also been detected in root tips in plants grown in aerated soil (Betz, 1957). Thus microaerobiosis and ethanol accumulation caused by actively respiring root meristems might support cellulase activity and penetration at these sites (Reinhold-Hurek et al., 1993a). Cellulases are also often found in phytopathogenic bacteria. *Pseudomonas solanacearum* (Huang et al., 1989) and *Erwinia chrysanthemi* (Chippaux, 1988) excrete endoglucanases into the culture supernatant. In contrast, exo- and endoglucanase of *Azoarcus* sp. BH72 are cell surface associated (Reinhold-Hurek et al., 1993a). Therefore, these enzymes might mediate a more localized digestion of plant polymers in comparison to phytopathogens, causing less damage to the host and allowing the systemic infection in healthy plants.

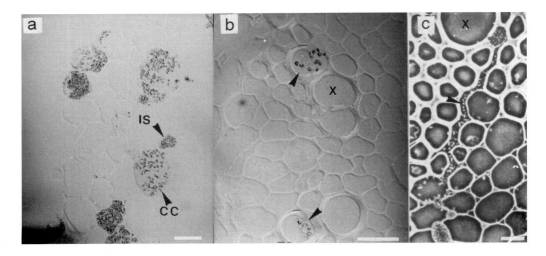

Figure 1. Colonization of the interior of grass roots by *Azoarcus* sp. BH72 in gnotobiotic culture. Kallar grass or rice (*Oryza sativa* IR 36) were grown for three weeks in cotton-stoppered glass tubes with agar medium. Roots embedded in resin were subjected to thin sectioning and light microscopic analyis. Bacteria were labeled with genus-specific antibodies by immunogold staining followed by silver enhancement (a, b). Root cortex of rice (**a**), and stele of rice (**b**) and Kallar grass (**c**). Bacteria are marked by arrowheads. IS, intercellular space; CC, cortex cell; X, xylem cell; bars, 10 μm. Photos from Hurek et al. (1991) and Hurek et al. (1994b), modified.

Table 4. Effects of inoculation with *Azoarcus sp.* strain BH72 on growth of rice seedlings

Treatment [a]	Inoculum	Dry weight (mg) [b]	Total protein (mg)[c]
Without malate	–	14.5 ± 1.9 (a)	0.55 ± 0.06 (a)
With malate	–	14.7 ± 2.1 (a)	0.54 ± 0.04 (a)
Without malate	strain BH72	16.8 ± 1.5 (b)	0.79 ± 0.07 (b)
With malate	strain BH72	10.9 ± 3.2 (c)	0.46 ± 0.05 (c)

[a] Rice seelings (*Oryza sativa* IR 36) were grown for three weeks in a gnotobiotic system in cotton-stoppered glass tubes with agar medium supplemented with mineral nutrients and 0.58 g L^{-1} of proline; 0.2 g L^{-1} of potassium-DL-malate was supplemented if indicated.

[b] Values represent means of 16 to 19 plants each. Results followed by different letters are significantly different from each other at the 5% level according to the Mann-Whitney test.

[c] Values represent means of six pools of three plants (each) measured three times each. Statistical analysis as in footnote [b].

Nitrogen fixation of Azoarcus *sp. BH72*

Bacteria of the genus *Azoarcus* fix nitrogen under microaerobic conditions (Hurek et al, 1987; Reinhold-Hurek et al., 1993b), such as *Azospirillum* spp. and *Herbaspirillum* sp. However, it has been shown for strain BH72 that it can reach a "hyperinduced" state under certain cultural conditions. When in the presence of proline cells shift into extremely low oxygen concentrations, nitrogen fixation becomes more active and efficient (Hurek et al., 1994a). In the course of hyperinduction cells form novel intracytoplasmic membrane stacks, "diazosomes" (Hurek et al., 1995). These stacked membranes are similar in structure to membranes with specialized functions, which are formed by phototrophic, nitrifying or methanotrophic bacteria. However, none of these functions can be assigned to the diazosomes. Most likely, they are involved in the process of efficient nitrogen fixation, due to the following line of evidence (Hurek et al., 1995): (i) The iron protein of nitrogenase is associated with diazosome membranes; in immunogold electron microscopic studies, antibodies to this protein detected it mainly on these membrane stacks, whereas nitrogenase was evenly distributed in the cytoplasma of the cells when the bacteria fixed nitrogen under standard conditions without membrane stacks; a second form of the iron protein having a slightly higher molecular weight (approx. 3 kDa) was detected in Western blot analysis when bacteria contained diazosomes, and it was associated with membrane fractions. Most likely, it is a covalently modified form of the iron protein. (Ii) Diazosome formation is tightly linked to the process of nitrogen fixation; when cells grew on ammonia, diazosome formation cannot be observed; moreover, a Nif⁻ mutant (deletion in *nifK* gene) was unable to develop membrane stacks.

Whether these membrane stacks are also formed in the natural habitat of *Azoarcus* sp., and whether they play any significant role in nitrogen fixation in situ, still awaits clarification.

Future prospects

Since *Azoarcus* spp. seem to occur naturally in roots of field-grown rice and are capable of infecting it in gnotobiotic culture, this raises the question whether the bacterial nitrogenase can be expressed inside rice roots. Therefore, we constructed a transcriptional reporter gene fusion of the *nifH* gene (coding for the iron protein of nitrogenase) with ß-glucuronidase, a reporter gene which had proven to be free of background staining in rice and Kallar grass previously (Hurek et al., 1994b), and stably integrated the construct into the *Azoarcus* chromosome. In the resulting *Azoarcus* mutant, nitrogenase activity and expression of ß-glucuronidase were in good correlation. In gnotobiotic cultures of rice seedlings grown for five days with the reporter strain, bacteria penetrated deeply into the cortex region at points of emergence of lateral roots. Blue precipitates in these bacterial cells after X-gluc staining indicated that nitrogenase genes were expressed when the bacteria were inside rice roots. Moreover, we demonstrated by immunogold staining with antibodies specific for the iron protein of nitrogenase, that the nitrogenase protein is actually present in these bacteria (Egener, Hurek and Reinhold-Hurek, manuscript in preparation).

This is the first report of nitrogenase expression by an endophytic bacterium in rice. This feature of *Azoarcus* sp. is very promising with respect to engineering a nitrogen-fixing plant-microbe interaction with rice; however, it is not yet clear if fixed nitrogen can be transferred to the plant, and if such a contribution is significant for rice growth. Our observations are encouraging us to study long-term effects of inocula-

tion on rice growth, but also to initiate comparative analyses on strain-cultivar interactions.

Acknowledgements

We would like to thank C Staubitz for help with photographic work. This work was supported by a grant of the Deutsche Forschungsgemeinschaft (Re 756/5-1) to B R-H

References

Amann R, Springer N, Ludwig L, Görtz H-D, and Schleifer KH 1991 Identification in situ and phylogeny of uncultured bacterial endosymbionts. Nature (London) 351, 161–164.

Armstrong W and Gaynard T J 1976 The critical oxygen pressures for respiration in intact plants. Physiol. Plant. 37, 200–206.

Anders H J, Kaetzke A, Kämpfer P, Ludwig W and Fuchs G 1995 Taxonomic position of aromatic-degrading denitrifying pseudomonad strains K 172 and KB 740 and their description as new members of the genera *Thauera*, as *Thauera aromatica* sp. nov., and *Azoarcus*, as *Azoarcus evansii* sp. nov., respectively, members of the beta subclass of the *Proteobacteria*. Int. J. Syst. Bacteriol. 45, 327–333.

Betz A 1957 Zur Atmung wachsender Wurzelspitzen. Planta 50, 122–143.

Bilal R and Malik K A 1987 Isolation and identification of a N₂-fixing zoogloea-for ming bacterium from Kallar grass histoplane. J. Appl. Bacteriol. 62, 289–294.

Bors J, Kloss M, Zelles I and Fendrik I 1982 Nitrogen fixation and nitrogen-fixing organisms in the rhizosphere of *Diplachne fusca* Linn. (Beauv.) J. Gen. Appl. Microbiol. 28, 11–118.

Chippaux M 1988 Genetics of cellulases in *Erwinia chrysanthemi*. *In* Biochemistry and genetics of cellulose degradation. Eds. J-P Aubert, P Béguin and J Millet. pp 219–234. Academic Press, London.

Crawford R M M and Baines M 1977 Tolerance of anoxia and ethanol metabolism in tree roots. New Phytol. 79, 519–526.

Döbereiner J and Pedrosa F O 1987 Nitrogen-fixing bacteria in non-leguminous crop plants. Springer Verlag, Berlin, Germany.

Döbereiner J, Reis V M, Paula M A and Olivares F 1993 Endophytic diazotrophs in sugar cane, cereals and tuber plants. *In* New horizons in nitrogen fixation. Eds. R Palacios, J Mora and W E Newton. pp 671–676. Kluwer Academic Publishers, Dordrecht.

Fries M R, Zhou J-Z, Chee-Sanford J and Tiedje J M 1994 Isolation, characterization, and distribution of denitrifying toluene degraders from a variety of habitats. Appl. Environ. Microbiol. 60, 2802–2810.

Giovannoni S J, Britschgi T B, Moyer C L, and Filed K G 1990 Genetic diversity in Sargasso sea bacterioplankton. Nature (London) 345, 60–63.

Huang J, Sukordhaman M and Schell M A 1989 Excretion of the *egl* gene product of *Pseudomonas solanacearum* J. Bacteriol. 171, 3767–3774.

Hurek T, Reinhold B, Fendrik I and Niemann E G 1987 Root-zone-specific oxygen tolerance of *Azospirillum* spp. and diazotrophic rods closely associated with Kallar grass. Appl. Environ. Microbiol. 53, 163–169.

64

Hurek T, Burggraf S, Woese C R and Reinhold-Hurek B 1993.16S rRNA-targeted polymerase chain reaction and oligonucleotide hybridization to screen for *Azoarcus* spp., grass-associated diazotrophs. Appl. Environ. Microbiol. 58, 3816–3824.

Hurek T, Reinhold-Hurek B, Turner G L and Bergersen F J 1994a Augmented rates of respiration and efficient nitrogen fixation at nanomolar concentrations of dissolved O_2 in hyperinduced *Azoarcus* sp. strain BH72. J. Bacteriol. 176, 4726–4733.

Hurek T, Reinhold-Hurek B, Van Montagu M and Kellenberger E 1994b Root colonization and systemic spreading of *Azoarcus* sp. strain BH72 in grasses. J. Bacteriol. 178, 1913–1923.

Hurek T and Reinhold-Hurek B 1995 Identification of grass-associated and toluene-degrading diazotrophs, *Azoarcus* spp., by analysis of partial 16S ribosomal DNA sequences. Appl. Environ. Microbiol. 61, 2257–2261.

Hurek T, Van Montagu M, Kellenberger E and Reinhold-Hurek B 1995 Induction of complex intracytoplasmic membranes related to nitrogen fixation in *Azoarcus* sp. BH72. Mol. Microbiol. 18, 225–236.

Khan M D 1966 "Kallar grass", a suitable grass for saline lands. Agric. Pak. 17, 375.

Malik K A, Zafar Y and Hussain A 1980 Nitrogenase activity in the rhizosphere of Kallar grass (*Diplachne fusca* (Linn.) Beauv.). Biologia 26, 107–112.

Mishra S, Béguin P and Aubert J-P 1991 Transcription of *Clostridium thermocellum* endoglucanase genes *celF* and *celD*. J. Bacteriol. 173, 80–85.

Patriquin D G, Döbereiner J and Jain D K 1983 Sites and processes of association between diazotrophs and grasses. Can. J. Microbiol. 29, 900–915.

Perry D A, Amaranthus M P, Borchers J G, Borchers S L and Brainerd N E 1989 Bootstrapping in ecosystems. BioScience 39, 230–237.

Reinhold B, Hurek T and Fendrik I 1985 Strain-specific chemotaxis of *Azospirillum* spp. J. Bacteriol. 162, 190–195.

Reinhold B, Hurek T and Fendrik I 1987a Cross-reaction of predominant nitrogen-fixing bacteria with enveloped round bodies in the root interior of Kallar grass. App. Environ. Microbiol. 53, 889–891.

Reinhold B, Hurek T, Fendrik I, Pot B, Gillis M, Kersters K, Thielemans S and De Ley J 1987b *Azospirillum halopraeferens* sp. nov., a nitrogen-fixing organism associated with the roots of Kallar grass (*Leptochloa fusca* (L.) Kunth). Int. J. Syst. Bacteriol. 37, 43–51.

Reinhold B, Hurek T, Niemann E G and Fendrik I 1986 Close association of *Azospirillum* and diazotrophic rods with different root zones of Kallar grass. Appl. Environ. Microbiol. 52, 520–526.

Reinhold-Hurek B, Hurek T, Claeyssens M and Van Montagu M 1993a Cloning, expression in *Escherichia coli*, and characterization of cellulolytic enzymes of *Azoarcus* sp., a root-invading diazotroph. J. Bacteriol. 175, 7056–7065.

Reinhold-Hurek, B, Hurek T, Gillis M, Hoste B, Vancanneyt M, Kersters K and DeLey J 1993b *Azoarcus* gen. nov., nitrogen-fixing proteobacteria associated with roots of Kallar grass (*Leptochloa fusca* (L.) Kunth), and description of two species, *Azoarcus indigens* sp. nov. and *Azoarcus communis* sp. nov.. Int. J. Syst. Bacteriol. 43, 574–584.

Reinhold-Hurek B and Shub D 1992 Self-splicing introns in tRNA genes of widely divergent bacteria. Nature (London) 357, 173–176.

Sandhu G R, Aslam Z, Salim M, Sattar A, Qureshi R H, Ahmad N and Wyn Jones R G 1981 The effect of salinity on the yield and composition of *Diplachne fusca* (Kallar grass). Plant Cell Environ. 4, 177–181.

Sandhu G R and Malik K A 1975 Plant succession - a key to the utilization of saline soils. Nucleus 12, 35–38.

Tadano T and Yoshida S 1978 Chemical changes in submerged soils and their effect on rice growth. *In* Soils and rice. Ed. International Rice Research Institute. pp 399–420. International Rice Reasearch Institute, Los Banos, Philippines.

Ueda T, Suga Y, Yahiro N, and Matsuguchi T 1995 Remarkable N_2-fixing bacterial diversity detected in rice roots by molecular evolutionary analysis of *nifH* gene sequences. J. Bacteriol. 177, 1414–1417.

Young J P W 1992 Phylogenetic classification of nitrogen-fixing organisms. *In* Biological nitrogen fixation. Eds. G Stacey, R H Burris and H J Evans. pp.43-86. Chapman and Hall, New York

Whitcomb R F and Hackett K J 1989 Why are there so many species of mollicutes? An assay on prokaryotic diversity. *In* Biotic diversity and germ plasm preservartion, global imperatives. Ed. K A Stoner. pp 205–240. Kluwer Academic Press, Amsterdam.

Zafar Y, Malik K A and Niemann E G 1987 Studies on N_2-fixing bacteria associated with the salt-tolerant grass, *Leptochloa fusca* (L.) Kunth. MIRCEN J. 3, 45–56.

Zhou J, Fries M R, Chee-Sanford J C and Tiedje J M 1995 Phylogenetic analyses of a new group of denitrifiers capable of anaerobic growth on toluene and description of *Azoarcus tolulyticus* sp. nov. Int. J. Syst. Bacteriol. 45, 500–506.

Guest editors: J K Ladha, F J de Bruijn and K A Malik

Plant and Soil **194**: 65–79, 1997.
65

Biological nitrogen fixation in non-leguminous field crops: Facilitating the evolution of an effective association between *Azospirillum* and wheat

Ivan R. Kennedy, Lily L. Pereg-Gerk, Craig Wood, Rosalind Deaker, Kate Gilchrist and Sunietha Katupitiya
SUNFix Centre for Nitrogen Fixation, Department of Agricultural Chemistry and Soil Science, University of Sydney, NSW 2006, Australia *

Key words: *Azospirillum brasilense*, evolution, *nifA-lacZ*, *nifH-lacZ*, nitrogen fixation, *para*-nodules, symbiosis, wheat, 2,4-D

Abstract

Recent advances towards achieving significant nitrogen fixation by diazotrophs in symbioses with cereals are reviewed, referring to the literature on the evolution of effective symbioses involving rhizobia and *Frankia* as microsymbionts. Data indicating that strains of *Acetobacter* and *Herbaspirillum* colonizing specific cultivars of sugarcane as endophytes make a significant contribution to the nitrogen economy of this crop improves the prospects that similar associative systems may be developed for other gramineous species such as rice and wheat. By contrast, the transfer of nodulation genes similar to those in legumes or *Parasponia* to achieve nodulation in crops like rice and wheat is considered to be a more ambitious and distant goal. Progress in developing an effective associative system for cereals has been materially assisted by the development of genetic tools based on the application of *lacZ* and *gusA* fusions with the promoters of genes associated with nitrogen fixation. These reporter genes have provided clear evidence that 'crack-entry' at the points of emergence of lateral roots or of 2,4-D induced *para*-nodules is the most significant route of endophytic colonization. Furthermore, using the laboratory model of *para*-nodulated wheat, there is now evidence that the ability of azospirilla and other nitrogen fixing bacteria to colonize extensively as endophytes can be genetically controlled. The most successful strain of *Azospirillum brasilense* (Sp7-S) for endophytic colonization and nitrogen fixation in wheat seedlings is a mutant with reduced exopolysaccharide production. Most other strains of azospirilla do not colonize as endophytes and it is concluded that though these are poorly adapted to providing nitrogen for the host plant, they are well adapted for survival and persistence in soil. A research program combining the study of endophytic colonization by azospirilla with an examination of the factors controlling the effectiveness of association (oxygen tolerance and nitrogen transfer) is now being pursued. It is proposed that a process of facilitated evolution of *para*-nodulated wheat involving the stepwise genetic improvement of both the prospective microsymbionts and the cereal host will eventually lead to effective nitrogen-fixing associations. In the attempt to achieve this goal, continued study of the endophytes occurring naturally in sugar cane and other grasses (e.g. *Azoarcus* sp.) should be of assistance.

Introduction

Globally, biological N$_2$ fixation, as a proportion of the total input of nitrogen to support the growth of crops, has probably been declining. It appears that a significant reduction in the relative use of fertiliser-N can only be achieved if biological nitrogen fixation is made directly available to cereal crops in an effective associative system with some of the characteristics of the legume symbiosis. Experience since the plant-associating diazotrophs were discovered by Döbereiner and her coworkers (see Baldani et al., 1983) has shown that this objective will be elusive and that achievement of nitrogen fixation with cereals is still some distance away although there may be other benefits from inoculation of field crops (Okon and Labandera-Gonzolez, 1994). Despite this, the probability of eventual success of nitrogen fixation with

_____
* FAX No: +612 9351 5108.
E-mail: kennedy@spiro.biz.usyd.edu.au

cereals should now be regarded as significant. Success will require a period of intensive developmental research, seeking a better understanding of the nature of plant-microbial associations while overcoming the obstacles to an effective symbiosis in a progressive fashion.

The thesis to be developed in this paper is that achieving this ambitious goal using microsymbionts such as *Azospirillum* may demand a process of *facilitated evolution,* designed to ultimately yield a spontaneously effective symbiosis. An essential set of desirable genetic characteristics need to be identified and then selected in new combinations of existing microbial and plant germ plasm or by deletions of genes that currently prevent desirable outcomes. In this review paper, it is intended to test this goal against recent research findings in the literature and to demonstrate that progress towards the goal has already been achieved.

Evolution of nitrogen-fixing symbiosis

In considering the possible evolution of a nitrogen-fixing symbiosis for cereals it is worth reviewing the information available regarding the evolution of the successful symbioses involving rhizobia (including bradyrhizobia and azorhizobia) and *Frankia* as microsymbionts.

Symbiosis may be defined as a mutual interdependence of dissimilar organisms. Its main characteristic is the complementary nature of the metabolic apparatus of each of the partners so that the overall task of satisfying nutritional needs for growth and reproduction are shared by both organisms. Although discussion of the conditions leading to the successful evolution of nitrogen fixing symbioses tend to be rather speculative it is significant that many species are involved as partners in the nitrogen-fixing symbioses. The apparent range of species involved is so diverse that several independent developments of nitrogen-fixing symbiosis in legumes have been proposed (Provorov, 1994). The advantage to the host plant of symbiotic nitrogen fixation is obvious but Udvardi and Kahn (1993) point to the key fact that the rhizobia have evolved a genetic regulatory system that requires nitrogen fixation to occur without associated ammonia assimilation in the bacteroids. The interactive signal transduction pathway induces nitrogenase in response to low oxygen concentration, even when bacterial cells have enough nitrogen. They pose and attempt to answer the ques-

tion of what the bacteroids gain by fixing nitrogen if they are not nitrogen-starved. However, attempting to assess the value of fixing nitrogen to bacteroids may be unnecessary in considering the phenomenon of symbiosis. Probably of much more significance to the rhizobia in terms of evolutionary selection pressure is the enhanced ability to eventually increase cell numbers and therefore to persist in soil.

Another hypothesis suggests (Sprent, 1994) that two separate nodulation events occurred in the humid tropics during the evolution of legumes in the late Cretaceous period. One of these involved an ancestor of *Rhizobium* and entry by root infection. This was initially parasitic and provided little benefit until bacteria were released from infection threads as in modern crop species. The other concerned a photosynthetic ancestor of *Bradyrhizobium* using a wound infection on stems, but never involved infection threads. The need for tolerance to stress by bacterial strains would also impose evolutionary constraints, where not all rhizobia (e.g. in arid regions) are capable of symbiosis, since symbiotic genes may be an expensive encumbrance in these conditions. Lateral transfer of material on megaplasmids could lead to a wide range of symbiotic and non-symbiotic forms in response to local pressures. When environmental constraints are superimposed on initial evolutionary developments, the result is an apparently chaotic situation where there is no obvious pattern of co-evolution between hosts and rhizobia. Evidence of such coevolution may be amenable to molecular analysis. Direct evidence of such interactive coevolution of host plant and microsymbiont was presented by Devine (1988) in the case of soybeans related to the production of rhizobitoxine by bradyrhizobia that can potentially interfere with chlorophyll synthesis leading to chlorotic plants.

Apart from the symbioses involving nodulated legumes (plus the non-legume *Parasponia*) and the rhizobia, a diverse range of non-legumes form nodular symbioses with *Frankia*. The taxonomic scheme of Cronquist (1981) classified plants with root nodule symbioses so that host plants of rhizobial symbionts were placed in subclasses Rosidae and Hamamelidae and those of *Frankia* in the widespread subclasses of Rosidae, Hamamelidae, Magnoliidae and Dilleniidae.

More recently, this broad phylogenetic distribution of nodulated plants has been challenged and angiosperm phylogenies based on DNA sequence comparisons reveals a more coherent group than previously thought (Chase et al., 1993; Soltis et al., 1995). This molecular data now indicates a single origin of the predisposition for root nodule symbiosis and also supports the occurrence of multiple origins of symbiosis within this more restricted group (Soltis et al., 1995; Swensen and Mullin, 1997). These findings indicate that only one small group of angiosperms in a single clade possesses the ability to host nitrogen-fixing microsymbionts.

There is still sufficient plant and microbial diversity involved to surmise that the initiation of nitrogen-fixing symbiosis was not an absolutely unique event. Possibly, some plant species or their ancestors may have periodically lost and then regained a symbiotic state in response to altered selection pressures. The complexity of the molecular dialogue now recognised as controlling the establishment of many legume-*Rhizobium* symbioses presumably represents a refinement of simpler relationships existing formerly. However, no mechanism can readily be proposed that would allow a sequential development of such a complex interactive process without initial advantages such as mutually beneficial nitrogen fixation. Thus we may assume that even legume symbiosis involved a prototype stage of development involving fewer genes but of lower stability. Alternatively, advantages such as improved mineral nutrition, stimulation of plant root growth or mild parasitism may have provided initial mutual benefits favouring acceptance of microbes by the plant host, with insertion of nitrogen fixation at a later stage. The genetic significance of the host plant in evolving an effective nitrogen-fixing symbiosis is obvious, supported by direct experimental evidence of heritable genetic variation for nitrogen fixation in pasture (Pinchbeck et al., 1980) and grain legumes (Smartt, 1986).

The reasons that cereals and most other Gramineae provide little clear evidence of effective symbioses with nitrogen-fixing microorganisms bear examination. The cereals do not fall within the clade of nodulating angiosperms mentioned above and it can be surmised that nodulation of a similar kind to that involving rhizobia and *Frankia* might be more difficult than previously thought (Swensen and Mullin, 1997). Therefore, a less developed nitrogen-fixing association between bacteria and Graminae might be be a preferable goal at this stage. Possibly, the relatively low nitrogen content of annual cereals does not provide sufficiently strong sink strength for nitrogen to elicit evolution involving as many genes as found in legume symbiosis. Although modern cereals bred for much higher yield than their genetic ancestors probably require more nitrogen, the deliberate choice of

fertile soil or its supplementation with fertiliser-N may have minimised selection pressures for developing or stabilizing nitrogen-fixing associations in cereal crops. For this reason, the search for germ plasm to facilitate nitrogen fixation in cereals should not ignore primitive cereals. The fact that cereals require much less nitrogen than legumes both simplifies the task in terms of the rates of nitrogen fixation required but possibly destabilizes any putative symbiosis that may develop since the needs for nitrogen may be more readily met from the soil.

Since our review of biological nitrogen fixation in non-leguminous field crops that included a discussion of *para*-nodulation (Kennedy and Tchan, 1992) considerable advances in knowledge have occurred. It is also striking that the focus is now on several nitrogen-fixing bacteria paid little attention before 1990. For example, strong credence is now granted to the significance of the naturally-occurring nitrogen-fixing association between sugarcane and *Acetobacter diazotrophicus* discussed below. Several other bacterial species potentially capable of nitrogen fixation with cereals are receiving more attention, including *Herbaspirillum* sp. and *Azoarcus* sp. Verification that some of these bacteria carry out symbiotic nitrogen fixation with crop plants and were already bringing previously unrecognised benefits to agricultural production is providing a welcome stimulus to attempts to engineer nitrogenfixation for the growth of wheat and rice. The availability of a range of genetic tools and reporter genes since 1992 is also providing strong impetus to the search for new information of value for this objective.

Models for development of N_2 fixing symbiosis in cereals

Key factors in establishing effective symbioses with cereals

Kennedy and Tchan (1992) discussed the key issues and problems likely to be encountered in any program of research seeking nitrogenfixation in cereals. In summary, these issues included:

1. the importance of adequate colonization and the probable ineffectiveness of diazotrophs located at the root surface (rhizoplane) in achieving significant nitrogenfixation. The main reasons for this included the inadequacy of carbon substrates diffusing from the root surface and competition

for these substrates from other microbes. Quispel (1991), in a discussion of the topic of achieving nitrogen fixation in cereals, drew attention to the probable significance of endophytic diazotrophs located within the root system in achieving such an objective. Recent research data to be described in this paper emphasises the validity of his proposal.

2. the oxygen paradox, posed by the fact that this gas is simultaneously an essential terminal electron acceptor for respiratory synthesis of ATP yet is extremely toxic to the nitrogenase enzyme. A large part of the physiological and morphological character of leguminous and non-leguminous nodules is devoted to regulating oxygen pressure within a satisfactory range.

3. the effective transfer of nitrogen to the host plant. This will be discussed later, but as indicated above in the discussion of evolution in legumes, the efficient transfer of newly-fixed nitrogen is one of the most salient features of the legume symbiosis. Even in the case of symbioses based on rhizobia and *Frankia,* there is obviously much to learn about the regulation of this process.

These issues regarding carbon substrates, oxygen and nitrogen transfer remain of central importance. However, as a result of new discoveries with both naturally occurring systems and with laboratory models, progress in our understanding of the solutions required is better defined. This review paper will illustrate this conclusion using some natural systems and then by focussing on the performance of a various diazotrophs in *para*-nodulated wheat.

Sugar cane

^{15}N and nitrogen balance studies conducted recently (see Boddey et al., 1995 for review) have shown that certain Brazilian cultivars of sugar cane can obtain over half their needs for nitrogen from BNF (>150 kg N ha^{-1} year^{-1}). The discovery of endophytic N$_2$-fixing bacteria within the roots, shoots and leaves of sugar cane is suggestive that these organisms are responsible, and that earlier scepticism about these claims was misplaced. It is suggested that the fact that sugar cane in Brazil has been bred over a long period for high yields with low fertilizer-N inputs has favoured the emergence of this naturally occurring symbiosis.

The key bacteria recognised so far in N$_2$-fixing sugar cane are *Acetobacter diazotrophicus* (Cavalcante and Döbereiner, 1988; Reis et al., 1994) and *Herbaspirillum* spp. (Baldani et al., 1986). Strains of *Herbaspirillum* also infect gramineous crops including rice in both the roots and the aerial tissue. *A. diazotrophicus* is a Gram-negative rod forming a rising pellicle in N-free medium with 100 g L^{-1} of sucrose. It is claimed to show a higher oxygen tolerance than *Azospirillum* species, continuing to fix up to 4 kPa (Reis et al.,1990). Its demonstrated capacity to directly transfer half the nitrogen fixed to an amylolytic yeast (*Lypomyces kononenkoae*) in mixed culture (Cojho et al., 1993) suggests that *A. diazotrophicus* would be capable of a similar transfer of fixed nitrogen to the plant tissues in sugar cane. Apparently, the high sugar requirement of this organism prevents it being found in significant numbers in soil but it has been recovered from a few other plant species. Studies on infection and colonization of micropropagated sugar cane seedlings using immunogold labelling indicated that it favoured crevices associated with lateral roots (James et al.,1994) and it is suggested that the bacteria are able to migrate within the xylem stream to the tops of the plants.

H. seropedicae with several other species have been isolated from sugarcane roots, stems and leaves but this species survives poorly in soil (Baldani et al., 1992). However, sorghum seedlings were able to stimulate growth of residual organisms in soil so that increased numbers of bacteria could be observed in the rhizosphere and roots (Olivares et al., 1993).

Döbereiner (1996) suggests that this organism is an obligate endophyte and that in sugar cane, it could be capable of conducting a complementary metabolism with *A. diazotrophicus* by consuming organic acids formed by the latter from sugar, thus maintaining homeostasis for pH within the plant tissues.

For future research with cereals, it will be of interest to obtain definition of the following aspects of this symbiosis:

1. The processes of infection and initial colonization of plants and the cell numbers per unit of host plant tissue needed for effective nitrogen fixation. Current data infers that the whole plant may be colonized by the nitrogen-fixing microsymbionts. If so, this raises some interesting questions regarding protection from oxygen produced in leaves, provision of carbon substrates at many such sites and the ability of bacteria and plant cells to achieve symbiosis.

2. The possible advantages of mixed communities of endophytic diazotrophs in achieving effective nitrogen fixation.

3. The robustness of the system under conditions of adequate nitrogen from soil. It is suggested that fertilizer nitrogen acts to severely reduce the ability of these diazotrophs to carry on nitrogen fixation. It will be important to establish the degree to which the system can tolerate periodic availability of fixed nitrogen and whether recovery of nitrogen-fixing ability can be achieved when required.

Azoarcus *spp. as nitrogen-fixing endophytes*

Originally isolated from Kallar grass (*Leptochloa fusa* Kunth) growing in saline-sodic soils of Pakistan, a strain of the gram-negative nitrogen-fixing bacterium *Azoarcus* sp. BH72 has been described (Reinhold-Hurek et al., 1993) that also colonizes grasses such as rice in laboratory experiments (Hurek et al., 1994). *Azoarcus indigens* is the type species and a second named species, *Azoarcus communis*, includes a strain obtained from French refinery oily sludge (Reinhold-Hurek et al., 1993). Bacteria of this genus have a strictly aerobic type of metabolism, fixed nitrogen microaerobically, and grew well on salts of organic acids but not on carbohydrates; the five species have now been identified as belonging to the beta subgroup of Proteobacteria (Hurek and Reinhold-Hurek, 1995). Related toluene-degrading bacterial isolates were investigated by 16S rRNA sequence analysis and a study of their biochemical and physiological features made (Zhou et al., 1995). Phylogenetic trees were constructed from 16S rRNA sequences and showed the isolates as belonging to a phylogenetically coherent cluster representing a sister lineage to *Azoarcus* species. A new species of the nitrogen-fixing genus *Azoarcus* was named *Azoarcus tolulyticus.*

A thorough and in-depth study of *Azoarcus* as an endophyte is yielding potential benefits, particularly the observation of extremely active form of nitrogenase activity associated with a characteristic morphology of the cells. Such high nitrogenase expression may be important in obtaining adequate rates of nitrogen fixation in associative systems with limited numbers of bacterial cells and the phenomenon deserves further study.

Ultrastructural analysis of cells in the course of hyperinduction revealed that complex stacks of intracytoplasmic membranes called diazosomes are formed; these are absent under standard nitrogen-fixing conditions (Hurek et al., 1995). The iron protein of nitrogenase was highly enriched on these membranes, as evidenced by immunohistochemical studies. Diazosome deficiency in NifH/K- mutants, a deletion mutant in the *nifK* gene and the character of NH_4^+-grown cells suggested, in concert with the membrane localization of nitrogenase, that these structures are specialized membranes related to nitrogen fixation. In batch cultures of *Azoarcus* sp. strain BH72, at nanomolar oxygen concentrations in the presence of proline, cells can shift into a state of higher activity and respiratory efficiency of nitrogen fixation in which diazosomes related to nitrogen fixation are formed (Karg and Reinhold-Hurek, 1996). Induction of intracytoplasmic membranes was most pronounced when *Azoarcus* sp. strain BH72 was cultured with an ascomycete originating from the same host plant, Kallar grass.

The invasive properties of *Azoarcus* sp. strain BH72, the endorhizospheric isolate of Kallar grass, on gnotobiotically grown seedlings of rice cv. IR36 and *L. fusca* were studied (Hurek et al., 1994). To facilitate localization and to assure identity of bacteria, genetically engineered microorganisms expressing β-glucuronidase were also used as inocula. β-glucuronidase staining indicated that the apical region of the root behind the meristem was the most intensively colonized and light and electron microscopy showed that strain BH72 penetrated the rhizoplane preferentially in the zones of elongation and differentiation and colonized the root interior both inter- and intracellularly. In addition to the root cortex, bacteria were also found in the xylem. No evidence was found for *Azoarcus* residing in living plant cells; plant cells were apparently destroyed after bacteria had penetrated the cell wall. A common pathogenicity test developed for tobacco leaves indicated that representative strains of *Azoarcus* spp. are not phytopathogenic. Compared with the non-inoculated controls, inoculation with strain BH72 significantly promoted growth of rice seedlings. However, this effect was reversed when the plant medium was supplemented with malate (0.2 g L^{-1}). Nitrogen fixation was apparently not involved, because the same response was obtained with a *nifK* mutant of strain BH72 with a Nif$^-$ phenotype. PCR and Western immunoblotting, using primers specific for eubacteria and antibodies recognizing type-specific antigens, respectively, indicated that strain BH72 could colonize rice plants systemically. When this data is considered with the observation that *Azoarcus indigens* expresses *nifH* and acetylene reduction activity to a maximum oxygen tension of 6.5% (Vande-Broek et al., 1996), this genus as a potential endophytic nitrogen-fixing microsymbiont for cereals merits continued study.

Hypertrophies on rice roots and their colonization by microbes

Recently, de Bruijn et al. (1995) have reviewed the potential and pitfalls of extending symbiotic interactions between nitrogen-fixing organisms and cereals such as rice. They considered chemical signalling between plant and microbe, nodulation and the prospects for symbiotic nitrogen fixation with such systems and the reader is advised to consult this critical review for this discussion which will not be covered here. The results of experiments carried out in China on the induction of "nodule-like" structures on rice roots by rhizobia were highlighted and data on attempts to extend these experiments in the USA given, providing some additional understanding of the induction of these structures by rhizobia and possibilities for their colonization.

They concluded that the formation of hypertrophies on rice roots infected and colonized with microbes had been confimed, but that little evidence supported their designation as "nodules" or even nodule-like. The frequency of formation of the root structures was low, but non-saprophytic colonization of the rice-root endorhizosphere had clearly been observed and deserved further study in this important area of research.

Rhizobia and azorhizobia as inoculants for cereals

There are potentially many approaches to the development of symbioses in cereals. The legume-*Rhizobium* or *Frankia* root nodule symbioses may be used directly as models, seeking to identify key features that would be regarded as essential for the development of analogous systems in cereals. The earlier approach used by Cocking and associates with wild *Parasponia* rhizobia and wheat and the recognition of a moderate degree of intercellular and intracellular infection associated with shortened, thickened lateral roots (Cocking et al., 1992) could be recognised as seeking a more primitive stage of legume nodule development, following their earlier work on more direct methods of inducing nodular structures with enzymes. More recently, this group has focussed their attention on associations between azorhizobia and cereals, also studied with some success by Chen et al. (1992), seeking evidence of possible chemical signalling between the bacteria and host plants that could influence success in establishing successful symbiosis based on colonization associated with the points of emergence of lateral roots (Webster et al., 1997). Using a *lacZ* marker gene, they showed that the flavanone naringenin at 10^{-5} M stimulated significantly the colonization of lateral root cracks and intercellular colonization of wheat roots by azorhizobia. Our experience with *Azorhizobium caulinodans* in *para*-nodulated wheat has been described (Kennedy et al., 1997; Yu and Kennedy, 1995). Acetylene reduction rates with *A. caulinodans* were found to be much less than with a strain of *Azospirillum brasilense* - mainly as a result of the small proportion of the *para*-nodules that were colonised (Yu and Kennedy, 1995). Recently Sabry et al. (1997) showed that wheat grown in pots and inoculated repeatedly with *A. caulinodans* colonised root tissues at the points of emergence of lateral roots and appeared to contribute significant amounts of nitrogen fixation to the plant. Acetylene reduction rates were similar to those reported with *para*-nodulated wheat. A potential advantage of *A. caulinodans* as an endophyte is that, like *Azospirillum*, it can more readily express nitrogenase activity than other members of the Rhizobiaceae which almost generally only fix nitrogen in legume nodules.

The para-*nodule induced with synthetic auxins such as 2,4-D*

During the past four years since we reviewed the status of biological nitrogen fixation in non-leguminous field crops (Kennedy and Tchan, 1992), several papers (see Katupitiya et al., 1995a, 1995b; Kennedy, 1994; Yu and Kennedy, 1995) have described our development of the laboratory model of associative N_2 fixation known as the *para*-nodule. The approach is based on the initial observations of Yanfu Nie at Shandong University that 2,4-D acting as an auxin could stimulate colonization by rhizobia in 'nodular structures' modified from the lateral roots of many plant species. We have emphasized that these structures, derived from the induction of the initials of lateral roots, are quite dissimilar to legume nodules particularly when colonized by azospirilla. In this case, the bacteria colonize intercellularly almost exclusively, usually in the basal zone of the *para*-nodules where plant cells appear loosely packed. Even when extensively infected with rhizobia (Nie et al., 1992), there is no evidence of intracellular colonization of living wheat root cells as occurs in legume nodules.

Although the use of synthetic auxin such as 2,4-D presents difficulties for field application, we consider that the use of this procedure is justified as a laboratory model, providing rates of improved colonization and nitrogenfixation that allow the controlling factors to be studied. It may prove possible to provide similar outcomes to the use of 2,4-D biologically, such as by using mutants producing higher amounts of indoleacetic acid. However, it should be recognised that many systems of agricultural monoculture already apply very stringent chemical treatments, sometimes with dramatic effects on plant development (e.g. cotton production using hormones and chemical defoliants).

We have not so far sought to demonstrate that *para*-nodulated wheat can fix nitrogen sufficiently well to become independent of soil nitrogen, preferring to concentrate on the factors controlling colonization in preliminary work. However, with improved experimental techniques we have now obtained enrichments using $^{15}N_2$ much higher than those reported previously (Yu et al., 1993), around 0.5 atom % ^{15}N in root total-N for overnight exposures, suggestive that most of the approximately 10^7 azospirilla in the root system of wheat seedlings (per 100 mg fresh weight) are actively fixing nitrogen (Wood and Kennedy, in preparation; see Kennedy et al., 1997).

Of extreme importance in this recent research has been the use of genetic markers and reporter genes based on fusions of the promoters of *nif* genes of *Azospirillum brasilense* with *lacZ* (Liang et al., 1991) and their application to the study of the colonization of wheat roots by azospirilla in collaborative work between our two laboratories (Arsène et al., 1994). These fusions have the facility to allow both visual and enzymic detection of bacterial cells in association with wheat. It is probable that analysis of β-galactosidase activity in the case of constitutively expressed fusions provides a better estimate of bacterial cell numbers than most probable number dilution counts. Similar genetic tools for the study of colonization of cereal roots by azospirilla based on the use of *gusA* fusions are also being applied for studies with cereals (Vande-Broek et al., 1993, 1996). These genetic tools provide much more reliability both in the data obtained and more confidence in its interpretation (see also Stoltzfus et al., 1997).

Key factors in colonization of para-nodules by *A. brasilense*

Our recent research using the genetic tools with *para*-nodulated wheat has led to the recognition of a number of factors that facilitate improved N_2 fixation under our specific conditions of assay.

Endophytic colonization by the flocculation$^-$ mutant of A. brasilense *Sp7*

A fortunate spontaneous mutation in our laboratory strain of *A. brasilense* Sp7, resulted in the selection of a new strain (Sp7-S) showing inability of colonies to bind Congo red with the loss of flocculating ability (Katupitiya et al., 1995b) from lowered exopolysaccharide (EPS) production and reduced swarming (Pereg et al., 1996). This reduction in EPS production was accompanied by loss of the ability of the bacterial cells to adhere to the wheat root surface when seedlings were inoculated. At the same time as the loss of EPS production there was an increased ability of Sp7-S to colonise wheat roots internally (endophytically), in an intercellular fashion, particularly in crevices at the point of emergence of lateral roots and in the basal region of the *para*-nodules (auxin-modified lateral roots). A simple interpretation is that the absence of EPS production provided a smooth-surfaced bacterial cell with less ability to adhere to the root cell surface of plants. The 'stickiness' of the wild-type Sp7 may tend to trap these cells on root surfaces and inhibits 'crack-entry' of cells into crevices at the root surface and near the points of emergence of lateral roots or *para*-nodules. The presence of such points of entry in the loosely structured monocots such as wheat was pointed out previously (Kennedy and Tchan, 1992). The associated qualities of absence of encystment and retention of the vegetative phenotype by Sp7-S both in pure culture and on wheat roots also seems likely to be of importance.

Upon prolonged incubation in semisolid or solid media many isolates of azospirilla form ovoid cysts which are distinct from the typical vibrioid or S-shaped cell morphology of this genus (Lamm and Neyra, 1981). Cyst-enriched cultures are much more resistant to desiccation than normal vegetative cultures (Lamm and Neyra, 1981; Papen and Werner, 1982) and their formation is presumed to represent a mechanism by which azospirilla can persist in soil during unfavourable environmental conditions. Encystment of *A. brasilense* Sp7 (ATCC 29145) is favoured in medium containing fructose (8 mM) and KNO_3 (0.5 mM) forming thick-walled structures filled with poly-β-hydroxybutyrate and associated with the production of a melanin-like pigment in aged cultures (Sadasivan and Neyra, 1987). Cyst formation in broth cultures is associated with flocculation and exopolysaccharide formation (Sadasivan and Neyra, 1985).

We have consistently observed apparent encystment of Sp7 on the wheat root surface after several days of colonization, shown by the ovoid shape of the cells (see Figure 1A); this observation contrasts with the consistently vibrioid shape of cells of Sp7-S when colonizing wheat seedlings (Katupitiya et al., 1995b), remaining in the vegetative form under all conditions we have observed on wheat roots. We have also found using lacZ and gusA fusions with Sp7-S and Sp7 simultaneously that colonization occurs preferentially at the same locations as observed when they colonize wheat separately (Katupitiya et al., 1997). Papen and Werner (1982) showed the inhibition of nitrogenase activity occurred when A. brasilense was encysted in broth culture so we assume that the cyst-like cells of Sp7 on the root surface could also be nitrogenase deficient although we have no direct demonstration of this. However, even these cyst-like cells still exhibit expression of nifH activity as the blue colour of β-galactosidase activity (Arsène et al., 1994; Figure 2), possibly indicative of a residual expression of nitrogenase activity, consistent with the lower activity acetylene reduction activity observed (Katupitiya et al., 1995b).

Colonization, carbon substrates and oxygen conditions suitable for nitrogenase expression in 2,4-D treated wheat seedlings

The effect of 2,4-D treatment on wheat seedlings is complex. Lateral root initials are activated en masse, providing many foci for endophytic infection of roots by 'crack-entry'. However, following initiation these lateral root initials fail to extend, thus providing the rounded nodular structures. This effect is concentration dependent, para-nodules being formed at a concentration of 2,4-D well below that at which herbicidal effects are exerted. Synthetic auxin also induces increased concentrations of soluble sugars (Feng, et al., 1997) and amino acids (Yu and Kennedy, unpublished). A glucose-utilizing mutant of A. brasilense Sp7-S (Sp7-Sg) had a modified pattern of colonization of wheat roots (Feng et al., 1997), with more bacteria occurring in intercellular channels established between rows of cortical cells. However, we have no evidence that this mutant is more capable of nitrogen fixation. In attempting to match the needs for carbon substrates by potential endophytes to those available in the plant it should be considered that cereals like wheat and rice are C_3 plants operating a Calvin photosynthetic cycle rather than the C_4 Hatch-Slack cycle involving organic acids as intermediates (Kennedy, 1992) that occurs in sugar cane.

Evidence has been obtained using nifH-lacZ that the FlcA⁻ mutant (Sp7-S) when colonising para-nodulated wheat roots (Deaker and Kennedy, 1996) shows greater β-galactosidase activity over a higher range of oxygen concentration (Figure 2), consistent with the endophytic mode of colonization by Sp7-S shown in Figure 1B. This result for nifH expression is in close agreement with the rates of acetylene reduction observed previously in para-nodulated wheat (Kennedy and Tchan, 1992), indicating that Sp7-S is more protected from the effects of oxygen than the Sp7 wild-type when colonizing the roots of wheat seedlings. Experiments using nifH-lacZ to monitor suitable oxygen conditions for nitrogen fixation with para-nodulated wheat seedlings growing in soils of different bulk density are currently in progress.

Ammonia excreting mutants

Previous attention has been paid to the selection of ammonia-excreting mutants of azospirilla as having potential benefits in associations with plants (Machado et al., 1991). Christiansen-Weniger (1992) employed an ammonia-excreting of A. brasilense in studies on para-nodulated wheat, but the degree to which this strain can function as an endophyte has not been determined. More recently, Christiansen-Weniger and Vanderleyden (1994) described the colonization of para-nodulated maize roots by the same ammonia-excreting mutant of Azospirillum, labelled with a gus fusion.

A double mutant of Sp7 selected to colonize endophytically and to facilitate ammonia transfer (Sp7-SA) is currently being studied in our laboratory to determine whether there can be a significant benefit from the association between Azospirillum and wheat (Wood and Kennedy, 1996). This research (Wood, Ritchie and Kennedy, unpublished) aims to establish the mechanism of ammonia excretion (see Figure 3) and its relationship with membrane potential, relating the potential to excrete ammonia to the physiological conditions actually existing in wheat roots.

Genetic control of EPS production and flocculating ability

The recognition using lacZ markers that the spontaneous mutant Sp7-S presented a distinctly different

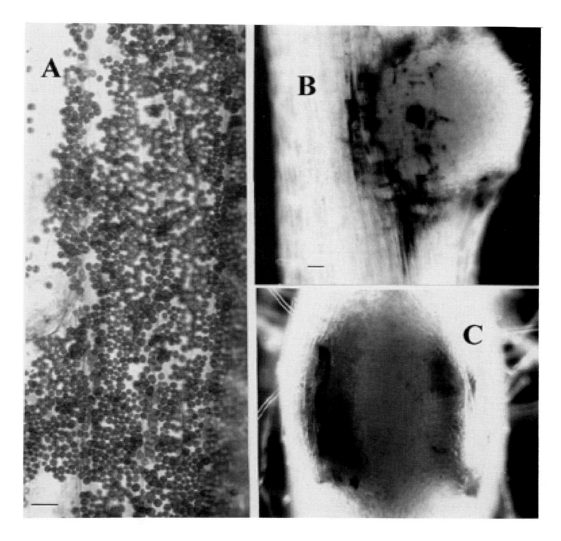

Figure 1. **A**) Cysts of *Azospirillum brasilense* Sp7 *nifA-lacZ* fusions at the rhizoplane of *para*-nodulated wheat seedlings. Plants were grown hydroponically for 10 days after inoculation (Zeman et al., 1992) and stained with X-gal (Arsène et al., 1994). *Bar* = 5 μm. **B**) Endophytic *A. brasilense* Sp7-S-*nifA-lacZ* fusions colonizing *para*-nodules on wheat seedlings stained with X-gal at 10 days growth after inoculation and treatment with 2,4-D. Cells of Sp7-S and Sp7-SA (ammonia-excreting mutant) with a FlcA⁻ phenotype (failing to bind Congo red and to flocculate and encyst in broth) are only found in crevices on the root surface, at the points of emergence of lateral roots and in the basal zone of *para*-nodules as shown in the photograph. **C**) Endophytic *Herbaspirillum seropedicae* (ATCC 35892) *nifA-lacZ* fusions stained with X-gal have a similar pattern of colonization as *A. brasilense* Sp7-S. *Bar* (B and C) = 50 μm.

phenotype to that of Sp7 on wheat roots that also profoundly affected the colonization pattern observed opened the way to genetic analysis of this character. It was found possible to complement Sp7-S with DNA fragments from a gene bank derived from the Sp7 wild-type held by C. Elmerich at the Institut Pasteur, restoring the property of binding of Congo red by colonies on agar plates and flocculation in liquid culture (Katupi-

tiya et al., 1995b). Subsequent studies (Pereg et al., 1995) have shown that site directed Tn*5* insertions in the complementing region of the chromosome of Sp7 produced mutants that did not flocculate or bind Congo red and that could be complemented by a shorter segment of DNA derived from that described previously (Katupitiya et al., 1995b) for a gene involved in the regulation of the ability to flocculate designated *flcA*

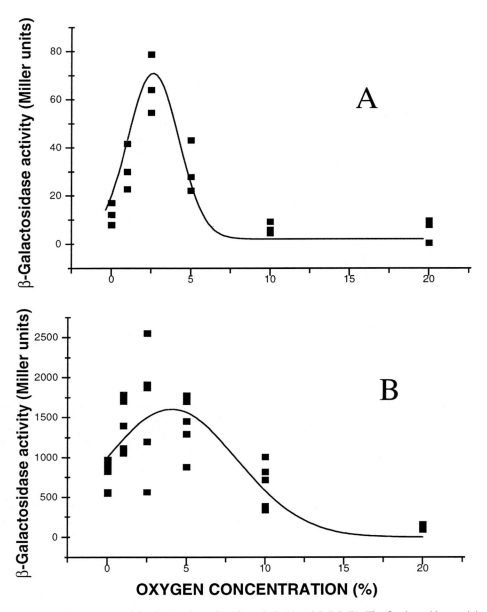

Figure 2. nifH-lacZ expression on *para*-nodulated wheat by *A. brasilense* Sp7 (**A**) and Sp7-S (**B**). The β-galactosidase activity (Miller Units per mg root protein) for separate seedlings is shown, following overnight incubation of seedlings at the oxygen pressure shown. Wheat seedlings were previously grown for several days in air-saturated hydroponic solution, completely repressing *nifH-lacZ* expression (Deaker and Kennedy, 1996 and in preparation).

(Pereg-Gerk et al., 1997). These Tn*5-flcA*⁻ mutants also had reduced surface colonization of wheat roots, similar to Sp7-S, but it was observed that Sp7-S lacked the ability to swarm, probably as a result of the absence of lateral flagellae while the Tn5 mutants retained the ability to swarm (Pereg et al., 1996). Thus the difference in the phenotype for colonization between Sp7

and Sp7-S involves more than one mutation. Sp7-S is still motile and thus probably retains its polar flagellae. Subsequently, the *flcA* gene has been identified, its nucleotide sequence established and the regulation of the gene studied (Pereg et al., 1996, 1997).

The origin of the original stable mutant Sp7-S is uncertain, but it may have originated from deliberate

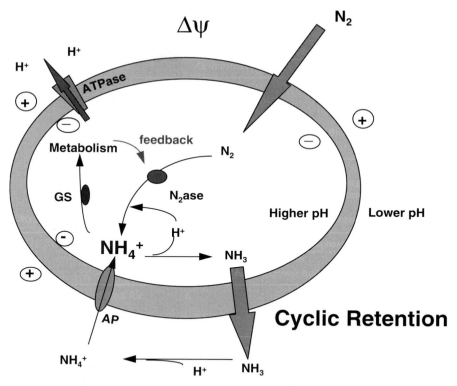

$\Delta\Psi$

Figure 3. Model for nitrogen transfer from diazotroph to host plant. The hypothesis of cyclic retention (Kleiner, 1985) proposes that ammonia is normally retained within bacterial cells for assimilation following nitrogen fixation by an effective transport mechanism (e.g. ammonia permease, AP). In mutants, a lesion in AP, or in the ammonia-assimilating enzymes (glutamine synthetase, GS or glutamate synthase) causing high internal concentrations of ammonia, can lead to leakiness. The mutant *A. brasilense* Sp7-SA has a lesion in GS expression (Wood and Kennedy, 1996) and the influence of membrane potential ($\Delta\Psi$) on possible ammonia leakage is being studied (Wood, Ritchie and Kennedy, unpublished).

attempts by Anita Zeman in our laboratory to count and isolate endophytic cells of *A. brasilense* using a culture of Sp7 from the CSIRO (Division of Plant Industry, Canberra, Australian Capital Territory) collection obtained by Y T Tchan from the late Dr Alan Gibson around 1989. A second example of the Canberra Sp7 culture obtained subsequently by the senior author from Dr Gibson was dissimilar to Sp7-S, binding Congo red and flocculating in broth, but still appeared intermediate in other phenotypic properties between Sp7 from the Institut Pasteur collection of C. Elmerich and Sp7-S, indicating that Sp7 (Paris) and Sp7-S differ in more than one gene. Possibly, sterilisation of the root surface using brief treatments with mercuric chloride or ethanol (Sriskandarajah et al., 1993; Zeman et al., 1992) preferentially selected Sp7-S since it is predominantly endophytic in 2,4-D treated wheat seedlings.

Subsequent work on the isolation of spontaneous mutants of Sp7 forming colonies that fail to bind Congo red and flocculate has shown that such mutations are relatively frequent. Some of these mutations proved unstable (L Pereg-Gerk, unpublished), but no instances of Sp7-S losing the mutant character has been observed during five years of laboratory culture.

It is noteworthy that the original source of both these cultures of Sp7 (Canberra and Paris) was directly from Johanna Döbereiner in Brazil, probably in the late 1970s. Obviously, a degree of evolutionary drift of genotypes has occurred. On the other hand, the significant phenotypic differences in colonization of wheat roots that have resulted produced fortunate results in our laboratory.

Performance of other nitrogen-fixing bacteria in para-*nodules*

It was of interest to determine the ability of different nitrogen-fixing bacteria to colonize wheat seedlings

Table 1. Colonization pattern and acetylene reduction activity by different strains of diazotrophs in *para*-nodulated wheat

Strain	Predominant colonization mode	Acetylene reduction assay	Oxygen sensitivity
A. brasilense Sp7	Rhizoplane	+	High
A. brasilense Sp7-S	Endophytic	+++++++	Medium
A. brasilense Sp7-SA	Endophytic	+++++++	Medium
A. lipoferum	Rhizoplane	++++	Medium
Herbaspirillum seropedicae	Endophytic	++++++	Medium
Acetobacter diazotrophicus	Rhizoplane	-	-
Derxia gummosa	Rhizoplane	++	Low
Azotobacter vinelandii	Rhizoplane	+++++	Low
Rhizobium spp.	Endophytic	-	-

Wheat seedlings were treated with 2,4-D and inoculated with diazotrophs as previously described (Zeman et al., 1992). Colonization mode was determined by light microscopy using *nifA-lacZ* transconjugants stained with X-gal (Arsène et al., 1994).

(Kennedy et al., 1997). The performance of seven different bacterial strains is summarized in Table 1. As shown in the table, only *A. brasilense* Sp7-S, *A. brasilense* Sp7-SA (an ammonia-excreting mutant isolated in this laboratory by C. Woods) and *Herbaspirillum seropedicae* performed as endophytes in a manner meeting the criteria for significant nitrogen fixation. The colonization pattern of *H. seropedicae* (ATCC 35892) on wheat seedlings is similar to that of *A. brasilense* Sp7-S (see Figure 1b) , the *flcA⁻* mutant. According to Döbereiner (1996), *Herbaspirillum* is an obligate endophyte and is not usually found in the exorhizosphere. This is borne out by our experiments using *lacZ* transconjugants (Kennedy et al., 1997). It will be of interest to see whether the EPS composition of its surface is similar to that observed with the Sp7-S mutant.

To identify associative nitrogen fixers that possess a more oxygen-tolerant nitrogen-fixation system, Vande-Broek et al. (1996) analyzed the expression of an *Azospirillum brasilense nifH-gusA* fusion and the acetylene reduction activity as a function of the oxygen concentration in eight aerobic associative diazotrophs (*Acetobacter diazotrophicus, Alcaligenes faecalis, Azoarcus indigens, Azorhizophilus paspali, Azospirillum brasilense, Azospirillum irakense, Burkholderia vietnamiensis* and *Herbaspirillum seropedicae*). Based on maximum oxygen concentration for activation of the *nifH* fusion and acetylene reduction, these organisms were classified into three groups of increasing oxygen tolerance. The groups were: (i) *Acetobacter diazotrophicus, Alcaligenes faecalis, Azospirillum brasilense, Azospirillum irak-*

ense, Burkholderia vietnamiensis and *Herbaspirillum seropedicae* (maximum oxygen tension for acetylene reduction between 2.0 and 3.0%); (ii) *Azoarcus indigens* (maximum oxygen tension for acetylene reduction 6.5%); and (iii) *Azorhizophilus paspali* (maximum oxygen tension for acetylene reduction at 8.5%).

However, our results with different strains of *Azospirillum* and other organisms such as *Derxia gummosa* indicates that the pattern of colonization of wheat seedlings and the degree of endophytic character may be of more significance than the inherent tolerance to oxygen. No doubt the availability of reporter genes such as *nifH-gusA* and *nifH-lacZ* will greatly facilitate future work designed to show the the effectiveness of colonization of cereal roots by diazotrophs.

Conclusion

The ambitious goal of achieving nitrogen fixation by cereals naturally arouses controversy. However, despite the biological obstacles, there are strong and obvious grounds for making an attempt to achieve it, based on environmental and economic considerations. This goal is in keeping with the trend towards a more organically based agriculture and would be welcomed by many sympathetic to this movement and by the large number of subsistence farmers world-wide for whom fertilizer costs are prohibitive.

It must be recognized that feeding the increasing world's population in the last two decades has only been possible because of a huge expansion in the application of nitrogenous fertilizers and the extra production from biological nitrogen fixation has probably been minimal. Some may even question the need for new systems capable of nitrogen fixation and there is no doubt that fertilizer-N must remain of predominant importance for the time being. In our opinion, the need for sustainable nitrogen-fixing systems is sufficiently great that there is an obligation on scientists to take some prudent risks in setting research goals.

We stress that even though the probability of success of obtaining an effective associative system with wheat soon is low (probably less than 0.5), this probability in the longer term is likely to approach unity. The progress since 1992 discussed in this short review suggests that the research is on schedule to deliver positive outcomes in the medium term of 5-15 years. This is particularly so if success is regarded as including reliable gains in crop production from associations with diazotrophs also involving other interactive factors influencing the nutrition and biological health of field crops.

Further development of the *para*-nodule so that it may be prepared for performance trials under field conditions is now our immediate goal. At this stage, however, the use of synthetic auxins and *para*-nodulation should be regarded as providing a re-iterative laboratory model for selecting desirable genetic features in azospirilla. Currently, there are challenges with respect to factors such as whether the high numbers of azospirilla found endophytically in wheat seedlings (ca. 10^8 cells g^{-1} fresh weight roots) can be maintained in larger plants and whether nitrogenase activity will be adequate with the oxygen concentrations found in soil. It seems likely that overcoming these obstacles will require significant genetic improvements of both host plant and bacteria as further steps in the process of facilitated evolution. However, these improvements may be hastened by selections made *in planta* aimed at generating superior genotypes favouring endophytic colonization of the plant.

Acknowledgements

This work has been supported by research grants from the Australian Research Council and the Australian Grains Research and Development Corporation. We are grateful to Dr Claudine Elmerich of the Institut Pasteur, Paris, without whose collaboration much of the research data reported in this paper would not have been possible.

References

Arsène F, Katupitiya S, Kennedy I R and Elmerich C 1994 Use of *lacZ* fusions to study the expression of *nif* genes of *Azospirillum brasilense* in association with plants. Mol. Plant-Microbe Interact. 7, 748–757.

Baldani V, Baldani J and Döbereiner J 1983 Effects of *Azospirillum* inoculation on root infection and nitrogen incorporation in wheat. Can. J. Microbiol. 29, 924–929.

Baldani J I, Baldani V L D, Seldin L and Döbereiner J 1986 Characterization of *Herbaspirillum seropedicae* gen. nov., sp. nov., a root-associated nitrogen-fixing bacterium. Int. J. Syst. Bacteriol. 36, 86–93.

Baldani V L D, Baldani J I, Olivares F L and Döbereiner J 1992 Identification and ecology of *Herbaspirillum seropediace* and the closely related *Pseudomonas rubrisubalbicans*. Symbiosis 13, 65–73.

Boddey R M, de Oliveira O C, Urquiaga S, Reis V M, de Olivares F L, Baldani V L D and Döbereiner J 1995 Biological nitrogen fixation associated with sugar cane and rice: Contributions and prospects for improvement. Plant Soil 174, 195–209.

Cavalcante V A and Döbereiner J 1988 A new acid-tolerant nitrogen-fixing bacterium associated with sugarcane. Plant Soil 108, 23–31.

Chase M and 37 others 1993 Phylogenetics of seed plants: an analysis of nucleotide sequences from the plastid gene *rbc*L. Ann. Mo. Bot. Gard. 80, 528–580.

Chen TW, Scherer S and Böger P 1992 Nitrogen fixation of *Azorhizobium* in artificially induced root para-nodules in wheat. Science China (B), 35, 1463–1470.

Christiansen-Weniger C 1992 N$_2$ fixation by ammonium-excreting *Azospirillum brasilense* in auxin-induced root tumours of wheat (*Triticum aestivum* L.). Biol. Fert. Soils 12, 100–106.

Christiansen-Weniger C and Vanderleyden J 1994 Ammonium-excreting *Azospirillum* sp. become intracellularly established in maize (*Zea mays*) para nodules. Biol. Fert. Soils 17, 1–8.

Cocking E C, Davey M R, Kothari S L, Srivastava J S, Jing Y, Ridge, R W and Rolfe B G 1992 Altering the specificity control of the interaction between rhizobia and plants. Symbiosis 14, 123–130.

Cojho E H, Reis V M, Schenberg A C G and Döbereiner J 1993 Interactions of *Acetobacter diazotrophicus* with an amylolytic yeast in nitrogen-free batch culture. FEMS Microbiol. Lett. 106, 341–346.

Cronquist A 1981 An Integrated System of Classification of Flowering Plants. Columbia University Press, New York.

Deaker R and Kennedy I R 1996 The use of *nifH-lacZ* fusions in the detection of nitrogen fixation in associations between *Azospirillum* spp. and wheat. Proc. 11th Aust. Nitrogen Fixation Conf. Perth, WA. pp. 34–35. Univ. of WA, Perth.

De Bruijn F J, Jing Y and Dazzo F B 1995 Potential and pitfalls of trying to extend symbiotic interactions of nitrogen-fixing organisms to presently non-nodulated plants, such as rice. Plant Soil 174, 225–240.

Devine T E 1988 Role of the nodulation restrictive allele Rj4 in soybean evolution. J. Plant Physiol. 132, 453–455.

Döbereiner J 1996 Biological N$_2$ fixation by endophytic diazotrophs in non-leguminous crops in the tropics. Abstr. 7th Int. Symp. BNF with Non-legumes, p 2.

Feng L, Copeland L and Kennedy I R 1997 Improved colonization of wheat roots by a glucose-utilising mutant of *Azospirillum brasilense* Sp7. Symbiosis. (*Submitted*).

Hurek T, Reinhold-Hurek B, Kellenberger E and Van Montagu M, 1994 Root colonization and systemic spreading of *Azoarcus* sp. strain BH72 in grasses. J. Bacteriol. 176, 1913–1923.

Hurek T, Van-Montagu M, Kellenberger E and Reinhold-Hurek B 1995 Induction of complex intracytoplasmic membranes related to nitrogen fixation in *Azoarcus* sp. BH72. Mol. Microbiol. 18, 225–236.

Hurek T and Reinhold-Hurek B 1995 Identification of grass-associated and toluene-degrading diazotrophs, *Azoarcus* spp., by analyses of partial 16S ribosomal DNA sequences. Appl. Environ. Microbiol. 61, 2257–2261.

James E K, Reis V M, Olivares F L, Baldani J I and Döbereiner J 1994 Infection of sugar cane by the nitrogen-fixing bacterium *Acetobacter diazotrophicus*. J. Exp. Bot. 45, 757–766.

Karg T and Reinhold-Hurek B 1996 Global changes in protein composition of N_2-fixing *Azoarcus* sp. strain BH72 upon diazosome formation. J. Bacteriol. 178, 5748–5754.

Katupitiya S, New P B, Elmerich C and Kennedy I R 1995a Improved nitrogen fixation in 2,4-D treated wheat roots associated with *Azospirillum lipoferum*: colonization using reporter genes. Soil Biol. Biochem. 27, 447–452.

Katupitiya S, Millet J, Vesk M, Viccars L, Zeman A, Zhao L, Elmerich C and Kennedy I R 1995b A mutant of *Azospirillum brasilense* Sp7 impared in flocculation with modified colonization and superior nitrogen fixation in association with wheat. Appl. Environ. Microbiol. 61, 1987–1995.

Katupitiya S, Wilson K and Kennedy I R 1997 Use of *lacZ* and *gusA* fusions to study effects of competition by azospirilla colonizing *para*-nodulated wheat. Symbiosis (*Submitted*).

Kennedy I R 1992 Acid Soil and Acid Rain, Research Studies Press/John Wiley and Sons, Taunton. 75 p.

Kennedy I R and Tchan Y T 1992 Biological nitrogen fixation in non-leguminous field crops: Recent advances. Plant Soil 141, 93–118.

Kennedy I R 1994 Auxin-induced N_2-fixing associations between *Azospirillum brasilense* and wheat. *In* Nitrogen Fixation with Non-legumes. Eds. N A Hegazi, M Fayez and M Monib. pp 513–523. American University in Cairo Press, Cairo.

Kennedy I R, Katupitiya S, Yu D, Deaker R, Gilchrist K, Pereg-Gerk L and Wood C 1997 Prospects for facilitated evolution of effective N_2-fixing associations with cereals: Comparative performance of *Azospirillum brasilense* Sp7-S with various free-living diazotrophs in *para*-nodulated wheat. Plant Soil. (*Submitted*).

Kleiner D 1985 Bacterial ammonium transport. FEMS Microbiol. Rev. 32, 87–100.

Lamm R B and Neyra C A 1981 Characterization and cyst production of azospirilla isolated from selected grasses growing in New Jersey and New York. Can. J. Microbiol. 27, 1320–1325.

Liang Y Y, Kaminski P A and Elmerich C 1991 Identification of a *nifA*-like regulatory gene of *Azospirillum brasilense* Sp7 expressed under conditions of nitrogen fixation and in air and ammonia. Mol. Microbiol. 5, 2735–2744.

Machado H B, Funayama S, Rigo L U and Pedrosa F O 1991 Excretion of ammonium by *Azospirillum brasilense* mutants resistant to ethylenediamine. Can. J. Microbiol. 37, 549–553.

Nie Y F, Vesk M, Kennedy I R, Sriskandarajah S and Tchan Y T 1992 Structure of 2,4-dichlorophenoxyacetic acid induced *para*-nodules on wheat roots. Phytochem. Life Sci. Adv. 11, 67–73.

Okon Y and Labandera-Gonzolez C A 1994 Agronomic applications of *Azospirilllum*: An evaluation of 20 years worldwide field inoculation. Soil Biol. Biochem. 26, 1591–1601.

Olivares F L, Baldani V L D, Baldani J I and Döbereiner J 1993 Ecology of *Herbaspirillum* spp. and ways of infection and colonization of cereals with these endophytic diazotrophs. *In* Nitro-gen Fixation with Non-legumes, Eds. N A Hegazi, M Fayez and M Monib. pp 357–358. Am. Univ. in Cairo Press, Cairo.

Papen M and Werner D 1982 Organic acid utilization, succinate excretion, encystation and oscillating nitrogenase activity in *Azospirillum brasilense* under microaerobic conditions. Arch. Microbiol. 132, 57–61.

Pereg L L, Millet J, Katupitiya S, Kennedy I R and Elmerich C 1995 Genetic analysis of an *Azospirillum brasilense* Sp7 mutant impaired in flocculation. *In* Nitrogen Fixation: Fumdamentals and Applications, Eds. I A Tikhhonovich, V I Romanov and W E Newton. p 345. Kluwer Academic Publishers, Dordrecht.

Pereg L L, Kennedy I R and Elmerich C 1996 Genetic factors controlling colonization of wheat roots by *Azospirillum brasilense* Sp7. Proc. 11th Aust. Nitrogen Fixation Conf. Perth, WA, pp. 122–123. Univ. of WA, Perth

Pereg-Gerk L, Paquelin A, Gounon P, Kennedy I R and Elmerich C 1997 A transcriptional regulator of the *LuxR-UhpA* family, *FlcA*, controls flocculation and wheat root surface colonization by *Azospirillum brasilense* Sp7. Mol. Plant-Microbe Interact. (*Submitted*).

Pinchbeck B R, Hardin R T, Cook P D and Kennedy I R 1980 Genetic studies of symbiotic nitrogen fixation in spanish clover. Can. J. Plant Sci. 60, 509–518.

Provorov N A 1994 The interdependence between taxonomy of legumes and specificity of their interaction with rhizobia in relation to evolution of the symbiosis. Symbiosis 17, 183–200.

Quispel A 1991 A critical evaluation of the prospects for nitrogen fixation with non-legumes. Plant Soil 137, 1–11.

Reinhold-Hurek B, Hurek T, Gillis M, Hoste B, Vancanneyt M, Kersters K and De Ley J 1993 *Azoarcus* gen. nov., a nitrogen fixing Proteobacteria associated with the roots of Kallar grass (*Leptochloa fusca* (L.) Kunth.), and description of two species *Azoarcus indigens* sp. nov. and *Azoarcus communis* sp. nov. Int. J. Syst. Bacteriol. 43, 574–588.

Reis V M, Olivares F L and Döbereiner J 1994 Improved methodology for isolation of *Acetobacter diazotrophicus* and confirmation of its endophytic habitat. World J. Microbiol. Technol. 10, 101–104.

Reis V M, Zang Y and Burris R H 1990 Regulation of nitrogenase activity by ammonium and oxygen in *Acetobacter diazotrophicus*. Ann. Acad. Bras. Cienc. 62, 317.

Sabry S R S, Saleh S A, Batchelor A, Jones J, Jotham J, Webster G, Kothari S L, Davey M R and Cocking E C 1997 Endophytic establishment of *Azorhizobium caulinodans* in wheat. Proc. Roy. Soc. Lond. B 264, 341–346.

Sadasivan L and Neyra C A 1985 Flocculation in *Azospirillum brasilense* and *Azospirillum lipoferum*: exoployssaccharides and cyst formation. J. Bacteriol. 163, 716–723.

Sadasivan L and Neyra C A 1987 Cyst production and brown pigment formation in aging cultures of *Azospirillum brasilense* ATCC 29145. J. Bacteriol. 169, 1670–1677.

Smartt J 1986 Evolution of grain legumes. VI. The future - the exploitation of evolutionary knowledge. Exp. Agric. 22, 39–58.

Soltis D E, Soltis P S, Morgan D R, Swensen S M, Mullin B C, Dowd J M and Martin P G 1995 Chloroplast gene sequence data suggest a single origin of the predisposition for symbiotic nitrogen fixation in angiosperms. Proc. Natl. Acad. Sci. 92, 2647–2651.

Sprent J I 1994 Evolution and diversity in the legume-rhizobium symbiosis: chaos theory? Plant Soil 161, 1–10.

Sriskandarajah S, Kennedy I R, Yu D and Tchan Y T 1993 Effects of plant growth regulators on acetylene-reducing associations between *Azospirillum brasilense* and wheat. Plant Soil 153, 165–178.

Stoltzfus J R, So R, Malarvithi P P, Ladha J K and de Bruijn F J 1997 Isolation of Endophytic bacteria from rice and assessment of their potential for supplying rice with biologically fixed nitrogen. Plant Soil 194, 25–36.

Swensen S M and Mullin B C 1997 Phylogenetic relationships among actinorhizal plants: the impact of molecular systematics and implications for the evolution of actinorhizal symbioses. Physiol. Plant. (In press).

Udvardi M K and Kahn M L 1993 Evolution of the (Brady) Rhizobium-legume symbiosis: why do bacteroids fix nitrogen? Symbiosis 14, 87–101.

Vande-Broek A, Michiels J, Van Gool A and Vanderleyden J 1993 Spatial-temporal colonization patterns of Azospirillum brasilense on the wheat root surface and expression of the bacterial nifH gene during association. Mol. Plant-Microbe Int. 6, 592–600.

Vande-Broek A, Keijers V and Vanderleyden J 1996 Effect of oxygen on the free-living nitrogen fixation activity and expression of the Azospirillum brasilense NifH gene in various plant-associated diazotrophs. Symbiosis 21, 25–40.

Webster G, Gough C, Vasse J, Batchelor C A, O'Callaghan K J, Kothari S L, Davey M R, Dénarié J and Cocking E C 1997 Interactions of rhizobia with rice and wheat. Plant Soil 194, 115–122.

Wood C and Kennedy I R 1996 Ammonia excreting mutants of Azospirillum. Proc. 11th Aust. Nitrogen Fixation Conf, Perth, WA. pp. 36–37. Univ. of WA, Perth.

Yu D, Kennedy I R and Tchan Y T 1993 Verification of nitrogenase activity (C_2H_2 reduction) in Azospirillum populated 2,4-dichloroacetic acid induced root structures of wheat. Aust. J. Plant Physiol. 20, 187–195.

Yu D and Kennedy I R 1995 Nitrogenase activity (C_2H_2 reduction) of Azorhizobium in 2,4-D-induced root structures of wheat. Soil Biol. Biochem. 27, 459–462.

Zeman A M M, Tchan Y T, Elmerich C and Kennedy I R 1992 Nitrogenase active wheat-root para-nodules formed by 2,4-dichlorophenoxyacetic acid (2,4-D)/Azospirillum. Res. Microbiol. 143, 847–855.

Zhou J Z, Fries M R, Chee-Sanford J C and Tiedje J M 1995 Phylogenetic analysis of a new group of denitrifiers capable of anaerobic growth on toluene and description of Azoarcus tolulyticus sp. nov. Int. J. Syst. Bacteriol. 45, 500–506.

Guest editors: J K Ladha, F J de Bruijn and K A Malik

Plant and Soil **194**: 81–98, 1997.

Rhizobial communication with rice roots: Induction of phenotypic changes, mode of invasion and extent of colonization

P.M. Reddy[1], J.K. Ladha[1,4], R.B. So[1], R.J. Hernandez[1], M.C. Ramos[1], O.R. Angeles[1], F.B. Dazzo and Frans J. de Bruijn[2,3]
[1]*International Rice Research Institute, P.O. Box 933, Manila 1099, Philippines*, [2]*Department of Microbiology and* [3]*DOE Plant Research Laboratory, Michigan State University, East Lansing, MI 48824, USA.* [4]*Corresponding author**

Key words: colonization, indole-3-acetic acid, invasion, Nod factors, *nod* gene induction, rhizobia, rhizobial attachment, rice, thick short lateral roots, *trans*-zeatin

Abstract

Legume-rhizobial interactions culminate in the formation of structures known as nodules. In this specialized niche, rhizobia are insulated from microbial competition and fix nitrogen which becomes directly available to the legume plant. It has been a long-standing goal in the field of biological nitrogen fixation to extend the nitrogen-fixing symbiosis to non-nodulated cereal plants, such as rice. To achieve this goal, extensive knowledge of the legume-rhizobia symbioses should help in formulating strategies for developing potential rice-rhizobia symbioses or endophytic interactions. As a first step to assess opportunities for developing a rice-rhizobia symbiosis, we evaluated certain aspects of rice-rhizobia associations to determine the extent of predisposition of rice roots for forming an intimate association with rhizobia. Our studies indicate that: **a.** Rice root exudates do not activate the expression of nodulation genes such as *nodY* of *Bradyrhizobium japonicum* USDA110, *nodA* of *R. leguminosarum* bv. *trifolii*, or *nodSU* of *Rhizobium*. sp. NGR234; **b.** Neither viable wild-type rhizobia, nor purified chitolipooligosaccharide (CLOS) Nod factors elicit root hair deformation or true nodule formation in rice; **c.** Rhizobia-produced indole-3-acetic acid, but neither trans-zeatin nor CLOS Nod factors, seem to promote the formation of thick, short lateral roots in rice; **d.** Rhizobia develop neither the symbiont-specific pattern of root hair attachment nor extensive cellulose microfibril production on the rice root epidermis; **e.** A primary mode of rhizobial invasion of rice roots is through cracks in the epidermis and fissures created during emergence of lateral roots; **f.** This infection process is *nod*-gene independent, nonspecific, and does not involve the formation of infection threads; **g.** Endophytic colonization observed so far is restricted to intercellular spaces or within host cells undergoing lysis. **h.** The cortical sclerenchymatous layer containing tightly packed, thick walled fibers appears to be a significant barrier that restricts rhizobial invasion into deeper layers of the root cortex. Therefore, we conclude that the molecular and cell biology of the *Rhizobium*-rice association differs in many respects from the biology underlying the development of root nodules in the *Rhizobium*-legume symbiosis.

Introduction

Nitrogen supply is critical for attaining yield potential in any crop. In rice, it takes 1 kg of nitrogen to produce 15-20 kg of grain. Lowland rice in the tropics can utilize the nitrogen that is naturally available in the soil through continuous biological nitrogen fixation (BNF) to produce 2-3 tons of grain per hectare. However, additional nitrogen must be applied to obtain higher yields.

Rice suffers from a mismatch of its nitrogen demand and its nitrogen supplied as chemical fertilizer, resulting in a 50-70% loss of the fertilizer applied. As pointed out in the introductory chapter, two approaches may be used to try to solve this problem. One is to regulate the timing of nitrogen application based on the plant's needs, thus increasing the efficiency of the plant's use of applied nitrogen (Cassman et al.,

---

* FAX No: +6328911292. E-mail: J.K.ladha@cgnet.com

1997). The other is to increase the efficiency of the use of available soil nitrogen, and meet the additional N-demand by making rice capable of "fixing its own nitrogen" either directly, or via a close interaction with diazotrophic bacteria (Ladha et al., 1997). Achieving the latter goal is a long-term strategy, but it potentially has a considerable payoff in terms of increasing rice production, helping resource-poor farmers and reducing environmental pollutants. If half of the nitrogen fertilizer applied to the 120 million hectares of lowland rice could be replaced with biologically fixed nitrogen, the equivalent of about 7.6 million tons of oil could be conserved annually.

Some free-living diazotrophic bacteria form natural associations with roots and submerged portions of the stem of rice plants. The amounts of N_2 fixed by associative diazotrophs are low and inefficiently utilized by the rice plant compared to the biologically fixed nitrogen provided by rhizobia to legumes under suitable conditions (see De Bruijn et al., 1995; Ladha and Reddy, 1995). Hence, associative BNF has only a limited capacity to render rice cultivation independent of external sources of N. If a BNF system could be assembled in the rice plant itself, then the fixed N would be directly available to the plant with little or no loss. This long-term goal could be achieved in a variety of ways, including the transfer to and expression of nitrogen fixation (nif, fix) genes in rice itself (Dixon et al., 1997). Alternatively, naturally occurring diazotrophic endophytic bacteria could be isolated and genetically modified to efficiently provide the plant with fixed nitrogen (Barraquio et al., 1997; Colnaghi et al., 1997; Kennedy et al., 1997; Kirchof et al., 1997; Stoltzfus et al., 1997). Yet another possibility is to try to establish a functional symbiotic interaction of rhizobia with rice via genetic manipulation of both plant and microbe. Extensive knowledge of the legume-rhizobia symbioses would need to be obtained to design strategies for extending this symbiosis to rice or other cereals (see De Bruijn et al., 1995). A first essential step in this process would be to critically evaluate and unders and the responses of rice plants to rhizobia, and vice versa, in comparison with the *Rhizobium*-legume symbiosis (De Bruijn et al., 1995; Gough et al., 1997; Kennedy et al., 1997; Reddy and Ladha, 1995; Stacey and Shibuya, 1997; Webster et al., 1997; Yanni et al., 1997). Moreover, a detailed assessmen of root morphogenesis in legumes versus cereals, and the distinct response of these systems to microbial infection, would need to be completed (Rolfe et al., 1997).

Parasponia is the only known non-legume nodulated by rhizobia (Trinick, 1973). Although the essential steps toward establishing a functional symbiosis in most legumes and *Parasponia* are remarkably similar, the latter involves a different mode of rhizobial infection. In *Parasponia*, rhizobia infect roots through cracks between epidermal cells or at the point of emergence of lateral roots. This mode of entry of rhizobia is interestingly shared by many legumes that live in an aquatic habitat similar to that where rice is typically grown (De Bruijn, 1995; Dreyfus et al., 1984; Subba-Rao et al., 1995; Tsien et al., 1983).

In the past, several researchers have examined rice-rhizobia interactions and reported a variety of responses, such as the ability of rhizobia to attach to rice roots (Terouchi and Syono, 1990), elicit the deformation of root hairs (Plazinski et al., 1985), and to form nodule-like structures/hypertrophies (Al-Mallah et al., 1989; Bender et al., 1990; De Bruijn et al., 1995; Jing et al., 1990, 1992; Li et al., 1991; Rolfe and Bender, 1990) or thick short lateral roots on rice plants (Cocking et al., 1993). Moreover, recently a report has appeared that examined a range of diverse rice and rhizobial genotypes to determine the variability of responses and interactions between the two partners. In this study, natural endophytic associations between rhizobia and rice were found and inoculation with certain endophytic rhizobial isolates was shown to promote rice growth under laboratory and field conditions (Yanni et al., 1997). Although some studies have reported the entry of rhizobia through the cracks at the point of emergence of lateral roots, microscopical details regarding the extent of invasion and patterns of colonization in relation to anatomical peculiarities of the rice root and the status of the cells of the interacting organisms during this interplay have been mostly lacking.

The present study forms a part of the international frontier project on assessing opportunities for nitrogen fixation in rice (see Introductory chapter of this volume). In order to create a foundation of information to genetically manipulate rice and/or nitrogen-fixing rhizobia so that a functional symbiosis could be achieved, we have conducted investigations on the following topics: a) Cellular and molecular aspects of the interactions between rice and rhizobia, b) Evaluation of rice genes similar to nodulin genes of legumes and elucidation of their functions, and c) Assessment of the expression of legume nodulin genes in a rice background and in response to rhizobial inoculation or Nod factors. Here we present recent results on the first topic, with the following specific objectives: to compare the ability of diverse rhizobia to interact with different rice genotypes; to determine if nodulation genes, CLOS Nod factors, and/or phytohormones are involved in the *Rhizobium*-rice association; and to document the patterns of rhizobial colonization and invasion of rice roots. Our findings on the other two topics will be published elsewhere.

Materials and methods

Rhizobial strains, culture methods and inoculum preparation

The rhizobial strains used are shown in Table 1. The bacteria were grown routinely on yeast extract mannitol agar medium (Vincent, 1970) at 30 °C under aerobic conditions in the dark. Single rhizobial colonies from agar plates were transferred to 50 mL of liquid media containing the appropriate antibiotics and incubated at 30 °C on a shaker. After 24 h, 5 mL of the culture was transferred to a 500 mL flask containing 95 mL of fresh liquid medium and allowed to grow for another 6-8 h. Subsequently, the bacterial cells were pelleted by centrifugation, washed with and resuspended in N-free Fahraeus (1957) or Jensen (1942) medium and the resuspension was adjusted to an OD_{650} of 0.25.

Surface sterilization of rice seeds

All operations for surface sterilization of seeds were performed at room temperature. Rice seeds, obtained from the International Rice Germplasm Center at IRRI, were gently dehulled, washed with sterile distilled water and immersed in either 70% ethanol for 4 min or 95% ethanol for 20 sec. Subsequently, the seeds were immediately washed with sterile distilled water (3×10 min) and incubated in 0.1% mercuric chloride solution for 4 min. Following this treatment, the seeds were washed repeatedly in excess amounts of sterile distilled water for 5-6 h on a shaker before seeding them in petri dishes containing either tryptone glucose yeast extract agar (Difco, USA) or potato dextrose agar (Difco, USA) medium, an incubated in the dark at 30 °C to test for possible contamination. Using this method, more than 98% of the seeds germinated, and 95% of these seedlings were found to be contamination-free. The seedlings devoid of any contamination were used in inoculation experiments.

Induction of nod gene expression

In vivo assays of *nod* gene expression were performed according to Redmond et al. (1986) and Peters and Long (1988) using *Bradyrhizobium japonicum* USDA110(ZB977) containing a *nodY::lacZ* reporter gene fusion, *Rhizobium leguminosarum* bv. *trifolii* ANU845(pRt032:218) harboring a *nodA::lacZ* fusion, or *Rhizobium* NGR234(pA27) harboring a *nodSU::lacZ* fusion and rice seedlings germinated from surface sterilized seeds. The induction of *nod* gene expression was determined by examining the blue color production due to ß-galactosidase activity using 5-bromo-4-chloro-3-indolyl-β-D-galactoside (X-gal, 40 μm L^{-1}) as indicator substrate.

Rice culture and inoculation

Two-three day old rice seedlings germinated from surface sterilized seeds were aseptically transferred to culture tubes (200×20 mm) containing 25–30 mL N-free Fahraeus or Jensen medium solidified with 0.3% agar and incubated for 2 days in the plant growth room (maintained at a 14 h light/10 h dark cycle, at temperatures of 27 °C/25 °C, respectively). One mL aliquots of mid-log rhizobial cultures (see above) were inoculated on rice seedlings on the 3rd day and the culture tubes were returned to the growth room for incubation. Rice seedlings inoculated with equal amounts of heat-killed bacteria served as controls. For the assessment of bacterial colonization, roots were sampled on the 15th and 30th day after inoculation (DAI).

Histochemical localization of β-galactoside activity

The histochemical staining method used to measure β-Gal activity is essentially the same as described by Boivin et al. (1990). Briefly, the entire root system of uninoculated rice plants or plants inoculated with *Azorhizobium caulinodans* strain ORS571 harboring plasmid pXLGD4 (a broad host-range plasmid harboring the reporter gene *lacZ* fused to the constitutively expressed *hemA* promoter of *R. meliloti* (Leong et al., 1985) was washed in 0.1 *M* Na-phosphate buffer (pH 7.2; 3×5 min) and fixed using 1.25% glutaraldehyde in 0.2 *M* Na-cacodylate buffer (pH 7.2) for 30 min under vacuum followed by 1 h at atmospheric pressure. The fixed roots were rinsed with 0.2 *M* Na-cacodylate buffer (pH 7.2; 2×15 min), transferred to a staining solution containing 800 mL of 0.2 *M* Na-cacodylate (pH 7.2), 50 mL of 100 m*M* K_3 [Fe(CN)$_6$], 50 mL

Table 1. Bacterial strains and plasmids used in the present study

Strain	Phenotype/Relevant characteristics	Legume host	Reference/Source[a]
Azorhizobium caulinodans		*Sesbania rostrata*	
IRBG315, IRBG366	Wild-type		IRRI
0RS571	Wild-type		ORSTOM
ORS571-3 *nodD*⁻	*nodD*::Tn5		Goethals et al. (1990)
ORS571*nodC*⁻	*nodC*::Tn5		Geelen et al.(1993)
ORS571-V44*nodA*⁻	*nodA*::Tn5		Geelen et al. (1993)
Bradyrhizobium spp.			
ORS322	Wild-type	*Aeschynomene afraspera*	ORSTOM
IRBG2	Wild-type	*Aeschynomene afraspera*	IRRI
IRBG87	Wild-type	*Aeschynomene scabra*	IRRI
IRBG91, IRBG92, IRBG276	Wild-type	*Aeschynomene indica*	IRRI
IRBG120	Wild-type	*Aeschynomene sensitiva*	IRRI
IRBG123, IRBG124, IRBG127,	Wild-type	*Aeschynomene pratensis*	IRRI
IRBG128	Wild-type	*Aeschynomene pratensis*	IRRI
IRBG229, IRBG233, IRBG234	Wild-type	*Aeschynomene evenia*	IRRI
IRBG230	Wild-type	*Aeschynomene nilotica*	IRRI
IRBG231	Wild-type	*Aeschynomene denticulata*	IRRI
IRBG270	Wild-type	*Aeschynomene fluminensis*	IRRI
Bradyrhizobium elkanii		*Glycine max*	
USDA94	Wild-type		USDA
USDA31	Wild-type		USDA
USDA94Δ*nod*	USDA94,*nodD₂D₁KABC*::del/ins *aph*		Yuhashi et al. (1995)
TN3	USDA31, IAA production-deficient mutant		Fukuhara et al. (1994)
Bradyrhizobium japonicum		*Glycine max*	
USDA110	Wild-type		Gary Stacey, UT, USA
USDA 110(ZB977)	*nodY*::*lacZ*		
Rhizobium trifolii		*Trifolium repens*	
ANU843	Wild-type		Djordjevic et al. (1987)
ANU845	Sym plasmid cured		Djordjevic et al. (1987)
ANU845(p032:218)	*nodA*::*lacZ*		
Rhizobium		Broad host range	
NGR234	Wild-type		Lewin et al. (1990)
NGR234(pA27)	*nodSU*::*lacZ*		
Rhizobium meliloti		*Medicago sativa*	
1021	Wild-type		Meade et al. (1982)
GMI225	250kb deletion of pSymA		Truchet et al. (1985)
Plasmids			
pTZS	Broad host range expression plasmid carrying a *trpR::tzs* fusion		Cooper and Long (1994)
pXLGD4	pGD499 prime (IncP) carrying a *hemA::lacZ* fusion		Leong et al. (1985)

[a] IRRI - International Rice Research Institute; ORSTOM - Institut Francais de Recherche Scientifique Pour le Developpement en Cooperation; USDA - United States Department of Agriculture; UT - University of Tennessee.

of 100 mM K$_4$ [Fe(CN)$_6$], 40 mL of 2% X-gal in N, N-dimethyl formamide and incubated overnight in the dark at 30 °C. Subsequently, the roots were rinsed with distilled water (5×10 min), and cleared using commercial chlorox solution for 40 sec, followed by a final rinse with distilled water. Stained whole roots and 30 mm transverse sections were examined by brightfield light microscopy.

Other microscopical methods

Rice roots colonized by bacteria, as visualized by X-gal staining, were further processed for detailed examination by brightfield light microscopy, scanning electron microscopy (SEM), laser scanning confocal microscopy (LSCM), and transmission electron microscopy (TEM). To prepare embedded sections, the stained roots were fixed again in a solution containing 4% glutaraldehyde, 1% paraformaldehyde and 1 mM CaCl$_2$ in 50 mM Na-cacodylate buffer (pH 6.8) at room temperature under vacuum for 2 h, washed with the same buffer (4×15 min), and post-fixed in 1% osmium tetroxide in 100 mM Na-cacodylate buffer (pH 6.8) for 2 h. Fixed roots were rinsed with distilled water (3×15 min) and suspended in 1% uranyl acetate for 2 h. Subsequently, the roots were rinsed again with distilled water (3×15 min), dehydrated through an acetone series, embedded in Spurr's firm epoxy medium. Three μm sections were stained with 0.1% (w/v) toluidine blue in 0.1% sodium tetraborate for about 1 min at 60 °C, or 15 min at room temperature, and examined by brightfield and phase contrast light microscopy. Ultrathin sections (70 nm) were restained with 2% uranyl acetate in 0.5% aqueous lead citrate solution and examined by TEM. For SEM, the root samples were fixed in glutaraldehyde and post-fixed in osmium tetroxide as described above, dehydrated through an ethanol series, critical point dried, mounted on stubs, and sputter-coated with 21 nm gold before examination. For LSCM, root segments were preserved in 1% Na-azide during transit to Michigan State University, stained with an aqueous solution of acridine orange (0.1 mg mL^{-1}), rinsed and mounted in 1% Na-pyrophosphate, and examined using the epifluorescence confocal mode (Subba-Rao et al., 1995).

Results and discussion

Nodule development in legumes results from a series of complex interactions between the host plant and rhizobia; including recognition and attachment of the bacteria, root hair curling, infection thread formation and the induction of cortical cell divisions (see Denarie and Cullimore, 1993; Hirsch, 1992; Nap and Bisseling, 1990; Vijn et al., 1993). The initial steps in this interplay are invoked by specific signal molecules, flavonoids and Nod factors, produced by legume plant and rhizobial symbionts, respectively. The process begins with the secretion of flavonoids from roots, and consequent flavonoid-triggered *nod* gene expression in the microsymbiont leading to the production of various Nod factors. Nod factors in turn induce nodule ontogeny, and activate nodulin genes governing the processes involved in root hair deformation/curling, initiation of infection threads and cortical cell divisions. To achieve the goal of establishing a symbiosis between rhizobia and rice, it is essential to systematically analyze the interactions of rice with rhizobia to identify similarities and differences in rice-rhizobial interplay vis-a-vis those occurring in legume-rhizobial interactions. Another important aspect to consider while comparing such interactions is the inherent differences between the root anatomy of rice and legumes (also Rolfe et al., 1997).

We examined the effects of diverse rhizobia on rice roots at 1–2 and 15–30 days after incubation. Examination at the first time point would reveal the extent to which rhizobia attach to the root epidermis and their ability to elicit root hair induction, deformation and/or curling. Examination at the second time point would reveal if rhizobia induce any other morphological changes in roots, including unique root-derived and nodule-like structures.

Morphological responses of rice root hairs to rhizobia and Nod factors

Most rhizobia were found to stimulate the formation of root hairs in the rice cultivars tested. However, rhizobia failed to induce deformation or curling of rice root hairs. This was the case for 70 different rice genotypes inoculated with 24 of the 25 different broad and narrow host-range strains of rhizobia (data not shown). The only exception was a marginal ability of *Bradyrhizobium* ORS322 to induce root hair deformation on rice variety Milyang 54 (Reddy et al., 1995). These results suggest that rhizobia may not produce Nod factors in the rice root environment and/or their Nod factors may be unable to promote root hair deformation in rice. We tested these hypotheses by two approaches. First, we used various *nod*::*lacZ* fusion reporter strains to examine whether axenically generated rice root exudate could activate expression of *nodY* in *B. japonicum* USDA110, *nodSU* in *Rhizobium* NGR234, and *nodA* in *R. leguminosarum* bv. *trifolii* ANU843. No such activation was observed (data not shown). Control experiments using the reporter strains and root exudates of the corresponding host legumes were found to be positive. These results suggest that rice root exudate either lacks the appropriate activators of *nod* gene expression and/or contains antagonistic substance(s) that inhibit(s) activation of the rhizobial *nod* genes.

In order to ascertain whether Nod factors were able to induce root hair deformation in rice at all, we examined more than 25 rice varieties for changes in root hair morphology at regular intervals during a 2-day incubation in the presence of a mixture of purified sulfated and nonsulfated CLOS Nod factors from *Rhizobium* NGR234 (at concentrations of 10^{-6} to 10^{-9} M). The results obtained showed that these Nod factors were unable to induce root hair deformation in rice under the experimental conditions tested (data not shown), and contrast with the induction of rice root hair deformation observed when using a recombinant strain of *R. leguminosarum* containing multiple copies of pSym-borne *nodDABC genes* from bv. *trifolii* strain ANU843 (Plazinski et al., 1985).

The apparent inability of rice root hairs to respond to CLOSs from NGR234 may be due to the absence of the appropriate CLOS structures required for induction, the putative plant receptor(s) that perceive them, or some other element(s) in the Nod factor signal transduction pathway required for induction of root hair deformation. It is also possible that the induction of these responses requires a critical concentration of other non-CLOS classes of Nod factors made by rhizobia (Orgambide et al., 1994; Philip Hollingsworth et al., 1991).

Induction of abnormal lateral root development in rice by rhizobia and IAA

None of the rhizobia tested induced genuine root nodules on any of the rice varieties examined. However, all rice varieties exhibited abnormal lateral root development within two weeks after inoculation with rhizobia, particularly those that nodulate aquatic legumes (Tables 2 and 3). The growth of newly emerging laterals in some main roots was highly retarded resulting in shorter and thicker modified lateral root structures (Figure 1A-D). More than 75% of the plants of all varieties tested exhibited this Thick Short Lateral Root (TSLR) phenotype by 30 DAI (Table 3). Only rice variety IR42 produced TSLRs without inoculation. Nevertheless, the frequency of TSLR formation in IR42 was significantly increased when inoculated with rhizobia. Although rhizobial inoculation induced TSLR formation in a very high percentage of plants, not all main roots in a plant developed TSLRs. Their formation was restricted to about 16-50% of the roots depending on the rice genotype and rhizobial strain combination used.

Nod factors (see Denarie and Cullimore, 1993; Nap and Bisseling, 1990; Vijn et al., 1993) produced by rhizobia are clearly involved in promoting nodule formation in legumes. Moreover, rhizobially produced indole acetic acid (IAA; Fukuhara et al., 1994; Yuhashi et al., 1995) has been shown to be involved in promoting nodule formation. Because wild-type strains of rhizobia that normally induce TSLR formation in rice have the ability under certain conditions to produce Nod factors as well as IAA, we examined whether rhizobia defective in the production of these bioactive molecules were capable of inducing any phenotypic changes in rice roots. Microscopical investigations showed that *R. meliloti* GMI255Δnod and *B. elkanii* USDA94 Δ*nod* induced TSLRs on IR42 with the same efficiency as their parent wild-type strains, *R. meliloti* 1021 and *B. elkanii* USDA94, respectively (Table 4). These results, indicating that *nod* genes required for CLOS production do not play a key role in the induction of TSLR in rice, were further confirmed by studies showing that addition of a 10^{-9} *M* mixture of sulfated and non-sulfated CLOSs from *Rhizobium* NGR234, supplied to the growth medium every alternate day for 15 days, did not induce TSLRs (data not shown). Contrary to the above results, *B. elkanii* TN3, a mutant of USDA31 deficient in IAA production but not in Nod factor synthesis, induced TSLRs with a lower frequency than the parent wild-type strain, suggesting a role for IAA in the promotion of TSLR formation in rice.

Along with IAA, rhizobia also produce cytokinins (see Taller and Sturtevant, 1991; Torrey, 1986). In alfalfa roots, localized *trans*-zeatin production by *R. meliloti* GMI2SS(pTZS) was found to induce the formation of nodule-like structures (Cooper and Long, 1994). In addition, certain early nodulin genes in legumes have been found to be induced by cytokinins (Dehio and de Bruijn, 1992; Hirsch and Fang, 1994; Silver et al., 1996), and the signal transduction pathway responsible for this phenomenon has been shown to be conserved in non-legumes, such as *Arabidopsis* (Silver et al., 1997). To explore the potential role of

Table 2. Induction of thick short lateral roots (TSLR) in IR42 by different strains of rhizobia after 15 days of inoculation

Strain	No. of plants examined	Percentage of plants showing TSLR	Percentage of roots with TSLR per plant
Control	19	7	1f
Azorhizobia			
ORS571	19	88ab	30a
IRBG315	18	78a-e	16bc
IRBG366	15	80a-e	13cd
Bradyrhizobia			
IRBG2	18	100a	8c-e
IRBGB7	6	100a	14b-d
IRBG91	7	71a-d	11c-e
IRBG92	7	43c-f	3de
IRBG120	10	40c-f	3de
IRBG123	14	50c-f	5c-e
IRBG124	9	33d-f	4c-e
IRBG127	17	83a-c	5c-e
IRBG128	8	50b-d	5c-e
IRBG229	5	80a-e	9c-e
IRBG230	11	73a-d	5c-e
IRBG231	7	43c-f	4c-e
IRBG233	8	75a-d	7c-e
IRBG234	8	25ef	2de
IRBG270	8	85b-e	10c-e
IRBG276	9	78a-d	7c-e

Values in a column followed by a common letter are not statistically different according to Duncan's Multiple-Range Test ($p = 0.05$).

cytokinins in the induction of TSLR formation or other changes in root morphology, we inoculated IR42 rice with the Nod mutant strain of *R. meliloti* GMI2SS with or without the plasmid pTZS, which constitutively expresses an isopentenyl transferase gene, enabling continuous *trans*-zeatin secretion (Cooper and Long, 1994). The results show that the presence or absence of this plasmid pTZS does not influence the frequency of TSLR formation by *R. meliloti* in rice, suggesting a probable lack of *trans*-zeatin involvement in TSLR formation (Table 4).

Mode of rhizobial colonization and invasion of rice roots

Since rhizobia from aquatic legumes induced TSLR formation more effectively than rhizobia from terrestrial legumes, *A. caulinodans* ORS571 was employed as the primary strain to study the mode of invasion and colonization patterns of rhizobia in the roots of IR42. For comparison, *nodA*::Tn5, *nodC*::Tn5, and *nodD*::Tr5 mutant derivatives of ORS571 were also examined. Rhizobial invasion and colonization of rice roots were visualized under light microscopy by X-gal staining of the plants inoculated with wild-type and nod mutants, each harboring pXLGD4, a broad-host-range plasmid containing the reporter gene *lacZ* fused to the constitutively expressed *hemA* promoter of *R. meliloti* (Leong et al., 1985). SEM, LSCM, and TEM studies were performed to obtain further details of the mode of invasion and patterns of colonization of rice tissues by the rhizobia. Wild type ORS571 and its Nod⁻ derivatives behaved similarly with regard to their attachment to rice root epidermal cells and patterns of invasion and colonization within the rice root, indicating a lack of *nod* gene involvement in these stages of the *Rhizobium*-rice association.

Table 3. Induction of thick short lateral roots (TSLR) in different rice varieties 15 and 30 days after inoculation (DAI) with rhizobial isolates from *Sesbania rostrata*

Strain	Variety	No. of plants examined	15 DAI		30 DAI	
			Percentage of plants showing TSLR	Percentage of roots with TSLR per plant	Percentage of plants showing TSLR	Percentage of roots with TSLR per plant
Control						
	Azucena	16	0	0	0	0
	IR42	58	15	0	15	4
	IRAT 110	16	0	0	0	0
	Kinandang Patong	16	0	0	0	0
	Moroberekan	16	0	0	0	0
	OS 4	16	0	0	0	0
ORS 571						
	Azucena	15	33c	7b	100a	36a
	IR 42	130	78a	9b	100a	54a
	IRAT 110	16	44bc	7bc	100a	28b
	Kinandang Patong	14	7d	2c	100a	27b
	Moroberekan	16	13cd	11ab	100a	35a
	OS4	14	71ab	13ab	100a	35a
IRBG 315						
	Azucena	12	42c	8b	100a	36a
	IR 42	124	60bc	11b	100a	33a
	IRAT 110	13	84ab	10b	93ab	32a
	Kinandang Patong	16	31c	7b	81b	16b
	Moroberekan	16	100a	21a	93ab	31a
	OS 4	14	64bc	11b	100a	30a
IRBG 366						
	Azucena	17	41b	8b	100a	37a
	IR42	132	81a	13a	100a	39a
	IRAT 110	17	83a	14a	100a	41a
	Kinandang Patong	16	6c	1b	75b	21b
	Moroberekan	16	88a	14a	100a	37ab
	OS 4	17	23bc	3b	100a	34ab

Values in a column followed by a common letter are not statistically different according to Duncan's Multiple-Range Test (*p*=0.05).

During the first 2 days of incubation, cells of ORS571 attached frequently in a polar orientation to rice root hairs (Figure 2A), in a fashion similar to that observed during rhizobial inoculant studies on rice by Terouchi and Syono 1990. No infection thread formation was observed in rice root hairs. Within 4-5 days after incubation, further colonization of the root epidermis occurred without a preferential cellular orientation of attachment or extensive elaboration of cellulose microfibrils (Figure 2B, 4A-D). Similar results were

observed with the nod mutant derivatives of ORS571. These early events in the *Rhizobium*-rice association contrast with the discrete *nodD*-dependent cell aggregation at root hair tips and extensive elaboration of cellulose microfibrils characteristic of Phase 1 attachment and Phase 2 adhesion in the *Rhizobium*-legume symbiosis, respectively (Dazzo et al., 1984, 1988; Mateos et al., 1995).

Rhizobial invasion of rice roots and further internal colonization were visualized by a combination of light,

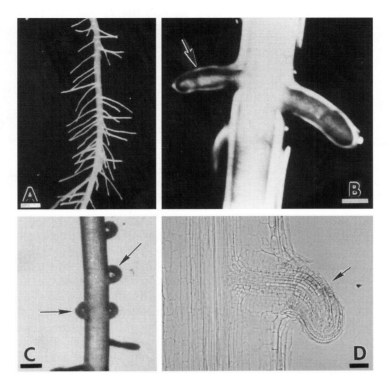

Figure 1. Morphology of (**A**) normal, and (**B-D**) thick short lateral roots (arrows) of rice. A, B, and D are inoculated with *Azorhizobium caulinodans* ORS571 whereas C is inoculated with *Bradyrhizobium elkanii* USDA94. *Bar* scales are 1 mm in A; 0.5 mm in B and C; and 150 μm in D.

Table 4. Effect of wild-type and mutant strains of rhizobia on the formation of thick short lateral roots (TSLR) in IR42 scored after 30 days of inoculation

Strain	No. of plants examined	Percentage of plants showing TSLR	Percentage of roots with TSLR per plant
Control	22	9c	3c
Bradyrhizobium elkanii			
USDA94, wild-type	22	82a	37b
USDA94, Δ*nod*	22	77a	47b
USDA31, wild-type	21	90a	53a
TN3, IAA deficient USDA31	22	41b	33b
Rhizobium meliloti			
1021, wild-type	10	30b	5c
GMI255, Δ*nod nif fix*	10	30b	6c
GMI255 (pTZS), Δ*nod nif fix*	10	40b	8c

Values in a column followed by a common letter are not statistically different according to Duncan's Multiple-Range Test ($p = 0.05$).

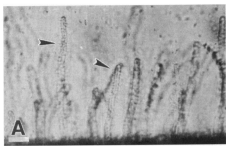 

Figure 2. Attachment of cells of *Azorhizobium caulinodans* ORS571 to the root epidermis of IR42 rice. **A** Light micrograph showing polar attachment of bacteria (arrowheads) to root hairs. **B.** Laser scanning confocal micrograph showing randomly oriented attachment of fluorescent bacteria to the non-root hair epidermis. *Bar* scales are 50 μm in A and 10 μm in B.

Table 5. Frequency of colonization of lateral root cracks in IR42 by wild-type and Nod- mutants of *Azorhizobium caulinodans* 15 days after inoculation

Strains	No. of plants examined	Percentage of lateral root cracks colonized
Azorhizobium caulinodans		
ORS571, wild-type	14	24a
ORS571-3 *nodD*⁻	20	25a
ORS571*nodC*⁻	20	18a
ORS571-V44 *nodA*⁻	15	34a

Values in a column followed by a common letter are not statistically different according to Duncan's Multiple-Range Test ($p = 0.01$).

All strains harbored pXLDG4 containing *lacZ* fused to constitutive hemA promoter. Bacterial colonization was scored as localized histochemical staining of β-galactosidase activity in whole roots. Blue product did not form in uninoculated controls, in controls inoculated with ORS571 without pXLDG4, or if the X-Gal substrate was omitted.

laser scanning confocal, scanning electron, and transmission electron microscopies. Brightfield microscopy showed localized X-gal staining of approximately one-fourth of the fissures at the point of lateral root emergence on plants inoculated with the *lacZ* reporter strain derivative of wild-type ORS571 (Table 5). The observed frequency of localized X-gal staining was similar (not statistically different) on plants inoculated with the corresponding *nod* mutants (Table 5), thereby indicating a lack of requirement for *nodD*, *nodA*, and *nodC* (hence Nod factors) in colonization of this site in rice roots.

Further examination by brightfield microscopy indicated that the bacteria had entered the roots on rare occasions through injured epidermal cells and more frequently through the natural wounds caused by splitting of the epidermis at the emergence of young lateral roots (Figures 3A-F). SEM and LSCM indicated that bacteria primarily entered the rice root through these fissures at lateral root emergence. SEM revealed numerous cells of ORS571 colonizing the area beneath the main root epidermis, within the fissures at lateral root emergence (Figures 4C-D), indicating that this environment seems to be favorable to the growth of ORS571. LSCM of non-dehydrated inoculated rice roots provided strong evidence that bacterial migration into the fissure cavity was not an artifact of dehydration during specimen preparation. Staining of bacteria with acridine orange followed by fluorescence LSCM confirmed the characteristic ring of heavy bacterial colonization surrounding the site of lateral root emergence (Figures 5A and C). Composite sets of serial Z optisections taken at higher magnification revealed bacterial colonization within the fissure cavity beneath the epidermal surface (Figures 5B and D). Further evidence of bacterial colonization within the fissure site beneath the epidermal root surface was provided by performing a 3-D computer reconstruction and rotation of a series of optisections followed by a vertical phi-Z section at an optimal location (Figures 5E and F).

Light microscopy and TEM revealed that the bacteria within the fissure disseminated deeper in the intercellular spaces (Figures 3F and 6A-D). The extent of intercellular colonization by ORS571 in the vicinity of normal rice lateral roots and TSLRs was similar, and it commonly penetrated from the fissure/epidermis lesions to a distance of 2 cell layers up to the cortical sclerenchymatous layer (Figures 3B and D). Less frequently, proliferating rhizobial cells surrounding the base of the lateral root migrated into deeper layers of the cortical zone even to the endodermis (Figure 3F). It

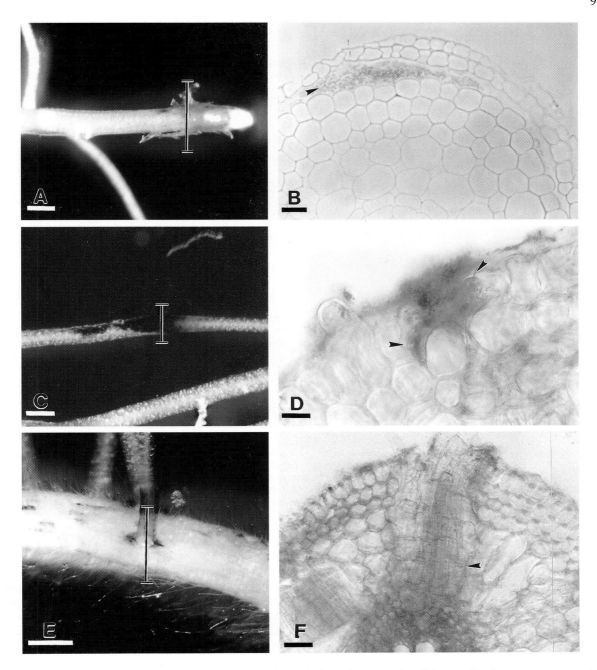

Figure 3. Light micrographs of roots of IR 42 rice showing invasion and colonization patterns by wild-type and Nod⁻ reporter mutant strains of *A. caulinodans* ORS571 harboring pXLGD4. Bacteria were localized by histochemical staining (blue) of β-galactosidase activity in whole roots (**A, C, E**) and semi-thin sections (**B, D, F**). A, bacterial colonization of the sub-apical immature elongation zone of the root. B, traverse section of the sub-apical elongation region of the root as indicated in A. Note bacteria in the enlarged intercellular space between the exodermis and the cortex. C, bacterial colonization at a mature region of the root. D, traverse section of the same mature region of the root as indicated in C. Note intercellular colonization of bacteria into the sclerenchymatous layer only. E, rhizobial colonization of a ring surrounding the site of lateral root emergence. F, cross-section of the main root at the site of lateral root emergence as indicated in E. Note that the bacteria have predominantly colonized the first two cell layers within the fissure created by the emerging lateral root. Occasionally, bacterial intercellular colonization was observed in the cortical zone as well (arrowhead) at the base of the emerging lateral root. *Bar* scales are 1 mm in A, C, and E; 10 μm in B; 15 μm in D; and 25 μm in F.

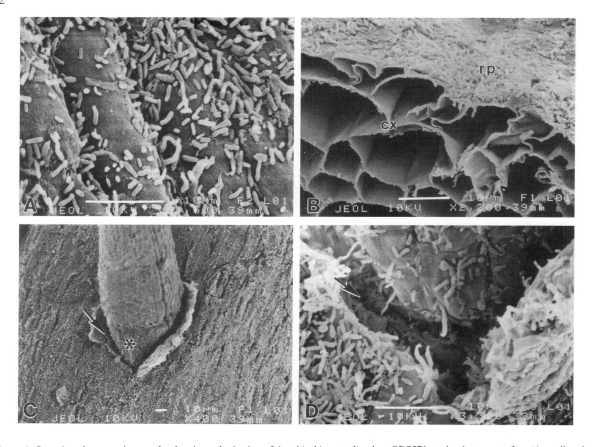

Figure 4. Scanning electron micrographs showing colonization of *Azorhizobium caulinodans* ORS571 on the rice root surface (**A**, undisturbed epidermis; **B**, rhizoplane [rp] at the edge of a free-hand cross-section [cx] of the root), and beneath the root surface within the fissure crevice created by lateral root emergence (C, D arrows). The absence of bacteria on the cut face of the root interior in micrograph B shows that fixed bacteria on the root epidermis are not redistributed to uncolonized surfaces during specimen preparation. In **C** and **D**, widening of the crevice during specimen dehydration reveals the in situ colonization of fixed bacteria within the fissure. Micrograph D is an enlargement of the starred region in C. *Bar* scales are 10 μm.

is worthwhile to mention here that, unlike in legumes, rice roots possess an additional cortical cell layer composed of thick walled fibers (Figure 7) which prevents the collapse of the root after cortical aerenchyma formation (Clark and Harris, 1981). It appears that this sclerenchymatous layer of tightly packed thick walled fibers is a major barrier for rhizobial invasion into intercellular spaces of the loosely packed parenchymatous cortical zone of rice roots.

Rhizobial cells were also found within rice epidermal and exodermal root cells. Clear indication that bacteria beneath the root surface had entered a rice cell (black arrow) adjacent to an uninvaded cell (white arrow) is presented in the LSCM fluorescence optisection presented in Figure 5G. TEM further showed rhizobia within cortical parenchyma cells, next to

aerenchymatous tissue (Figures 6E and F). No evidence of infection threads was observed. This contrasts with the mode of rhizobial infection even in aquatic legumes, where *bona fide* infection threads ultimately develop as a route for rhizobial intracellular dissemination (De Bruijn, 1995; Dreyfus et al., 1984; Napoli et al., 1975; Subba-Rao et al., 1995; Tsien et al., 1983). We predict that this transition from the intercellular to the intracellular status of rhizobia inside rice occurs via local thinning and solubilization of the plant fibrillar wall due to the action of hydrolytic enzymes secreted by the bacteria (Chalifour and Benhamou, 1989; Mateos et al., 1992). Figures 6E and F further exemplify the finding that host cells invaded by rhizobia exhibit several changes in integrity, including loss in electron density of their cytoplasm and cell wall, lack

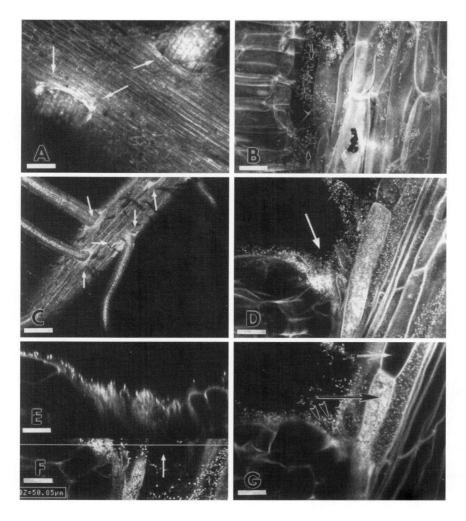

Figure 5. Laser scanning confocal micrographs of *Azorhizobium caulinodans* ORS571 (**A**, **B**, **E**, **F**) and *Bradyrhizobium* sp. ORS322 (**C**, **D**, **G**) colonizing the fissure crevice (white arrows) that encircles the site of lateral root emergence. Fluorescent bacteria are stained with acridine orange. A-D are reconstructed composite Z serial optisections. E is a computer rendered 50 μm thick vertical phi-Z section of the composite optisections in F, made at the position of the horizontal white line showing the fluorescent bacteria beneath the root surface within the fissure crevice. G is a single confocal optisection showing fluorescent bacteria within the fissure crevice at lateral root emergence (arrowheads) and a rice cell (black arrow) located 2 cell layers below the root surface adjacent to an uninfected rice cell (white arrow) in the same confocal plane. *Bar* scales are 250 μm in A; 500 μm in C; and 10 μm in B, D, E, F and G.

of intact membranous organelles, and disruption of the cytoplasmic membrane. All of these changes within the invaded cell can occur adjacent to an intact host cell (Figures 6E and F). Unlike in the *Rhizobium*-legume symbiosis, no membrane-enclosed endosymbiotic rhizobia within symbiosomes of intact host cells were found in our studies of the *Rhizobium*-rice association. In the initial stages of invasion of rice roots, ORS571 cells had a normal rod shape. However rhizobia that had colonized within the rice root were somewhat larger and abound with electron-transparent intracellular granules (presumably poly-β-hydroxybutyrate), and more frequently plasmolyzed (Figure 6). All of these ultrastructural features indicate that invading cells of ORS571 were able to obtain excess carbon and energy sources from the rice root, but nevertheless were unable to establish a compatible intracellular association with the plant cells, further restricting their development.

94

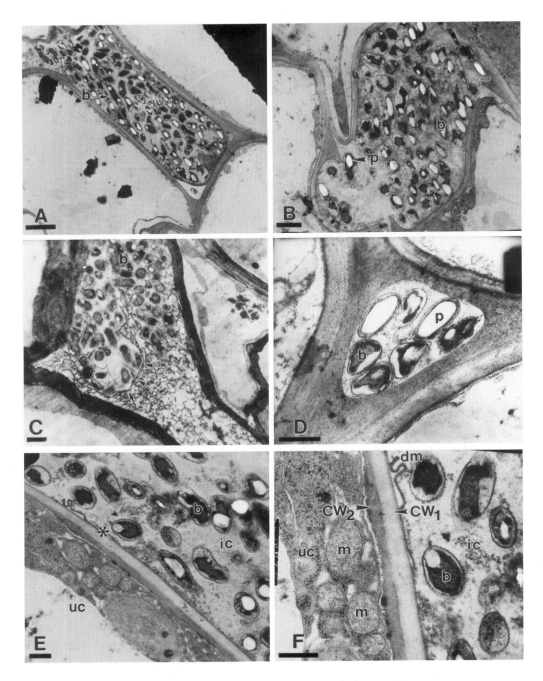

Figure 6. Transmission electron micrographs showing intercellular (**A-D**) and intracellular (**E-F**) invasion of IR42 roots by *Azorhizobium caulinodans* ORS571. Note that most bacteria contain electron-transparent storage granules and some bacteria are pleiomorphic. In C, some bacteria are embedded in extracellular electron-transparent material (possibly capsule) with a partially defined border (arrows) that has displaced the fibrillar lamellar matrix. In E and F, bacteria have entered a rice cell within the cortical zone. In contrast to the adjacent uninfected host cell, the infected host cell has a less electron-dense cytoplasm, lacks intact membraneous organelles, has a disrupted cytoplasmic membrane, and a marked reduction in electron-density of its fibrillar cell wall. Also, some of the bacteria appear degenerated. Micrograph F is an enlargement of the starred region within micrograph E. **b**-bacterial cells; **cw1**-cell wall of the invaded cell; **cw2**-cell wall of the adjacent uninfected cell; **dm**-disrupted cell membrane; **ic**-invaded cell; **m**-mitochondria; **p**-intracellular bacterial lipid granules (presumably poly-β-hydroxybutyrate); **uc**-uninfected cell. *Bar* scales are 2 μm in A; 1 μm in B, C; and 0.5 μm in D-F.

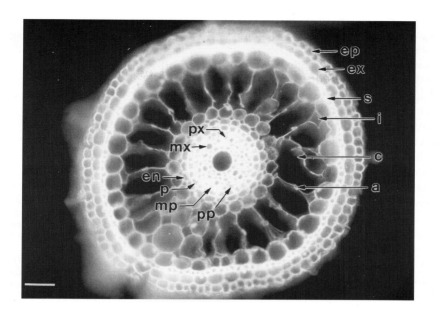

Figure 7. Epifluorescence micrograph of a free-hand cross-section of a rice (IR42) root made 5-6 cm from the apex and stained with acridine orange. Note the collapse of cortical parenchyma leading to the formation of aerenchymatous tissue with large intercellular spaces. **a**-aerenchyma; **c**-cortex; **en**-endodermis; **ep**-epidermis; **ex**-exodermis; **i**-intercellular spaces; **mp**-metaphloem; **p**-pericycle; **pp**-protophloem; **s**-thick-walled sclerenchymatous cell layer.

Conclusions

This study has provided useful information about the responses of rice towards selected rhizobia. Most of the events found to occur in this *Rhizobium*-rice association are different from the characteristic sequence of developmental stages in the *Rhizobium*-legume symbiosis. These variations in response exhibited by rice vis-a-vis legumes may not be that surprising, since there are considerable differences in root morphology (also Rolfe et al., 1997), response to microbes, and general physiological and developmental processes between these widely divergent genera of plants.

Although *A. caulinodans* ORS571 has been found to enter the rice roots (as found also by Webster et al., 1997) the magnitude of its invasion appears to be limited. Our studies further showed that rice root exudates are unable to activate *nod* gene expression and that exogenously added chitolipooligosaccharide Nod factors do not elicit obvious morphological responses in rice roots. Furthermore, evidence of an incompatible intracellular association was found. ORS571 infection of rice and wheat roots has also been studied by Webster et al. (1997). Interestingly, these investigators report that in the wheat system, the addition of the *Nod* gene inducer naringenin stimulates infection

by ORS571, but that the infection process is *nod* gene independent, as observed in our studies. These results warrant a continued screening of more rice germplasm with diverse rhizobia in order to identify compatible rice-*Rhizobium* combinations that may lead to more extensive root colonization and responsive rice genotypes. Of major importance to this plan is the discovery of natural beneficial associations and endophytic interactions of rhizobia and rice in fields rotated with legumes since antiquity (Yanni et al., 1997) or in more primitive rice varieties not bred to efficiently respond to N fertilizer (Barraquio et al., 1997; Stoltzfus et al., 1997), since it offers excellent opportunities to identify superior combinations of symbionts that may potentially impact agricultural rice production.

The research on the *Rhizobium*-legume symbiosis has revealed that the host plant possesses a genetic program for the development of root nodules that is activated by signal molecules such as Nod factors, produced by the microsymbiont. It is unlikely that a monocot plant such as rice would possess the complete complement of genes involved in the nodule ontogeny program, and that rhizobial strains could induce the formation of genuine nodules on these plants. Therefore, rice would need to be genetically modified to respond to the appropriate rhizobial morphogenic triggers and subsequent *Rhizobium*-modulated nodule ontogeny requirements. Recent studies have shown that rice possesses homologues of some of the (nodulin) genes specifically expressed in legume nodules during early events in infection and nodule formation (Reddy et al., 1996a,1996b, unpublished). These results suggest that rice may have at least part of the genetic program involved in a functional symbiosis with *Rhizobium*. Other studies have shown that CLOS Nod factors enhance the promoter activity of Mt*ENOD12* within rice roots, indicating that a portion of the signal transduction machinery important for legume nodulation may also already exist in rice (Reddy et al., 1996c).

Another reason for optimism may be that rice is able to enter into symbiotic associations with mycorrhizal fungi (Secilia and Bagyaraj, 1992). Genetic links between the processes involved in nodulation and arbuscular mycorrhiza have been found in legumes (Cook et al., 1997; Gianinazzi-Pearson, 1996). Thus, rice may possess part of the genetic program necessary for entering into mutually beneficial, endosymbiotic associations with other soil microorganisms.

In conclusion, many differences exist between the association of rice and rhizobia relative to the root- or stem-nodule symbiosis with legumes. However, some of the molecular interactions that occur in these plant-microbe associations may be similar. It is therefore essential that studies be further extended at the cellular and molecular levels to identify why such responses do not fully occur in rice, in order to contemplate genetically engineering this major cereal crop to form a more intimate endosymbiotic association with rhizobia.

Acknowledgements

We are thankful to the following scientists for kindly providing us plasmids, chitolipooligosaccharide Nod factors, and/or rhizobial strains: J Denarie for pXL-GD4 and purified Nod factors; W J Broughton for *Rhizobium* NGR234 and NGR234(pA27); C Gough for *A. caulinodans* ORS571-3 nodD⁻ (pXLDG4), ORS571 nodC⁻ (pXLGD4) and ORS571-V44 nodA⁻; S R Long for *R. meliloti* 1021, GM1225 ad GMI225 (pTZS); K Minamisawa for *B. elkanii* USDA94, USDA31, USDA94Dnod and TN3; B G Rolfe for *R. leguminosarum* bv. *trifolii* ANU843, ANU845 and ANU845(pRt032:218); G Stacey for *B. japonicum* USDA11O and USDA11O(ZB977). This work was supported by grants from the Danish International Development Agency (DANIDA) to IRRI and MSU, the MSU-NSF Center for Microbial Ecology (NSF Grant No DEB 9120006) and the DOE (De-FGO2-91ER20021) to MSU.

References

Al-Mallah M K, Davey M R and Cocking E C 1989 Formation of nodular structures on rice seedlings by rhizobia. J. Exp. Bot. 40, 473–478.

Barraquio W L, Revilla L and Ladha J K 1997 Isolation of endophytic diazotrophic bacteria from wetland rice. Plant Soil 194, 15–24.

Bender G L, Preston L, Barnard D and Rolfe B G 1990 Formation of nodule-like structures on the roots of the non-legumes rice and wheat. *In* Nitrogen Fixation: Achievements and Objectives. Eds. P M Gresshoff, L E Roth, G Stacey and W E Newton. p 825. Chapman and Hall, London.

Boivin C, Camut S, Malpica C A, Truchet G and Rosenberg C 1990 *Rhizobium meliloti* genes encoding catabolism of trigonellin are induced under symbiotic conditions. Plant Cell 2, 1157–1170.

Cassman K G, Peng S, Olk D C, Ladha J K, Reichardt W, Dobermann A and Singh U 1997 Opportunities for increased nitrogen use efficiency from improved resource management in irrigated rice systems. Field Crops Res. (*In press*).

Chalifour F-P and Benhamou N 1989 Indirect evidence for cellulase production by *Rhizobium* in pea root nodules during bacteroid differentiation: Cytochemical aspects of cellulose breakdown in rhizobial droplets. Can. J. Microbiol. 35, 821–829.

Clark L H and Harris W H 1981 Observations on the root anatomy of rice (*Oryza sativa* L.). Am. J. Bot. 68, 154–161.

Cocking E C, Srivastava J S, Cook J M, Kothari S L and Davey M R 1993 Studies on nodulation of maize, wheat, rice and oilseed rape: interactions of rhizobia with emerging lateral roots. *In* Biological Nitrogen Fixation - Novel Associations with Non-legume Crops. Eds. N Yanfu, I R Kennedy and C Tingwei. pp 53–58. Qingdao Ocean University Press, Qingdao.

Colnaghi R, Green A, He L, Rudnick P and Kennedy C 1997 Strategies for increased ammonium production in free-living or plant associated nitrogen fixing bacteria. Plant Soil 194, 145–154.

Cook D R, VandenBosch K, de Bruijn F J and Huguet T 1997 Model legumes get the nod. Plant Cell 3, 275–281.

Cooper J B and Long S R 1994 Morphogenetic rescue of *Rhizobium meliloti* nodulation mutants by *trans*-zeatin secretion. Plant Cell 6, 215–225.

Dazzo F B, Hollingsworth R, Philip-Hollingsworth S, Robeles M, Olen T, Salzwedel J, Djordjevic M and Rolfe B 1988 Recognition process in the *Rhizobium trifolii*-white clover symbiosis. *In* Nitrogen Fixation: Hundred Years After. Eds. H Bothe and F J de Bruijn. pp 431–435. Gustav Fischer, Stuttgart.

Dazzo F B, Truchet G L, Sherwood J, Hrabak E M, Abe M and Pankratz H S 1984 Specific phases of root hair attachment in the *Rhizobium trifolii*-clover symbiosis. Appl. Environ. Microbiol. 48, 1140–1150.

De Bruijn F J, Jing Y and Dazzo F B 1995 Potential and pitfalls of trying to extend symbiotic interactions of nitrogen-fixing organisms to presently non-nodulated plants, such as rice. Plant Soil 172, 207–219.

Dehio C and de Bruijn F J 1992 The early nodulin gene SrEnod2 from *Sesbania rostrata* is inducible by cytokinin. Plant J. 2, 117–128.

Denarie J and Cullimore J 1993 Lipo-oligosaccharide nodulation factors: A minireview. New class of signaling molecules mediating recognition and morphogenesis. Cell 74, 951–954.

Dixon R, Cheng Q, Shen G-F, Day A and Dowson-Day M 1997 Nif gene transfer and expression in chloroplasts: prospects and problems. Plant Soil 194, 193–203.

Djordjevic M A, Redmond J W, Batley M and Rolfe B G 1987 Clovers secrete specific phenolic compounds which either stimulate or repress *nod* gene expression in *Rhizobium trifolii*. EMBO J. 6, 1173–1179.

Dreyfus B L, Alazard D and Dommergues Y R 1984 New and unusual microorganisms and their niches. *In* Current Perspectives in microbial Ecology. Eds. M J Klug and C A Reddy. pp 161–169. American Society of Microbiology, Washington DC.

Fahraeus G 1957 The infection of clover root hairs by nodule bacteria studied by a simple glass slide technique. J. Gen. Microbiol. 16, 374–381.

Fukuhara H, Minakawa Y, Akao S and Minarnisawa K 1994 The involvement of indole-3-acetic acid produced by *Bradyrhizobium elkanii* in nodule formation. Plant Cell Physiol. 35, 1261–1265.

Geelen D, Mergaert P, Geremia R A, Goormachtig S, Van Montagu M and Holsters M 1993 Identification of *nodSUIJ* genes in Nod locus 1 of *Azorhizobium caulinodans*.: evidence that *nodS* encodes a methyltransferase involved in Nod factor modification. Mol. Microbiol. 9, 145–154.

Gianinazzi-Pearson V 1996 Plant cell responses to arbuscular mycorrhizal fungi: Getting to the roots of the symbiosis. Plant Cell 8, 1871–1883.

Goethals K, Van den Eede G, Van Montague M and Holsters M 1990 Identification and characterization of a *nodD* gene in *Azorhizobium caulinodans* ORS571. J. Bacteriol. 172, 2658–2666.

Gough C, Vasse J, Galera C, Webster G, Cocking E and Dénarié J 1997 Interactions between bacterial diazotrophs and non-legume dicots: *Arabidopsis thaliana* as a model plant. Plant Soil 194, 123–130.

Hirsch A M 1992 Developmental biology of legume nodulation. New Phytol. 122, 211–237.

Hirsch A M and Fang Y 1994 Plant hormones and nodulation: what's the connection? Plant Mol. Biol. 26, 5–9.

Jensen H L 1942 Nitrogen fixation in leguminous plants. I. General characters of root nodule bacteria isolated from species of *Medicago* and *Trifolium* in Australia. Proc. Linn. Soc. N.S.W. 66, 98–108.

Jing Y, Li G, Jin G, Shan X, Zhang, B Guan C and Li J 1990 Rice root nodules with acetylene reduction activity. *In* Nitrogen Fixation Achievements and Objectives. Eds. P M Gresshoff, L E Roth, G Stacey and W E Newton. p 829. Chapman and Hall, London.

Jing Y, Li G and Shan X 1992 Development of nodule-like structure on rice roots. *In* Nodulation and Nitrogen Fixation in Rice. Eds. G S Khush and J Bennett. pp 123–126. International Rice Research Institute, Manila.

Kennedy I R, Pereg-Gerk L, Wood C, Deaker R, Gilchrist K and Katupitiya S 1997 Biological nitrogen fixation in non-leguminous field crops: Facilitating the evolution of an effective association between *Azospirillum* and wheat. Plant Soil 194, 65–79.

Kirchhof G, Reis V M, Baldani J I, Eckert B, Döbereiner J and Hartmann A 1997 Occurence, physiological and and molecular analysis of endophytic diazotrophic bacteria in gramineous energy plants. Plant Soil 194, 45–55.

Ladha J K, Kirk G, Bennett J, Peng S, Reddy C K, Reddy P M and Singh U 1997 Opportunities for increased nitrogen use efficiency from improved lowland rice germplasm. Field Crops Res. (*In press*).

Ladha J K and Reddy P M 1995 Extension of nitrogen fixation to rice-necessity and possibilities. Geojournal 35, 363–372.

Leong S A, Williams P H and Ditta G S 1985 Analysis of the $5'$ regulatory region of the gene for d-amino levulinic acid synthetase of *Rhizobium meliloti*. Nucl. Acids Res. 13, 5965–5976.

Lewin A, Cervantes E, Chee-Hoong W and Broughton W J 1990 *nodSU*, two new nod genes of the broad host range *Rhizobium* strain NGR234 encode host-specific nodulation of the tropical tree *Leucaena leucocephala*. Mol. Plant-Microbe Interact. 3, 317–326.

Li G, Jing Y, Shan X, Wang H and Guan C 1991 Identification of rice nodules that contain *Rhizobium* bacteria. Chin. J. Bot. 3, 8–17.

Mateos P, Baker D, Philip-Hollingsworth S, Squartini A, Peruffo A, Nuti M and Dazzo F B 1995 Direct in situ identification of cellulose microfibrils associated with *Rhizobium leguminosarum* bv. *trifolii* attached to the root epidermis of white clover. Can. J. Microbiol. 41, 202–207.

Mateos P, Jiminez-Zurdo, J Chen, A Squartini, S Haack, E Martinez-Molina, D Hubbell and Dazzo F B 1992 Cell-associated pectinolytic and cellulolytic enzymes in *Rhizobium leguminosarum* bv. *trifolii*. Appl. Environ. Microbiol. 58, 1816–1822.

Meade H, Long S R, Ruvkun G B, Brown S E and Ausubel F M 1982 Physical and genetic characterization of symbiotic and auxotrophic mutants of *Rhizobium meliloti*. J. Bacteriol. 149, 114–122.

Nap J-P and Bisseling T 1990 Developmental biology of a plant-prokaryote symbiosis: The legume root nodule. Science 250, 948–954.

Napoli C A, Dazzo F B and Hubbell D H 1975 Ultrastucture of infection and common antigen relationships in the *Rhizobium-Aeschynomene* symbiosis. *In* Proc. 5th Australian Legume Nodulation Conference. Ed. J Vincent. pp 35–37. Brisbane.

Orgambide G G, Philip Hollingsworth S, Hollingsworth R T and Dazzo F B 1994 Flavone-enhanced accumulation and symbiosis-related activity of a diglycosyl diacylglycerol membrane glycolipid from *Rhizobium leguminosarum* bv. *trifolii*. J. Bacteriol. 176, 4338–4347.

Peters N K and Long S R 1988 Alfalfa root exudates and compounds which promote or inhibit induction of *Rhizobium meliloti* nodulation genes. Plant Physiol. 88, 396–400.

Philip-Hollingsworth S, Hollingsworth R I and Dazzo F B 1991 N-acetylglutamic acid: an extracellular Nod signal of *Rhizobium trifolii* ANU843 which induces root hair deformation and nodule-like primordia in white clover roots. J. Biol. Chem. 266, 16854–16858.

Plazinski J, Innes R W and Rolfe B G 1985 Expression of *Rhizobium trifolii* early nodulation genes on maize and rice plants. J Bacteriol. 163, 812–815.

Reddy P M, Kouchi H, Hata S and Ladha J K 1996a Homologs of *GmENOD93* from rice. 8th International Congress on Molecular Plant-Microbe Interactions, Knoxville, TN.

Reddy P M, Kouchi H, Hata S and Ladha J K 1996b Identification, cloning and expression of rice homologs of *GmN93* 7th International Symposium on BNF with Non-Legumes, Faisalabad.

Reddy P M and Ladha J K 1995 Can symbiotic nitrogen fixation be extended to rice? *In* Nitrogen fixation: Fundamentals and Applications. Eds. I A Tikhonovich, N A Provorov, V I Romanov and W E Newton. pp 629–633. Kluwer Academic Publishers, Dordrecht.

Reddy P M, Ramos M C, Hernandez R J and Ladha J K 1995 Rice-rhizobial interactions. 15th North American Conference on Symbiotic Nitrogen Fixation, Raleigh, NC.

Reddy P M, Torrizo L, Ramos M C, Datta S K and Ladha J K 1996c Expression of *MtENOD12* promoter driven GUS in transformed rice. 8th International Congress on Molecular Plant-Microbe Interactions, Knoxville, TN.

Redmond J W, Batley M, Djordjevic M A, Innes R W, Kuempel P L and Rolfe B G 1986 Flavones induce expression of nodulation genes in *Rhizobium*. Nature 323, 632–635.

Rolfe B G and Bender G L 1990 Evolving a *Rhizobium* for non-legume nodulation. *In* Nitrogen Fixation:Achievements and Objectives. Eds. P M Gresshoff, L E Roth, G Stacey and W E Newton. pp 779–786. Chapman and Hall, London.

Rolfe B G, Djordjevic M A, Weinman J J, Mathesius U, Pittock C, Gärtner E, Dong Z, McCully M and McIver J 1997 Root morphogenesis in legumes and cereals and the effect of bacterial inoculation on root development. Plant Soil 194, 131–144.

Secilia j and Bagyaraj D J 1992 Selection of efficient vesicular-arbuscular mycorrhizal fungi for wetland rice (*Oryza sativa* L.). Biol. Fert. Soils 13, 108–111.

Silver D L, Pinaev A, Chen R and de Bruijn F J 1996 Post-transcriptional regulation of the *Sesbania rostrata* early nodulin gene *SrEnod2* by cytokinin. Plant Physiol. 112, 559–567.

Silver D L, Deikman J, Chen R and de Bruijn F J 1997 The *SrEnod2* gene is contnolled by a conserved cytokinin signal transduction pathway. Plant Physiol. (*In press*).

Stacey G and Shibuya N 1997 Chitin recognition in rice and legumes. Plant Soil 194, 161–169.

Stoltzfus J R, So R, Malarvizhi P P, Ladha J K and de Bruijn F J 1997 Isolation of endophytic bacteria from rice and assessment of their potential for supplying rice with biologically fixed nitrogen. Plant Soil. 194, 25–36.

Subba-Rao N S, Mateos P F, Baker D, Pankratz H S, Palma J, Dazzo F B and Sprent J I 1995 The unique root nodule symbiosis between *Rhizobium* and the aquatic legume, *Neptunia natans*. (L.F.) Druce. Planta 196, 311–320.

Taller B J and Sturtevant D B 1991 Cytokinin production by rhizobia. *In* Advances in Molecular Genetics of Plant-Microbe Interactions. Vol. 1. Eds. H Hennecke and D P S Verma. pp 215–221. Kluwer Academic Publishers, Dordrecht.

Terouchi N and Syono K 1990 *Rhizobium* attachment and curling in asparagus, rice and oat plants. Plant Cell Physiol. 31, 119–127.

Torrey J G 1986 Endogenous and exogenous influences on the regulation of lateral root formation. *In* New Root Formation in Plants and Cuttings. Eds. M B Jackson. pp 31–66. Martinus Nijhoff Publishers, Dordrecht.

Trinick M J 1973 Symbiosis between *Rhizobium* and the non-legume, *Trema aspera*. Nature 244, 459–460.

Truchet G, Debelle F, Vasse J, Terzaghi B, Garnerone A M, Rosenberg C, Batut J, Maillet F and Denarie J 1985 Identification of *Rhizobium meliloti* Sym2011 region controlling the host specificity of root hair curling and nodulation. J. Bacteriol. 164, 1200–1210.

Tsien H C, Dreyfus B L and Schmidt E L 1983 Initial stages in the morphogenesis of nitrogen-fixing stem nodules of *Sesbania rostrata*. J. Bacteriol. 156, 888–897.

Vijn I, das Neves L, van Kammen A, Franssen H and Bisseling T 1993 Nod factors and nodulation in plants. Science 260, 1764–1765.

Vincent J M 1970 A manual for the study of root nodule bacteria. IBP Handbook 15, Blackwell Scientific Publ. London. 164 p.

Webster G, Gough C, Vasse J, Batchelor C A, O'Callaghan K J, Kothari S L, Davey M R, Dinari J and Cocking E C 1997 Interactions of rhizobia with rice and wheat. Plant Soil 194, 115–122.

Yanni Y G, Rizk E Y, Corich V, Squartini A, Ninke K, Philip-Hollingsworth S, Orgambide G G, de Bruijn F J, Stoltzfus J, Buckley D, Schmidt T M, Mateos P F, Ladha J K and Dazzo F B 1997 Natural endophytic association between *Rhizobium leguminosarum* bv. *trifolii* and rice roots and assessment of its potential to promote rice growth. Plant Soil 194, 99–114.

Yuhashi, K-I, Akao S, Fukuhara H, Tateno E, Chun J-Y, Stacey G, Hara H, Kubota M, Asami T and Minamisawa K 1995 *Bradyrhizobium elkanii* induces outer cortical root swelling in soybean. Plant Cell Physiol. 36, 1571–1577.

Guest editors: J K Ladha, F J de Bruijn and K A Malik

Plant and Soil **194**: 99–114, 1997.
© 1997 *Kluwer Academic Publishers. Printed in the Netherlands.*

Natural endophytic association between *Rhizobium leguminosarum* bv. *trifolii* and rice roots and assessment of its potential to promote rice growth

Youssef G. Yanni[1], R.Y. Rizk[1], V. Corich[2], A. Squartini[2], K. Ninke[3,4],
S. Philip-Hollingsworth[3], G. Orgambide[3], F. de Bruijn[3,4,5], J. Stoltzfus[5], D. Buckley[3,4],
T.M. Schmidt[3,4], P.F. Mateos[6], J.K. Ladha[7] and Frank B. Dazzo[3,4,8]
[1]*Sakha Agricultural Research Station, Kafr El-Sheikh, 33717 A. R. Egypt;* [2]*Dipt. di Biotecnologie Agrarie,
Universita Degli Studi di Padova, Padova, Italy;* [3]*Dept. of Microbiology,* [4]*Center for Microbial Ecology and*
[5]*Plant Research Laboratory, Michigan State University, East Lansing, MI 48824, U.S.A.;* [6]*Dept. de
Microbiologia y Genetica, Universidad de Salamanca, Salamanca, Spain and* [7]*International Rice Research
Institute, P.O. Box 933, 1099 Manila, Philippines.* [8]*Address all correspondence to Dr. Frank B. Dazzo, Dept. of
Microbiology, Michigan State University, East Lansing, MI 48824, USA**

Key words: association, clover, endophyte, PGPR, *Rhizobium leguminosarum* bv. *trifolii,* rice, root, symbiosis

Abstract

For over 7 centuries, production of rice (*Oryza sativa* L.) in Egypt has benefited from rotation with Egyptian berseem clover (*Trifolium alexandrinum*). The nitrogen supplied by this rotation replaces 25- 33% of the recommended rate of fertilizer-N application for rice production. This benefit to the rice cannot be explained solely by an increased availability of fixed N through mineralization of N- rich clover crop residues. Since rice normally supports a diverse microbial community of internal root colonists, we have examined the possibility that the clover symbiont, *Rhizobium leguminosarum* bv. *trifolii* colonizes rice roots endophytically in fields where these crops are rotated, and if so, whether this novel plant-microbe association benefits rice growth. MPN plant infection studies were performed on macerates of surface-sterilized rice roots inoculated on *T. alexandrinum* as the legume trap host. The results indicated that the root interior of rice grown in fields rotated with clover in the Nile Delta contained $\sim 10^6$ clover-nodulating rhizobial endophytes g^{-1} fresh weight of root. Plant tests plus microscopical, cultural, biochemical, and molecular structure studies indicated that the numerically dominant isolates of clover-nodulating rice endophytes represent 3 – 4 authentic strains of *R. leguminosarum* bv. *trifolii* that were Nod$^+$ Fix$^+$ on berseem clover. Pure cultures of selected strains were able to colonize the interior of rice roots grown under gnotobiotic conditions. These rice endophytes were reisolated from surface-sterilized roots and shown by molecular methods to be the same as the original inoculant strains, thus verifying Koch's postulates. Two endophytic strains of *R. leguminosarum* bv. *trifolii* significantly increased shoot and root growth of rice in growth chamber experiments, and grain yield plus agronomic fertilizer N-use efficiency of Giza-175 hybrid rice in a field inoculation experiment conducted in the Nile Delta. Thus, fields where rice has been grown in rotation with clover since antiquity contain Fix$^+$ strains of *R. leguminosarum* bv. *trifolii* that naturally colonize the rice root interior, and these true rhizobial endophytes have the potential to promote rice growth and productivity under laboratory and field conditions.

Abbreviations: GC/MS – gas chromatography-mass spectrometry, ^{1}H-NMR – proton nuclear magnetic resonance spectroscopy, IRRI – International Rice Research Institute, LSD – least significant difference, MPN – most probable number, PCR – polymerase chain reaction, PGPR – plant growth promoting rhizobacteria, RFLP – restriction fragment length polymorphism, RDP – ribosomal database project, SDS-PAGE – sodium dodecylsulfate-polyacrylamide gel electrophoresis, YEM – yeast extract mannitol

---

* FAX No: +15173538953. E-mail: 23249mgr@msu.edu

Introduction

Cereals are the world's major source of food for human nutrition. Among these, rice (*Oryza sativa* L.) is very prominent and represents the staple diet for more than two-fifth's (2.4 billion) of the world's population, making it the most important food crop of the developing world (IRRI, 1996). Production of rice, and hence global food security, depends on reaching even higher levels of sustainable grain production, which is not possible without additional nutrient input. Indirectly, rice is able to utilize a basal level of fixed-N as a source of its N nutrition from the N_2-fixing activities of diazotrophs in its agronomic ecosystem (Ladha, 1986; Roger and Ladha, 1992; Yanni, 1991). If rice were able to establish a more direct and efficient symbiotic association with N_2-fixing organisms, serious economic and ecological problems associated with the use of inorganic and organic fertilizers to enhance rice production could be mitigated.

It is well known that a remarkable diversity of N_2-fixing bacteria naturally associate with field-grown rice (Bally et al., 1983; Ladha, 1986; Ladha, 1993; Natalia et al., 1994; Roger and Watanabe, 1986; Ueda et al., 1995; Yanni, 1991). During the last few years, there has been an increased interest in exploring the possibility of extending the beneficial interactions between rice and some of these N_2-fixing bacteria. This line of investigation came into full focus in 1992, when the International Rice Research Institute hosted an international workshop to assess knowledge on the potential for nodulation and nitrogen fixation in rice associated with symbiotic bacteria (Khush and Bennett, 1992). One of the future research directions recommended at that workshop was to determine if rhizobia naturally colonize the interior of rice roots when this cereal is grown in rotation with a legume crop, and if so, to assess the potential impact of this novel plant-microbe association on rice production. This idea is derived from the general concept that roots of healthy plants grown in natural soil eventually develop a continuum of root-associated microorganisms extending from the rhizosphere to the rhizoplane, and even deeper into the epidermis, cortex, endodermis, and vascular system (Balandreau and Knowles, 1978; Klein et al., 1990; Old and Nicolson, 1975; Old and Nicolson, 1978). Typically, the presence of these microorganisms within roots does not induce obvious symptoms of disease. Although originally described as *endorhizosphere microorganisms*, it has been proposed that the microflora that colonize this specialized habitat inside roots should instead be referred to as *endophytes* or *internal root colonists* (Kloepper et al., 1992). This habitat has already been identified as an important reservoir for isolation of N_2-fixing plant growth-promoting rhizobacteria (PGPR). Examples that illustrate this point are the isolation of *Azospirillum* strains from "inside" host roots (after surface-sterilization) which efficiently promote yield when inoculated on that homologous host (Boddey and Dobereiner, 1988), and the diazotrophic endophytes of *Azoarcus* inside Kallar grass (Bilal and Malik, 1987; Hurek et al., 1994) and *Acetobacter diazotrophicus* inside sugar cane (Dobereiner et al., 1993). Presumably, nature selects endophytes that are competitively fit to occupy compatible niches within this nutritionally enriched and protected habitat of the root interior without causing pathological stress on the host plant.

Our interest has been to assess the possible existence and agronomic importance of naturally occurring rhizobial endophytes within rice roots, particularly in regions of the world where rice production *is significantly benefited* by rotation with a legume crop that could sustain the populations of the corresponding rhizobial symbiont at a high inoculum potential for the next rice growing season. One of the regions ideally suited to address these questions is in the Nile Delta of Egypt, the major producer of rice in Northern Africa. For more than 7 centuries, most of the rice cultivated in this region has been rotated with the legume, Egyptian berseem clover (*Trifolium alexandrinum* L.). Currently, about 60–70% of the $\sim 546,000$ ha of land area used in Egypt for rice production is in rice-clover rotation. This clover species is well adapted to the Middle-East where it is believed to originate, and its high yields, protein content, and N_2-fixing capacity enhance its use as a forage and green manure plant (Graves et al., 1987). In the Nile Delta region, rice is cultivated by transplantation in irrigated lowlands and includes both Indica and Japonica cultivars. Typically, irrigation from the Nile River is stopped 15–20 days before harvest to partially dry the soil. After the rice grain and straw are harvested, berseem clover seed is broadcast (most often without tillage) and becomes naturally nodulated by indigenous rhizobia (*Rhizobium leguminosarum* bv. *trifolii*) in the soil. Vegetative regrowth of the rice roots produces so-called "ratoon rice" intermingled among the clover plants.

The benefit of clover rotation replaces 25-33% of the recommended rate of fertilizer-N application for optimal rice production in the Nile Delta, but this benefit cannot be explained solely by an increased availability of fixed N through mineralization of the N-rich clover crop residues. Because of this benefit of clover rotation, and the many years during which native clover rhizobia have had the opportunity to interact with rice, we chose this agronomic system of the Egyptian Nile Delta for investigation. In this study, we examined the extent to which rice supports a natural endophytic association with clover rhizobia in fields where these two crops have been rotated continuously since antiquity, and assessed the potential of this endophytic association in promoting rice growth under laboratory conditions, and both rice productivity and agronomic fertilizer N-use efficiency under field conditions. Natural associations of endophytic diazotrophs in rice roots under rice-*Sesbania* rotation in the Philippines are also being studied by some of us (Ladha et al., 1989; Ladha et al., 1996a; Ladha et al., 1996b).

Materials and methods

Microscopical examination of endophytes within rice roots

Rice seedlings were grown for 30 days in non-sterile potted soil in a growth chamber as previously described (de Bruijn et al., 1995). Roots were sampled, cleaned with running water, freehand sectioned, and processed for examination by scanning electron microscopy (Umali-Garcia et al., 1980). Other seedlings were grown under microbiologically controlled conditions in enclosed tubes inoculated with pure cultures of rice endophytes. Freehand sections were stained with 0.01% acridine orange, washed and mounted in 1% sodium pyrophosphate, and examined by laser scanning microscopy in the epifluorescence confocal mode using computer-enhanced reconstruction of serial section overlays and digital image processing (Subba-Rao et al., 1995).

Enumeration and isolation of clover-nodulating rhizobia in rhizosphere soil and within roots of field-grown rice

Rhizosphere soil and roots of the Japonica rice cultivar Giza-172 were collected from fields at Sakha Kafr El-Sheikh, Egypt, near the middle Nile Delta region where rice has been rotated with Egyptian berseem clover for several hundred years. The roots were washed with running water, blotted and weighed, surface-sterilized with 70% ethanol followed by 10% sodium hypochlorite solution, rolled over yeast extract mannitol (YEM)

agar plates to verify surface-sterilization, and macerated in sterile 5 mM Na-phosphate buffer (pH 7.0). Enumeration of *R. leguminosarum* bv. *trifolii* in rhizosphere soil and surface-sterilized / macerated rice roots was performed by the five-tube most probable number (MPN) - plant infection test using berseem clover as the legume trap host (Somasegaran and Hoben, 1985). Tubes containing Vincent's nitrogen-free agar medium were planted with surface-sterilized clover seeds and germinated for 3 days before inoculation. Seedlings were incubated for one month and then scored for root nodulation. Root nodules that developed on clover plants receiving the highest dilutions were excised, surface-sterilized, and the nodule occupants isolated into pure culture by plating on YEM agar followed by restreaking isolated colonies on defined BIII agar (Dazzo, 1982).

Analyses of the symbiotic properties of selected isolates in pure culture were performed on Egyptian berseem clover seedlings (8 replicates per treatment) grown on agar slopes of N-free Fahraeus medium under microbiologically controlled conditions (Dazzo, 1982). Seedling roots were inoculated with 10^6 cells of a 5 day-old inoculum and incubated in a growth chamber under 16 hr day^{-1} of light, 70% relative humidity, and 22 °C day / 20 °C night cycle. Nodulation kinetics were assessed by periodically inspecting plants under the stereomicroscope for emergence of root nodules. Plants were harvested at 41 days after inoculation and evaluated for effectiveness in symbiotic N fixation by comparison of their dry weight and N-content by way of the micro-Kjeldahl steam distillation method (Black et al., 1965) to that of the uninoculated control plants.

Analysis of strain diversity and identification of endophytic Rhizobium *isolates by molecular methods*

Plasmid profiles were analyzed by the method of Eckhardt (1978) as modified by Espuny et al. (1987). Genomic restriction fragment length polymorphism (RFLP) of *Xba*1 digests was analyzed by pulsed-field gel electrophoresis as described by Corich et al. (1991). BOX-PCR amplification fragment length polymorphism was analyzed as described by Versalovic et al. (1994). Cells were also boiled in SDS gel buffer (Laemmli, 1970) and the profiles of their total cellular proteins were compared by SDS-PAGE in 12% running gels stained with Coomassie blue. The phylogenetic relationships of isolates E11 and E12 were analyzed by sequencing their total 16S rDNA. The 16S ribosomal RNA-encoding genes were PCR amplified from genomic DNA using conserved eubacterial primers 8F and 1540R. The amplified product was sequenced using dye terminators on the Applied Biosystems' DNA Sequencing System (Foster City, CA). Sequences were aligned against those in the Ribosomal Database Project (Larsen et al., 1993) on the basis of conserved regions of sequence and secondary structure. Phylogenetic relationships were inferred using regions of unambiguous alignment and the distance method of DeSoete (1984).

Further phenotypic characterization of selected clover-nodulating rice endophytes

Additional studies were performed on selected isolates to determine if they share phenotypic traits with typical wild-type strains of *R. leguminosarum* bv. *trifolii.* These include their growth characteristics on defined BIII agar, Gram's reaction and cellular morphology, production of well-defined extracellular capsules and intracellular lipid granules, structural features of their acidic exopolysaccharide produced during growth on BIII agar, production of cell-bound cellulase, and nodulation host range (Dazzo, 1982; Mateos et al., 1992; Philip-Hollingsworth et al., 1989). For the latter test, isolates were inoculated on 4 replicate seedlings of white clover (*T. repens* var. Dutch), Egyptian berseem clover, and alfalfa (*Medicago sativa* var. Gemini) in enclosed tube cultures and scored for root nodulation after 30 days of incubation in the growth chamber.

Cultivation of rice under gnotobiotic conditions

Laboratory studies of the interaction between rice and clover-nodulating rhizobial endophytes were performed using an enclosed tube culture system to exclude microbial contaminants from the roots under climate-controlled growth chamber conditions. Japonica rice Giza-171 and Indica rice IR-28, both currently cultivated in the Nile Delta, were used for this study. Separate tests established that seeds of these two rice cultivars harbored no endophytic clover-nodulating rhizobia that would survive surface-sterilization. Seeds weighing approximately 30 mg each were surface-sterilized by treatment with 70% ethanol for 1 min followed by 10% sodium hypochlorite solution for 4 min, and then washed for 4×1 min with sterile water. Surface-sterilized seeds were transferred to 25 $\times$ 200 mm tubes, each enclosed with foam plugs and containing 20 ml of Hoagland's #2 plant growth medium (Sigma Chem. Co., St. Louis, MO) solidified with 1% purified agar (United States Biochemical, Cleveland, OH), above which was layered ca. 5 g of sterile acid-washed quartz sand and 4 mL sterile Hoagland's #2 liquid medium. Tubes were incubated for 2 days in the dark at 30 °C for seed germination. The bacterial inocula were grown separately on BIII agar for 5 days at 30 °C, suspended in sterile Hoagland's medium, and adjusted to a density of 10^7 cells mL^{-1}. Each seedling root was inoculated with 10^6 cells (6 plant replicates per treatment) and incubated in the growth chamber as described above. When grown to sufficient length, the stem was repositioned through a slit on the edge of the foam plug to allow continuous growth while preventing microbial contamination of the root system. Tubes were irrigated with sterile water alternating with sterile Hoagland's solution as needed.

Endophytic colonization of rice by various strains of R. leguminosarum *bv.* trifolii *and assessment of their potential to promote plant growth*

Rice plants in tube culture were gently uprooted 32 days after inoculation, and then excised at the stem base. Shoot biomass (stem plus leaves) was measured as dry weight. Roots were rinsed free of agar and sand, blotted, weighed, surface-sterilized with 70% ethanol followed by 10% sodium hypochlorite solution, rolled over plates of BIII agar and trypticase soy agar to check for surface sterility, and then macerated in 5 m*M* Na-phosphate buffer as described above. Viable plate counts of the rice endophyte populations were made after 5 days incubation of diluted root macerates plated on YEM agar. Then colonies were picked, restreaked on BIII plates, and stocked in pure culture. Authenticity of these endophyte reisolates was evaluated by a comparison of their plasmid profiles and BOX-PCR patterns to those of the original inoculant strains. In similar experiments, rice roots were assayed for N_2-fixing activity by incubating the entire root system of each plant replicate for 2 hr at 22 °C in 14 ml serum vials containing 10% acetylene in air, followed by flame-ionization gas chromatography of 1 ml gas samples to detect acetylene-dependent ethylene production. Gross morphological responses to inoculation were then evaluated by preparing calibrated photocopies of shoots and roots, followed by computer-assisted image analysis of photocopied imprints using a constant threshold setting (Dazzo and Petersen, 1989; Smucker, 1993). Shoots were then dried and their N-content measured.

Field evaluation of rice growth responses to inoculation with rice endophytes of R. leguminosarum *bv.* trifolii

A field inoculation experiment was conducted in a lowland, irrigated field at the Sakha Agricultural Research Station using the short duration (135 days) rice cultivar Giza-175 (a hybrid of the Japonica cultivar Giza-14 and the Indica cultivar IR-28). This rice cultivar is characterized by its short stature and grain, blast resistance, early maturing, high N-response, high yielding, and good hasking qualities. The field soil was an alluvial clay-loam that had originated from the annual Nile flood sediments. Its characteristics were 50–55% clay, 20–25% silt, 20–25% coarse + fine sand; pH 8.0; ~2% organic matter, 0.11% total N; 4 ppm available P, and a CEC of 40–45 meq 100 g^{-1} soil. This field soil had previously been cultivated in rice-clover rotation for many years and contained an indigenous population of *R. leguminosarum* bv. *trifolii* with a MPN of ~3.8 x 10^4 g^{-1} soil at the start of the experiment. Calcium superphosphate (15% P_2O_5) was added at 36 kg P_2O_5 ha^{-1} before tillage. One-month old rice seedlings were transplanted at a density of 500 seedlings per 20 m^2 (4 × 5 m) subplot with a spacing of 20 × 20 cm between plants. Main plot treatments were fertilized with urea (46% N) at 0, 48, 96, or 144 kg N ha^{-1} added in two equal doses, 25 days after transplanting and at the mid-tillering stage. Plots were irrigated to maintain a 5–7 cm waterhead above the soil surface. Inoculant cultures of rice endophyte strains E11 and E12 were grown in YEM broth for 96 hrs at 30 °C and adjusted to a density of 10^9 CFU mL^{-1}. Each sub-plot of 20 m^2 was inoculated with a 100 ml suspension of one of the two isolates (according to the experimental design) 5 days after transplanting. Some sub-plot treatments received no inoculum as control. The main and sub-plot treatments were distributed at random within each of four replications. Various crop and agronomic responses to inoculation were evaluated from rice plants 135 days after sowing (105 days after transplanting). These parameters included grain size (1000-grain weight), grain yield and N-content, straw yield and N-content, harvest index (% grain yield / grain + straw yields), and the agronomic fertilizer N-use efficiency (kg grain per kg fertilizer-N). All rice plants were harvested and evaluated to obtain the yield data. The collected data were analyzed as a split-plot design experiment, with the N-fertilizer doses as main plot treatments and endophyte inoculation as sub-plot treat-ments. The mean differences were compared to their corresponding least significant differences.

Results and discussion

Direct microscopy reveals a diverse natural community of endophytes within the root interior of rice grown in soil

The first step in this project was to document that the root interior of healthy rice plants is colonized by microorganisms when grown in non-sterile soil. Scanning electron microscopy of the freehand cut face of rice roots grown in potted soil provided direct evidence that this is indeed the normal case, particularly where emergence of lateral roots forms open wounds in the epidermis and cortex, providing a portal of entry of the rhizoplane microflora. Figures 1A-D reveal a diverse community of bacteria exhibiting various morphotypes that colonized this habitat.

Enumeration and characterization of clover-nodulating rhizobia in rhizosphere soil and within roots of field-grown rice

The next step was to measure the extent of natural association between rhizobia and rice roots in fields rotated with legumes. Figure 2A shows the field sampling site 25 days after harvest of the rice grain with regrowth of rice ratoon intermingled among berseem clover. The MPN plant infection test using berseem clover as the trap host indicated a population density of ~1.7 × 10^6 indigenous clover-nodulating rhizobia per gram of rhizosphere soil surrounding rice roots in this field (Figure 2B). By strategically combining surface-sterilization and the use of the appropriate trap legume host, the MPN plant infection test indicated that clover-nodulating rhizobia naturally invade rice roots and achieve an internal population density of ~1.1 × 10^6 endophytes per gram fresh weight of rice roots (Figure 2B).

Twelve isolates of nodule occupants representing the numerically dominant clover-nodulating rice endophytes were established in pure culture. Each isolate produced large, mucoid, pearl white colonies on defined BIII agar and were Gram negative rods containing intracellular lipid granules, typical of other wild type *R. leguminosarum* bv. *trifolii* strains (figures not shown). Eight isolates of these rice endophytes were retested on berseem clover in N-free tube culture, and

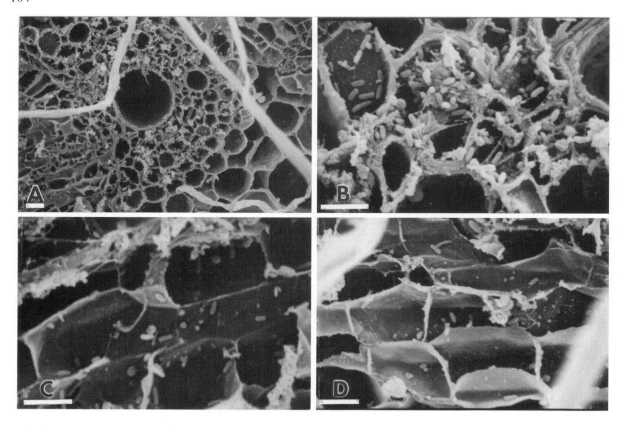

Figure 1. Scanning electron micrographs of the natural community of microbial endophytes that colonized the interior of rice roots grown for 30 days in potted soil. (**A**) Low magnification view of the cut face of the root. (**B-D**) Higher magnification showing the density and variety of bacterial morphotypes. Bar scales are 10 μm in A, 5 μm in B-D.

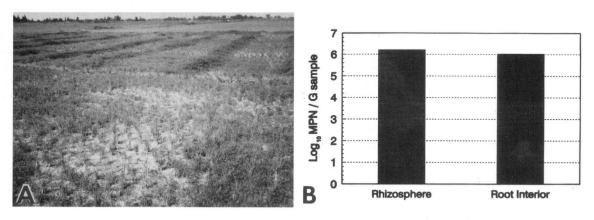

Figure 2. (**A**) Example of a field sampling site in the Nile Delta where rice is rotated with berseem clover. (**B**) MPN populations of clover-nodulating rhizobia in the rhizosphere soil and root interior of field-grown rice.

each isolate was able to nodulate and fix N_2 symbiotically with this host under these microbiologically controlled conditions (Figure 3, Table 1). Isolates E11 and E12 were most noteworthy, since they nodulated berseem clover very rapidly and were very effective in symbiotic N_2 fixation on this host.

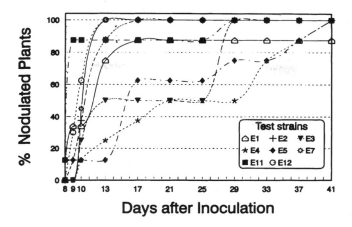

Figure 3. Nodulation kinetics of berseem clover inoculated with various test strains of rice endophytes and grown in gnotobiotic tube cultures.

Table 1. Symbiotic performance of clover-nodulating rice endophytes inoculated on Egyptian berseem clover (*Trifolium alexandrinum*) and incubated 41 days in tube culture

Endophyte Inoculum Strain	Nodules per plant	Plant dry weight (mg)	Plant N-content (mg)
E1	3.6	8.2[b]	0.24[a]
E2	3.5	7.7	0.21[a]
E3	4.5	7.5	0.23[a]
E4	3.6	7.6	0.24[a]
E5	4.5	8.3[b]	0.24[a]
E7	4.6	11.3[a]	0.33[a]
E11	3.6	13.7[a]	0.40[a]
E12	5.3	12.4[a]	0.36[a]
Control	0	6.4	0.13
LSD		**0.05 0.01**	**0.05 0.01**
		1.5 2.0	0.01 0.05

Mean values followed by the letter a or b are significantly different from that of the uninoculated control at the 99% and 95% confidence levels, respectively.

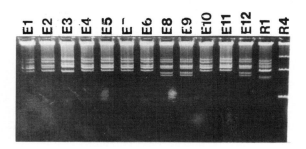

Figure 4. Plasmid profiles of various clover-nodulating rice endophytes (E strains) and rice rhizosphere isolates (R strains). Lanes are labeled with strain numbers.

The diversity of these Fix[+] clover-nodulating rice endophytes was determined by various molecular analyses. The evaluation of their plasmid profiles indicated several groups; one group contained only isolate E3, a second group contained isolates E8, E9, and E12, and a third group contained E1, E4, E5, E6, E7, E10, and E11 (Figure 4). The plasmid profile of isolate E2 was very similar to the latter isolates comprising the third group, except that the intensity of one of its bands was decreased. Additional genomic DNA analyses of these endophyte isolates using pulsed-field RFLP and BOX-PCR, and total protein profiles using SDS-PAGE indicated the same 3 groupings with retention of isolate E2 in group 3 (figures not shown). These results indicate that a diversity of at least 3 (possibly 4) different groups of clover-nodulating rice endophytes were isolated from the root interior of field-grown rice sampled in the Sakha region of Egypt. The plasmid profiles of these clover-nodulating rice endophytes differed from 2 clover-nodulating rhizobia isolated from the rice rhizosphere soil in the same field (Figure 4), and from *R. leguminosarum* bv. *trifolii* strains ARC100 and ARC101 used for inoculant production in other areas of Egypt (Hashem and Kuykendall, 1994).

Because of their superior symbiotic competence on berseem clover and their distinct differences in the above tests of strain diversity, endophyte strains E11 and E12 were examined further to verify their authenticity as *R. leguminosarum* bv. *trifolii*. Phylogenetic analysis of their total 16S rRNA sequences indicate that they both belong to the α proteobacteria within the cluster of *Rhizobium leguminosarum* that is clearly distinguished from other rhizobia and rhi-

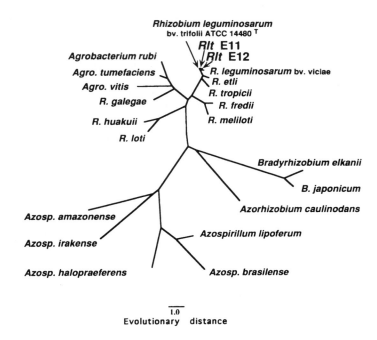

Figure 5. Phylogenetic tree showing the relatedness of clover-nodulating rice endophyte strains E11 and E12 to *R. leguminosarum* bv. *trifolii* and other endophytic rhizobacteria. Data are based on a comparison of the total 16S rRNA sequences of strains E11 and E12 to other official type strains in the Ribosomal Project Database. The calibrated scale of evolutionary distance is indicated.

zobacteria in the RDP database (Figure 5). Interestingly, their 16S rRNA sequences differ from the official type strain of *R. leguminosarum* bv. *trifolii* (ATCC 14480) **at only one base position** (*Escherichia coli* position #1137) (Willems and Collins, 1993; Young and Haukka, 1996). The GenBank accession numbers for the 16S rDNA sequences of strains E11 and E12 are designated as U73208 and U73209, respectively. In Jensen tube cultures grown on N-free medium, strains E11 and E12 nodulated both Egyptian berseem clover and Dutch white clover in the clover cross-inoculation group [inducing an average of 6 and 7 nodules per plant by 41 days after inoculation, respectively], but neither strain nodulated alfalfa under the same test conditions. Like other wild type *R. leguminosarum* bv. *trifolii* (Dazzo, 1982; Mateos et al., 1992), both strains were positive in the plate assay for cell-bound cellulase (figure not shown) and produced distinctive capsules surrounding the bacterial cells (Figure 6 insert). Both strains produced an extracellular acidic heteropolysaccharide composed of glucose, glucuronic acid, and galactose residues, and pyruvate, acetate, and 3-hydroxybutyrate substitutions. Although the GLC/MS (figure not shown) and ^{1}H-NMR spectra (Figure 6) of the acidic exopolysaccharide of both strains are typical

of other wild type strains of *R. leguminosarum* bv. *trifolii* (Philip-Hollingsworth et al., 1989), their ^{1}H-NMR spectra differ from one another in resonances for 3-hydroxybutyrate substitutions (Figure 6). Considered collectively, all of these molecular, cellular, cultural, and symbiotic characteristics provide compelling evidence that endophytes E11 and E12 isolated from the root interior of field-grown rice are two authentic strains of *Rhizobium leguminosarum* bv. *trifolii* capable of forming an effective N_2-fixing root-nodule symbiosis with *Trifolium alexandrinum*. The data argue against the alternative possibility that they are some other bacterial species that acquired the Sym plasmid of indigenous clover-nodulating rhizobia within the same soil.

Colonization of rice plants by rice root endophytes of R. leguminosarum *bv.* trifolii *under gnotobiotic conditions*

We next developed a gnotobiotic tube culture system to study the interactions of endophyte strains and rice under climate-controlled growth chamber conditions while excluding microbial contaminants. In this system, seedlings developed from surface-sterilized

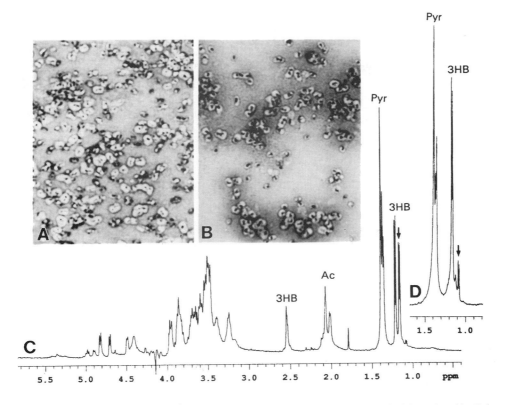

Figure 6. Photomicrographs of encapsulated cells and ^{1}H-NMR spectra of isolated acidic heteropolysaccharide produced by *R. leguminosarum* bv. *trifolii* strains E11 (**A**, **C**) and E12 (**B**, **D**). Differences in 3-hydroxybutyrate resonances are indicated by arrows.

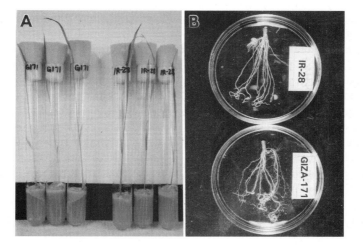

Figure 7. (**A**) Examples of the Japonica rice Giza-171 and the Indica rice IR-28 grown in enclosed tube cultures using a submerged sand overlay above semi-solid agar. (**B**) Morphology of uninoculated roots after 32 days of growth in axenic tube culture.

seeds will grow roots through a layer of quartz sand submerged in the plant growth medium above soft agar containing the same growth medium, while their shoots protrude through the side of a foam plug closure into open humid air (Figure 7A). Under these growth conditions, uninoculated roots of the Japonica

Table 2. Populations of *R. leguminosarum* bv. *trifolii* rice endophytes colonizing the root interior of rice grown in gnotobiotic tube culture for 32 days

Endophyte inoculum strain	Plant growth medium	Rice endophyte population (Log_{10} CFU / g root fresh wt)	
		Giza-171	IR-28
E2	Fahraeus -N	6.34	6.34
E7	Fahraeus -N	7.82	9.94
E11	Fahraeus -N	8.43	9.68
E12	Fahraeus -N	6.44	5.36
E2	Hoagland +N	7.94	8.85
E7	Hoagland +N	8.36	8.72
E11	Hoagland +N	8.23	9.52
E12	Hoagland +N	7.95	9.81

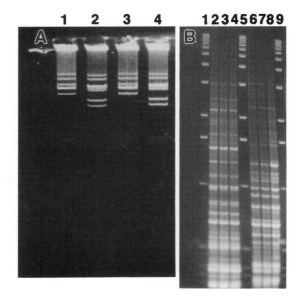

Figure 8. Analyses of plasmid profiles (**A**) and BOX-PCR patterns (**B**) which confirm that the *R. leguminosarum* bv. *trifolii* endophytes reisolated from rice roots are the same as the inoculant strains. Lanes A1 and B2 are inoculant strain E11; A2 and B6 are inoculant strain E12. Lanes A3 , B3 and B4 are endophyte reisolates of E11; A4, B7 and B8 are endophyte reisolates of E12. Lanes B1, B5, and B9 represent a standard 1 kb ladder.

rice Giza-171 and the Indica rice IR-28 remain axenic and produce similar biomasses, but Giza-171 inherently produces more extensive branching of lateral roots (Figure 7B). Plating experiments of macerates from surface-sterilized roots of the inoculated plants indicated substantial populations of the endophytic bacteria, the magnitude of which varied with the inoculant strain, the rice cultivar, and the plant growth medium (Table 2). For most strains, higher endophytic populations were established inside rice roots of Indica IR-28 grown in both plant growth media. (Table 2). These roots appeared healthy without development of nodule-like hypertrophies or obvious symptoms of disease, e.g., localized brown discolorations as previously described (de Bruijn et al. 1995). Under these experimental conditions, internal root colonization by the rhizobial endophytes was not suppressed in Hoagland's No. 2 plant growth medium, which contains both NH_4^+ and NO_3^- as sources of combined N. This contrasts with a distinct suppression of rhizobial invasion of the legume host root via root hair infection when the symbionts are cultured together in media containing either of these two sources of N (Abdel Wahab et al., 1996; Dazzo and Brill, 1978).

Colonies formed on defined BIII agar in these plating experiments had the typical pearl white and mucoid appearance of the inoculum. Well-isolated colonies picked from plates containing macerated samples of surface-sterilized roots originally inoculated with strain E11 or E12 had the same plasmid profiles and genomic BOX-PCR patterns as the corresponding inoculant strains (Figure 8A and 8B). These results indicate that endophytic colonization of rice roots by selected strains of clover-nodulating rhizobia can be

reliably established under microbiologically controlled conditions, thus fulfilling Koch's postulates for this newly described plant-microbe association. Obviously, these strains of *R. leguminosarum* bv. *trifolii* are fully capable of invading rice and colonizing their root interior without needing other soil microorganisms to assist their entry or endophytic multiplication.

Laser scanning confocal microscopy of rice grown under these gnotobiotic conditions revealed that the inoculant strains had colonized the root epidermal surface (figure not shown), consistent with previous reports that rhizobia can colonize other cereal rhizoplanes (Chabot et al., 1996; Hoflich et al., 1995; Ladha et al., 1989; Shimshick and Herbert, 1979). Other studies have indicated that rhizobia presumably access the root interior of rice and other non-legumes by crack entry at emergence of lateral root primordia or between epidermal cells, and subsequently they colonize the root cortex intercellularly and within dead host cells adjacent to living host cells in a non-structured way (Cocking et al., 1992; de Bruijn et al., 1995; Gough et al., 1996; Ladha et al., 1996a; Reddy et al., 1997; Spencer et al., 1994). Although some rhizobial symbionts use the same portal of entry into roots of their aquatic legume hosts, they ultimately disseminate within well-structured nodules through *bona-fide* tubular infection threads (Ndoye et al., 1994; Subba-Rao et al., 1995). Further examination of rice plants grown with strain E11 in gnotobiotic tube culture revealed numerous bacteria within leaf whirls at the stem base above the taproot (Figures 9A-9C). This latter finding identifies a second protected site of internal bacterial colonization that has potential importance to the rice-rhizobia association since it locates these endophytic diazotrophs in proximity to photosynthetically active host cells, and suggests a possible ascending migration of the bacteria within the rice plant. Whether the natural endophytic colonization of rice by clover rhizobia is restricted to these two specific habitats or also includes a truly endosymbiotic state within intact host cells remains an open question to be addressed in future investigations.

Growth promotion of rice by selected rice endophytes of R. leguminosarum *bv.* trifolii *under gnotobiotic and field conditions*

Growth stimulation of wheat, corn, radish, and mustard shoots following seed inoculation with a strain of *R. leguminosarum* bv. *trifolii* in open pot experiments has been previously reported (Hoflich et al., 1995). Our finding that selected strains of *R. leguminosarum* bv. *trifolii* can colonize the surface and interior of rice plants under gnotobiotic and field conditions prompted us to examine the impact of this close association on rice growth and development. We therefore used the tube culture system in quantitative bioassays of short-term PGP responses of rice to microbial inoculation during the period in which rice growth and development was not restricted, while avoiding uncontrolled effects of airborne microbial contaminants in the growth chamber environment. Quantitative measurements of plant growth responses to inoculation revealed that certain rhizobial endophytes significantly promoted growth of rice shoots and roots, the extent of which was influenced by the rice cultivar, the inoculant strain, the plant growth medium, and the growth parameter measured (Figures 10A and 10B, Table 3). Growth responses to inoculation were generally higher using the Japonica rice Giza-171 grown in Hoagland's (+N) No. 2 plant growth medium and inoculated with strain E11 or E12. These optimal experimental conditions resulted in significantly higher shoot dry wt, shoot plus leaf area, root fresh weight, average length of crown roots, and cumulative length of crown roots on

inoculated as compared to uninoculated plants. Interestingly, although the endophytic rhizobia generally developed higher populations within roots of Indica rice (Table 2), they elicited higher short-term PGP responses on the Japonica rice (Figures 10A and 10B). This result indicates that identification of superior combinations of rhizobia and rice genotypes for optimal growth responses will likely require PGP bioassays rather than just an assessment of the bacterial endophyte's ability to colonize the root interior, and manipulations to increase the endophyte population above the natural level achievable within rice per se may not necessarily improve the resultant growth promotion response. At the time of harvest, none of the plants used to obtain the data reported in Table 3 were active in acetylene-dependent ethylene production as a measure of N_2 fixation.

The ultimate assessment of the potential importance of this newly described plant-microbe association on rice productivity requires experimentation under field conditions. Therefore, we have conducted our first field inoculation experiment to evaluate the growth responses of hybrid rice cultivar Giza-175 to inoculation with strains E11 and E12 under N-limited and fertilizer-N supplemented conditions on experimental farms at the Sakha Agricultural Research Station. Both inoculant strains performed very well in this field experiment and clearly demonstrated their potential to enhance rice productivity. Inoculation without added fertilizer-N resulted in statistically significant increases (95% confidence level) in straw N-content and statistically highly significant increases (99% confidence level) in grain yield, grain N-content, and the harvest index (Tables 4 and 5). Statistically significant interactions between inoculation plus certain doses of N-fertilizer application were detected which increased the level of certain agronomic parameters of rice productivity more than did either treatment alone. Most noteworthy is the highly significant response of this rice cultivar to application of 1/3 the recommended dose of N-fertilization (48 kg N ha^{-1}) plus inoculation with strain E11, which increased the grain yield to a level exceeding that obtained by application of the full recommended fertilizer dose (144 kg N ha^{-1}) alone. This important result is also reflected in the very high agronomic N-use efficiency in the corresponding inoculated rice plots, indicating a significant return in rice grain production per unit of chemical fertilizer-N input. The requirement of added N-fertilizer in order for inoculation to significantly increase straw yield (Table 4) indicates that the quantity of available N derived direct-

110

Table 3. Evaluation of various morphological responses of Giza-171 rice in tube culture 32 days after inoculation with *R. leguminosarum* bv. *trifolii* rice endophytes

Plant growth medium	Endophyte inoculum strain	Shoot+ leaf area (cm^2)	Avg. crown root length (cm)	Cumulative length of crown roots (cm)
Fahraeus (-N)	E11	5.4	8.8[a]	90.3[b]
"	E12	5.0	7.0[a]	70.2
"	None	4.7	5.5	42.7
Hoagland (+N)	E11	15.9[b]	8.7[a]	96.2[b]
"	E12	15.3[a]	9.6[a]	128.3[a]
"	None	8.5	5.3	40.5

Mean values followed by the letter a or b are significantly different from the corresponding uninoculated control at the 99% and 95% confidence levels, respectively.

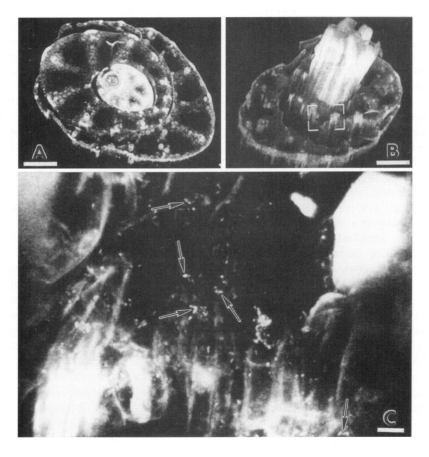

Figure 9. Epifluorescence laser scanning confocal micrographs (reconstructed serial section overlays) of fresh rice tissue cut at the stem base above the root and stained with acridine orange. The rice plant was grown in gnotobiotic tube culture with *R. leguminosarum* bv. *trifolii* endophyte strain E11. (**A**) Low magnification top view, showing the leaf whirls beneath the stem surface. (**B**) Low magnification tilted view. (**C**) Higher magnification of the corresponding boxed area in (B), revealing numerous fluorescent endophytic bacteria (arrows) within the interior of the stem base. Bar scales are 0.3 mm in A and B, and 10 μm in C.

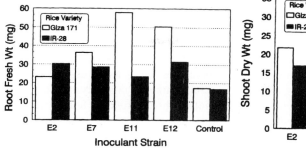
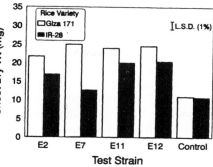

Figure 10. Growth responses of Japonica and Indica rice following inoculation with various endophytic strains of *R. leguminosarum* bv. *trifolii* in gnotobiotic tube culture (**A**) Root fresh weight [computed average from composite samples]. (**B**) Shoot dry weight [means of replicate samples measured individually, least significant differences at the 1% level is shown].

Table 4. Effect of N-fertilization and inoculation with *R. leguminosarum* bv. *trifolii* rice endophytes on production of Giza-175 rice under field conditions[a]

Fertilization (kg N ha^{-1})	1000-grain weight (g)			Straw yield (tons ha^{-1})			Grain yield (tons ha^{-1})		
	Cont	E11	E12	Cont	E11	E12	Cont	E11	E12
0 N	19.3	19.7	18.8	10.20	11.71	11.25	4.33	6.31	6.15
48 N	20.5	19.6	19.7	13.66	11.94	11.11	5.84	7.93	6.96
96 N	20.0	19.4	19.7	11.62	12.34	13.35	5.41	7.11	7.85
144 N	19.2	18.8	20.2	13.93	15.82	16.27	6.38	7.00	6.96
LSD	0.05	0.01		0.05	0.01		0.05	0.01	
Fertilization (N)	n.s.	n.s.		1.25	1.80		0.31	0.44	
Inoculation (E)	n.s.	n.s.		n.s.	n.s.		0.32	0.44	
N × E	0.6	0.8		1.54	2.09		0.65	0.88	

[a]LSD (0.05 and 0.01) are the least significant differences at the 95% and 99% confidence level, respectively. n.s.; statistically not significant.

Table 5. Effect of N-fertilization and inoculation with *R. leguminosarum* bv. *trifolii* rice endophytes on straw and grain N-contents, the harvest index, and the agronomic fertilizer N-use efficiency of Giza-175 rice under field conditions[a]

Fertilization (kg N ha^{-1})	Straw N-content (kg ha^{-1})			Grain N-content (kg ha^{-1})			Harvest index			Agronomic fertilizer N-use efficiency		
	Cont	E11	E12	Cont	E11	E12	Cont	E11	E12	Cont	E11	E12
0 N	34.8	48.3	48.3	36.4	55.7	48.9	30.1	35.0	35.4	–	–	–
48 N	49.1	47.8	42.4	54.9	70.9	63.1	30.0	40.0	38.5	121.7	165.2	145.0
96 N	47.6	61.7	62.3	52.9	73.7	75.3	31.8	36.7	37.4	56.4	74.1	81.8
144 N	65.8	75.0	74.3	58.9	68.5	70.2	31.6	30.6	30.0	44.3	48.6	48.4
LSD	0.05	0.01		0.05	0.01		0.05	0.01		0.05	0.01	
Fertilization (N)	4.5	6.4		3.8	5.5		2.5	3.7		3.4	4.9	
Inoculation (E)	4.9	n.s.		3.0	4.1		1.9	2.6		4.4	6.0	
N × E	n.s.	n.s.		6.0	n.s.		3.9	5.3		8.8	11.9	

[a]The harvest index is the % grain yield / grain + straw yields. The agronomic fertilizer N-use efficiency is calculated as kg grain yield kg^{-1} fertilizer-N applied. LSD (0.05 and 0.01) are the least significant differences at the 95% and 99% confidence levels, respectively. n.s., statistically not significant.

ly from mineralization of the inoculum itself (bacterial cells plus diluted culture medium) was not a significant factor influencing the outcome of this field experiment.

The mechanism(s) responsible for these inoculation-induced changes in rice growth and yield under gnotobiotic and field conditions are unknown. In particular, more studies will be necessary to fully assess whether the rhizobia fix N_2 in association with rice under conditions in which these bacteria promote rice growth, and if so, what portion of the plant-N content can be derived from this biological N_2-fixing activity. The ability of these rice endophytic strains of *R. leguminosarum* bv. *trifolii* to produce Fix$^+$ nodules and significantly increase plant N-content on berseem clover clearly demonstrates their capacity for symbiotic N_2 fixation and release of fixed N at significant levels that benefit the growth of the appropriate legume host. However, rice plants grown under N-free conditions in gnotobiotic tube culture were not consistently increased in N-content nor did they have detectable acetylene reduction activity when examined 32 days after inoculation with these rhizobia (data not shown). These results plus the ability of selected inoculant strains to significantly promote rice growth in the presence of adequate sources of combined-N in tube culture and in N-fertilized field soil raise questions about the potential importance of biological N_2 fixation in this particular plant-microbe association, and will require further investigation (including studies using ^{15}N labeling and Nif-minus mutant derivatives). An alternative working hypothesis is that endophytic colonization by these native rhizobia modulates growth physiology of rice (possibly by hormone action) enabling the plant root system to utilize the existing resources of available nutrients and water more efficiently in ways that may be independent of biological N_2 fixation. Such PGP mechanisms appear to be largely responsible for the ability of the endophytic diazotroph *Azospirillum brasilense* to enhance growth of various cereal crops in approximately 70% of world field trials (Tien et al., 1979; Umali-Garcia et al., 1980; Okon and Labandera-Gonzalez, 1994 and references therein). Although flavone induction of pSym *nod* genes in wild type *R. leguminosarum* bv. *trifolii* leads to production of a wide variety of low molecular weight metabolites that are bioactive at very low hormonal concentrations in eliciting growth responses on clover (Dazzo et al., 1996; Hollingsworth et al., 1989; Orgambide et al., 1994; Orgambide et al., 1996; Philip-Hollingsworth et al., 1991), tests for natural inducers of *nodA* expression in axenically produced root exudate of Giza-171

or IR-28 rice using an appropriate *nodA::lacZ* reporter strain of *R. leguminosarum* bv. *trifolii* (provided by M. Djordjevic, Australian National University) have been negative (data not shown). An important future direction will be to determine if rice produces signal molecules that activate expression of novel genes in certain rhizobia, uniquely enabling them to invade rice roots without activating host defense responses and to establish an endophytic association that promotes these beneficial growth responses.

In nature, *Rhizobium* is normally viewed as a microbe that survives saprophytically in soil between periods in which the host legume is absent. However, our studies have shown that clover rhizobia also occupy another endophytic niche *inside rice* plants, and can tap in on this third niche in ways that can benefit this cereal host. The outcome of such a mutually beneficial alternate symbiosis (dissimilar organisms living together) would help to perpetuate this natural plant-microbe association and can potentially improve sustainable agriculture, thus enhancing world food production. Our understanding of the natural ability of rhizobia to inhabit cereals and enhance their growth is only beginning to be explored. Nevertheless, the novel findings described here represent a major step forward in achieving the technically challenging goal of increasing rice productivity by reducing its dependence of the need for fertilizer-N through enhancement of its natural association with rhizobia without requiring as highly developed a system as the root nodule *Rhizobium*-legume symbiosis.

Acknowledgements

We thank A Trinh and J Whallon for technical assistance and Deborah Dodge and Douglas Smith from the Applied Biosystems Division of Perkin Elmer for assistance with sequencing. Portions of this project were funded by the Egyptian National Agricultural Research Project through the Ministry of Agriculture and Land Reclamation (Arab Republic of Egypt), the Danish International Development Agency through the International Rice Research Institute (The Philippines), the Michigan State University Center for Microbial Ecology (National Science Foundation Grant BIR 91-20006), and the Michigan Agricultural Experiment Station. This study is dedicated in memory of Professor David H Hubbell, whose teachings to combine basic and applied microbial ecology in the design of research on the *Rhizobium*-legume symbiosis and to follow Leonardo da Vinci's wisdom to look first to nature for the best design before invention were inspirational throughout this study.

References

Abdel Wahab A M, Zahran H H, Abd-Alla M H 1996 Root-hair infection and nodulation of four grain legumes as affected by the form and the application time of nitrogen fertilizer. Folia Microbiol. 41, 303–308.

Balandreau J and Knowles R 1978 The rhizosphere. In Interactions Between Non-Pathogenic Soil Microorganisms and Plants. Eds. Y R Dommergues, S V Krupa. pp 243–268, Elsevier, Amsterdam.

Bally R, Thomas-Bauzon D, Heulin T, Balandreau J, Richard C and Ley J D 1983 Determination of the most frequent N_2-fixing bacteria in a rice rhizosphere. Can. J. Microbiol. 29, 881–887.

Bilal R and Malik K A 1987 Isolation and identification of a N_2-fixing zoogloea-forming bacterium from kallar grass histoplane. J. Appl. Bacteriol. 62, 289–294.

Black A D, Evans F, Ensminger J, White F, Clar J and Dinaver R 1965 Methods of Soil Analysis - II. Chemical and Microbiological Properties. No. 9 in the Series of Agronomy, American Society for Agronomy, Madison, WI.

Boddey R and Dobereiner J 1988 Nitrogen fixation associated with grasses and cereals: recent results and perspectives for future research. Plant Soil 108, 53–65.

Chabot R H, Antoun H, Kloepper J and Beauchamp C 1996 Root colonization of maize and lettuce by bioluminescent Rhizobium leguminosarum biovar phaseoli. Appl. Environ. Microbiol. 62, 2767–2772.

Cocking E C, Davey M R, Kothari S L, Srivastava J S, Jing Y, Ridge R W and Rolfe B G 1992 Altering the specificity control of the interaction between rhizobia and plants. Symbiosis 14, 123–130.

Corich V, Giacomini A, Ollero F J, Squartini A and Nuti M P 1991 Pulsed-field electrophoresis in contour-clamped homogeneous electric fields (CHEF) for the fingerprinting of Rhizobium spp. FEMS Microbiol. Lett. 83, 193–198.

Dazzo F B 1982 Leguminous root nodules. In Experimental Microbial Ecology. Ed. J Slater and R Burns. pp 431–446. Blackwell Scientific Publications, Oxford, UK.

Dazzo F B and Brill W J 1978 Regulation by fixed nitrogen of host-symbiont recognition in the Rhizobium-clover symbiosis. Plant Physiol. 62, 18–21.

Dazzo F B, Orgambide G G, Philip-Hollingsworth S, Hollingsworth R I, Ninke K O and Salzwedel J L 1996 Modulation of development, growth dynamics, wall crystallinity, and infection sites in white clover root hairs by membrane chitolipooligosaccharides from Rhizobium leguminosarum biovar trifolii. J. Bacteriol. 178, 3621–3627.

Dazzo F B and Petersen J 1989 Applications of computer-assisted image analysis for microscopical studies of the Rhizobium-legume symbiosis. Symbiosis 7, 193–210.

de Bruijn F J, Jing Y and Dazzo F B 1995 Potentials and pitfalls of trying to extend symbiotic interactions of nitrogen-fixing organisms to presently non-nodulated plants, such as rice. Plant Soil 174, 225–240.

DeSoete G 1984 Additive-tree representations of incomplete dissimilarity data. Quality Quantity 18, 387–393.

Dobereiner J, Reis V, Paula M and Olivares F 1993 Endophytic diazotrophs in sugar cane, cereals, and tuber plants. In New Horizons in Nitrogen Fixation. Eds. R Palacios, J Mora and W E Newton. pp 671–679. Kluwer Academic Publishers, Dordrecht, The Netherlands.

Eckhardt T 1978 A rapid method for the identification of plasmid deoxyribonucleic acid in bacteria. Plasmid 1, 584–588.

Espuny M R, Ollero F J, Bellogin R A, Ruiz-Sainz J E and Perez-Silva J 1987 Transfer of the Rhizobium leguminosarum biovar trifolii symbiotic plasmid pRtr5a to a strain of Rhizobium sp. that nodulates Hedysarum coronarium. J. Appl. Bacteriol. 63, 13–20.

Graves W L, Williams W A, Wegrzyn V A, Calderon D, George M R and Sullins J L 1987 Berseem clover is getting a second chance. California Agriculture September-October 1987, pp 15–18.

Gough C, Webster G, Vasse J, Galera C, Batchelor C, O'Callaghan K, Davey M, Denarie J and Cocking E 1996 Intercellular infection of wheat and Arabidopsis by rhizobial strains. Abstr. S-45, 8th International Congress on Molecular Plant-Microbe Interactions, Knoxville, TN.

Hashem F M and Kuykendall D 1994 Plasmid DNA content of several agronomically important Rhizobium species that nodulate alfalfa, berseem clover, or Leucaena. In Symbiotic Nitrogen Fixation. Eds. P Graham, M Sadowsky and C Vance. pp 181–188. Kluwer Academic Publishers, Dordrecht, The Netherlands.

Holflich G, Wiehe W and Hecht-Bucholz C 1995 Rhizosphere colonization of different crops with growth promoting Pseudomonas and Rhizobium bacteria. Microbiol. Res. 150, 139–147.

Hollingsworth R I, Squartini A, Philip-Hollingsworth S and Dazzo F B 1989 Root hair deforming and nodule initiating factors from Rhizobium trifolii. In Signal Molecules in Plants and Plant-Microbe Interactions. Ed. B Lugtenberg. pp 387–393. Springer-Verlag, Berlin, Germany.

Hurek T, Reinhold-Hurek B, van Montague M, and Kellenberger E 1994 Root colonization and systemic spreading of Azoarcus sp. strain BH72 in grasses. J. Bacteriol. 176, 1913–1923.

International Rice Research Institute 1996 IRRI towards 2020. International Rice Research Institute. P.O. Box 933, Manila, Philippines. 43 p.

Khush G S and Bennett J 1992 Nodulation and Nitrogen fixation in rice: Potential and prospects. International Rice Research Institute Press, Manila, Philippines. 136 p.

Klein D A, Salzwedel J L and Dazzo F B 1990 Microbial colonization of plant roots. In Biotechnology of Plant-Microbe Interactions. Eds. J P Nakas and C Hagedorn. pp 189–225. McGraw-Hill Publishing Company, NY.

Kloepper J W, Schippers B and Bakker P A 1992 Proposed elimination of the term endorhizosphere. Phytopathol. 82, 726–727.

Ladha J K 1986 Studies on nitrogen fixation by free-living and rice-plant associated bacteria in wetland rice field. Bionature 6, 47–58.

Ladha J K, Tirol-Padre A, Reddy K and Ventura W 1993 Prospects and problems of biological nitrogen fixation in rice production: a critical assessment. In New Horizons in Nitrogen Fixation. Eds. R Palacios, J Mora and W E Newton. pp 677–682. Kluwer Academic Publishers, Dordrecht, The Netherlands.

Ladha J K, Garcia M, Miyan S, Padre A T and Watanabe I 1989 Survival of Azorhizobium caulinodans in the soil and rhizosphere of wetland rice under Sesbania rostrata-rice rotation. Appl. Environ. Microbiol. 55, 454–460.

Ladha J K, So R, Hernandez R, Dazzo F B, Reddy P M, Angeles O R, Ramos M C, de Bruijn F J and Stoltzfus J 1996a Rhizobial invasion and induction of phenotypic changes in rice roots are independent of Nod factors. Proceedings 8th International Congress on Molecular Plant-Microbe Interactions (Abstr. #L-11), Knoxville, TN.

Laemmli U 1970 Cleavage of structural proteins during the assembly of the head of bacteriophage T4. Nature (London) 227, 680–685.

114

Larsen N, Olsen G, Maidak B, McCaughey M, Overbeek R, Macke T, Marsch T and Woese C R 1993 The ribosomal database project. Nucl. Acid Res. 21, 3021–3032.

Mateos P, Jiminez J, Chen J, Squartini A, Martinez-Molina E, Hubbell D H and Dazzo F B 1992 Cell-associated pectinolytic and cellulolytic enzymes in *Rhizobium trifolii*. Appl. Environ. Microbiol. 58, 1816–1822.

Natalia K, Gennady K and Vitaly K 1994 Endophytic association of nitrogen-fixing bacteria with plants. Proceedings 1st European N_2-Fixation Conference (Abstr. P104), Szeged, Hungary.

Ndoye I, DeBilly F, Vasse J, Dreyfus B and Truchet G 1994 Root nodulation of *Sesbania rostrada*. J. Bacteriol. 176, 1060–1068.

Old K and Nicolson T 1975 Electron microscopical studies of the microflora of roots of sand dune grasses. New Phytol. 74, 51–58.

Old K and Nicolson T 1978 The root cortex as part of a microbial continuum. *In* Microbial Ecology. Eds. M Loutit and J Miles. pp 291–294. Springer-Verlag, NY.

Okon Y and Labandera-Gonzalez C A 1994 Agronomic applications of *Azospirillum*: an evaluation of 20 years worldwide field inoculation. Soil Biol. Biochem. 26, 1591–1601.

Orgambide G, Lee J, Hollingsworth R and Dazzo F B 1995 Structurally diverse chitolipooligosaccharide Nod factors accumulate primarily in membranes of wild type *Rhizobium leguminosarum* bv. *trifolii*. Biochemistry 34, 3832–3840.

Orgambide G, Philip-Hollingsworth S, Hollingsworth R I and Dazzo F B 1994 Flavone-enhanced accumulation and symbiosis-related biological activity of a diglycosyl diacylglycerol membrane glycolipid from *Rhizobium leguminosarum* bv. *trifolii*. J. Bacteriol. 176, 4338–4347.

Orgambide G, Philip-Hollingsworth S, Mateos P F, Hollingsworth R I and Dazzo F B 1996 Subnanomolar concentrations of membrane chitolipooligosaccharides from *Rhizobium leguminosarum* bv. *trifolii* are fully capable of eliciting symbiosis-related responses on white clover. Plant Soil, 186, 93–98.

Philip-Hollingsworth S, Hollingsworth R I and Dazzo F B 1989 Host-range related structural features of the acidic extracellular polysaccharides of *Rhizobium trifolii* and *Rhizobium leguminosarum*. J. Biol. Chem. 264, 1461–1466.

Philip-Hollingsworth S, Hollingsworth R I and Dazzo F B 1991 N-acetylglutamic acid: an extracellular Nod signal of *Rhizobium trifolii* ANU843 which induces root hair branching and nodule-like primordia in white clover roots. J. Biol. Chem. 266, 16854–16858.

Reddy P M, Ladha J K, So R, Hernandez R, Dazzo F B, Angeles O R, Ramos M C and de Bruijn F J 1997 Rhizobial communication with rice: induction of phenotypic changes, mode of invasion and extent of colonization. Plant Soil 194, 81–98.

Roger P and Ladha J K 1992 Biological nitrogen fixation in wetland rice fields: estimation and contribution to nitrogen balance. Plant Soil 141, 41–55.

Roger P A and Watanabe I W 1986 Technologies for utilizing biological nitrogen fixation in wetland rice: potentialities, current usage, and limiting factors. Fert. Res. 9, 39–77.

Shimshick E J and Herbert R R 1979 Binding characteristics of N_2-fixing bacteria to cereal roots. Appl. Environ. Microbiol. 38, 447–453.

Smucker A J 1993 Soil environmental modifications of root dynamics and measurement. Annu. Rev. Phytopathol. 31, 191–216.

Somasegaran P and Holben H J 1985 Methods in Legume-*Rhizobium* Technology. NifTAL Project, University of Hawaii, Maui, Hawaii. 367 p.

Spencer D, James E K, Ellis G J, Shaw J E and Sprent J I 1994 Interaction between rhizobia and potato tissues. J. Exp. Bot. 45, 1475–1482.

Subba-Rao N S, Mateos P F, Baker D, Pankratz H S, Palma J, Dazzo F B and Sprent J I 1995 The unique root-nodule symbiosis between *Rhizobium* and the aquatic legume, *Neptunia natans* (L. f.) Druce. Planta 196, 311–320.

Tien T, Gaskins M H and Hubbell D H 1979 Plant growth substances produced by *Azospirillum brasilense* and their effect on the growth of pearl millet (*Pennisetum americanum* L.). Appl. Environ. Microbiol. 37, 1016–1024.

Ueda T, Suga Y, Yahiro N and Matsuguchi T 1995 Remarkable N_2-fixing bacterial diversity detected in rice roots by molecular evolutionary analysis of *nifH* gene sequences. J. Bacteriol. 177, 1414–1417.

Umali-Garcia M, Hubbell D H, Gaskins M H and Dazzo F B 1980 Association of *Azospirillum* with grass roots. Appl. Environ. Microbiol. 39, 219–226.

Versalovic J, Schneider M, de Bruijn F J and Lupski J R 1994 Genomic fingerprinting of bacteria using repetitive sequence-based polymerase chain reaction. Methods Molec Cell. Biol 5, 25–40.

Willems A and Collins M D 1993 Phylogenetic analysis of rhizobia and agrobacteria based on 16S rRNA gene sequences. Int. J. Syst. Bacteriol. 43, 305–313.

Yanni Y G 1991 Potential of indigenous cyanobacteria to contribute to rice performance under different schedules of nitrogen application. World J. Microbiol. Biotech. 7, 48–52.

Young J P and Haukka K E 1996 Diversity and phylogeny of rhizobia. New Phytol. 133, 87–94.

Guest editors: J K Ladha, F J de Bruijn and K A Malik

Plant and Soil **194**: 115–122, 1997.
© 1997 *Kluwer Academic Publishers. Printed in the Netherlands.*

115

Interactions of rhizobia with rice and wheat

G. Webster[1,2], C. Gough[2], J. Vasse[2], C.A. Batchelor[1], K.J. O'Callaghan[1], S.L. Kothari[1], M.R. Davey[1], J. Dénarié[2] and E.C. Cocking[1,3]

[1]*Plant Genetic Manipulation Group, Department of Life Science, University of Nottingham, Nottingham NG7 2RD, U.K. and* [2]*Laboratoire de Biologie Moléculaire des Relations Plantes-Microorganismes, INRA-CNRS, BP27, 31326 Castanet-Tolosan, France.* [3]*Corresponding author**

Key words: Azorhizobium caulinodans, crack entry, intercellular colonization, naringenin, rice, wheat

Abstract

Recently, evidence has been obtained that naturally occurring rhizobia, isolated from the nodules of non-legume *Parasponia* species and from some tropical legumes, are able to enter the roots of rice, wheat and maize at emerging lateral roots by crack entry. We have now investigated whether *Azorhizobium caulinodans* strain ORS571, which induces root and stem nodules on the tropical legume *Sesbania rostrata* as a result of crack entry invasion of emerging lateral roots, might also enter rice and wheat by a similar route. Following inoculation with ORS571 carrying a *lacZ* reporter gene, azorhizobia were observed microscopically within the cracks associated with emerging lateral roots of rice and wheat. A high proportion of inoculated rice and wheat plants had colonized lateral root cracks. The flavanone naringenin at 10^{-4} and 10^{-5} M stimulated significantly the colonization of lateral root cracks and also intercellular colonization of wheat roots. Naringenin does not appear to be acting as a carbon source and may act as a signal molecule for intercellular colonization of rice and wheat by ORS571 by a mechanism which is *nod* gene-independent, unlike nodule formation in *Sesbania rostrata*. The opportunity now arises to compare and to contrast the ability of *Azorhizobium caulinodans* with that of other rhizobia, such as *Parasponia* rhizobia, to intercellularly colonize the roots of non-legume crops.

Introduction

The inoculation of non-leguminous crops with diazotrophic bacteria has been studied for many years with the expectation that these bacteria would fix dinitrogen gas and provide combined nitrogen to the plant for enhanced crop production (Sloger and Van Berkum, 1992). One major limitation of such associative nitrogen fixation is that diazotrophic bacteria in the rhizosphere of plants utilise the products of nitrogen fixation for their own growth, but release little while they are alive (Van Berkum and Bohlool, 1980). Measurements of nitrogen fixation in rice and wheat using ^{15}N have confirmed that the majority of the fixed nitrogen remains in the bacteria within the root environment (Okon, 1985). Another major limitation of associative nitrogen fixation is that, in most instances, bacteria colonize only the surface of roots and remain vulnerable to competition from other rhizosphere microorganisms. In contrast, a major advantage of symbiotic systems is that the nitrogen-fixing bacteria colonize the plant internally and become endophytic, thus being protected from competition with rhizosphere microorganisms and having the possibility of more intimate metabolic exchange with host plants. Quispel (1991) has suggested that only in endophytic systems are the prerequisites for effective nitrogen fixation likely to be fulfilled in interactions between non-legumes and diazotrophic bacteria. In an endophytic situation, there would need to be a reliable supply of metabolic substrates from host plant photosynthesis to provide sufficient energy and reducing conditions, protection against too high oxygen concentrations and transport of the nitrogen fixation products to the host plant. One approach is to seek naturally occurring endophytic diazotrophs in cereals, such as rice and wheat, and to attempt to optimise such endophytic nitrogen fixation. Currently, this approach is being exploited with sugar

* FAX No: +441159513240.
E-mail: Edward Cocking@nottingham.ac.uk

cane (Boddey et al., 1995) and with rice (Ladha and Reddy, 1995; Yanni et al., 1997).

An alternative approach, which we are investigating, is to determine whether any rhizobial strains can colonize non-legumes internally. This approach has been developed in the light of our previous observations that some strains of rhizobia are able to induce, at low frequency, nodule-like structures on the roots of rice, wheat and oilseed rape seedlings after treatment of the seedling roots with a mixture of the cell wall degrading enzymes, cellulase and pectinase (Al-Mallah et al., 1989, 1990). Enzymatic treatment of the root hairs of white clover has also been shown to remove the barrier to host-specificity and to increase the range of rhizobia able to nodulate this legume (Al-Mallah et al., 1987). These findings led to the discovery that some naturally occurring rhizobia can invade the emerging lateral roots of rice, wheat, maize and oilseed rape without the need for any enzyme treatment, although at low frequency (Cocking et al., 1990, 1992, 1994). The rhizobia used in the latter work all enter their host plant root systems by crack entry infection, that is intercellularly between adjacent plant cells, and not by the formation of infection threads at the tips of root hairs (Sprent and Raven, 1992). Rhizobia that are capable of symbiotic nitrogen fixation in non-legume *Parasponia* species (Webster et al., 1995a) were also tested. These latter rhizobia gain entry into the roots of *Parasponia* through ruptures in the root epidermis. A small percentage of lateral roots of inoculated rice and wheat plants were observed to be shorter and thicker than lateral roots of uninoculated plants. Sections of such lateral roots, cut for both light and electron microscopy, revealed that rhizobia were within, and between, cells of the root cortex. Among the rhizobia used in the previous work, *Azorhizobium caulinodans*, which induces root and stem nodules on the tropical legume *Sesbania rostrata*, is especially interesting, since in addition to forming nodules after crack entry invasion of emerging lateral roots (Tsien et al., 1983; Ndoye et al., 1994) it is able to fix nitrogen in the free-living state in up to 3% (v/v) oxygen, and without differentiation into bacteroids (Kitts and Ludwig, 1994). We have therefore concentrated our investigation on *A. caulinodans* strain ORS571 and studied, in more detail, how it can colonize the roots of the non-legumes rice and wheat. Endophytic establishment of *A. caulinodans* in the root systems of rice and wheat would present a novel situation in these non-legume crops and could provide opportunities for endophytic nitrogen fixation.

Recently, studies of plant-microbe interactions have benefited from the use of genetically modified micro-organisms harbouring reporter genes under the control of constitutively expressed promoters, since expression of such genes facilitates visualisation and identification of the micro-organisms associated with plant tissues (Boivin et al., 1990; Vasse et al., 1995). We have adopted an approach, combining bacterial genetics and cytology, using strain ORS571 and its derivatives containing a constitutively expressed *lacZ* reporter gene which can be assayed histochemically, to localise bacteria within the roots of rice and wheat. A similar study has been performed on the model dicot *Arabidopsis thaliana* (Gough et al., 1997). The effect of flavonoids on bacterial colonization of wheat and rice roots was also investigated by the addition of specific flavonoids to the plant growth medium. Our recent work has demonstrated that the flavonoid, naringenin, is able to stimulate *Azorhizobium* colonization of non-legume root systems (Gough et al., 1996).

Materials and methods

Plant culture and bacterial inoculation

Rice (*Oryza sativa* cv. Lemont) and wheat (*Triticum aestivum* cv. Canon) were grown aseptically in tubes (25×150 mm, 60 mL; 25×200 mm, 70 mL capacity respectively) containing either 20 mL (rice) or 25 mL (wheat) of nitrogen-free Fåhraeus medium (Fåhraeus, 1957) semi-solidified with 0.8% (w/v) agar (Sigma) and inoculated after 2 days with ORS571 or ORS571 (pXLGD4). The latter strain contained a constitutively-expressed *lacZ* reporter gene (Leong et al., 1985). Wheat plants were also inoculated with strain ORS571 (pPR3408) carrying a translational fusion between *nifD* of ORS571 and the *lacZ* reporter gene (Pawlowski et al., 1987).

β-galactosidase activity associated with plant roots and microscopic techniques

Bacterial colonization of the roots of inoculated plants was visualised, after 2 weeks, by light microscopy of the dark blue precipitate resulting from the degradation of X-gal by β-galactosidase, essentially according to Boivin et al. (1990), but with the sodium cacodylate reduced from 0.2 to 0.12 M. A scoring system was developed to evaluate bacterial colonization of root systems. Prior to counting, the samples were coded and randomised to ensure a blind assay. The number of colonized lateral root cracks (LRCs) was recorded and the percentage of LRCs which were colonized per plant was calculated. When appropriate, sections cut to approximately 2 μm in thickness for light microscopy were stained with 0.5% (w/v) toluidine blue in 0.1% (w/v) sodium tetraborate for 3 min at 60 °C (Davey et al., 1993; Webster et al., 1995b).

Re-isolation of bacterial endophytes from plant roots

Bacteria were re-isolated from colonized roots by surface sterilisation of excised roots in 95% (v/v) ethanol for 10 sec, followed by 8% (v/v) "Domestos" bleach (Lever Industrial Ltd., Runcorn, U.K.) for 4 min. Roots were rinsed thoroughly with sterile distilled water. Sterilised root pieces were crushed in liquid TY medium (Somasegaran and Hoben, 1994) and the resultant mixture streaked onto 20 mL of 1.5% (w/v) agar-solidified TY medium in 9 cm Petri dishes. Cultures were incubated at 25°C in the dark for 5 days.

Results and discussion

Intercellular colonization at lateral root cracks of rice and wheat by Azorhizobium caulinodans

In wheat and rice, following inoculation with ORS571 (pXLGD4), bacteria were observed on the root surface, at root tips and inside LRCs. The latter result from lateral root emergence and are present at the bases of lateral roots. Sections of blue-staining LRCs of wheat showed intercellular pockets of bacteria in the cortex of the main root in which rhizobia had multiplied extensively, as in the first stages of nodulation of *Sesbania rostrata* (Ndoye et al., 1994). β-galactosidase activity was still detected at colonization sites of wheat roots up to 4 weeks after inoculation, suggesting the persistent presence of viable bacteria. No obvious signs of hypersensitivity were observed in either wheat or rice after inoculation with *A. caulinodans* strain ORS571. In addition, limited expression of a *nif* promoter-*lacZ* chimaeric gene (Pawlowski et al., 1987) was also observed at colonization sites in wheat, when roots of wheat seedlings were inoculated with ORS571(pPR3408).

In experiments where the *lacZ* reporter gene was not used, bacteria re-isolated from surface-sterilised roots of wheat were shown to be *A. caulinodans* strain ORS571 by RFLP finger printing, coupled with DNA hybridisation using a *nifH* gene probe from *Rhizobium phaseoli* strain CFN-42 (R Palacios, personal communication). In addition, re-isolated strains of *A. caulinodans* were shown to re-colonize LRCs of wheat and to induce root and stem nodules on the legume *Sesbania rostrata*, thus confirming Koch's postulates.

A high proportion of plants (83% of rice and 75-100% of wheat) had colonized LRCs, including both crack colonization and further intercellular colonization. This high level of colonization enabled us to determine the percentage of colonized LRCs per plant, which were 7.1% for rice and 10.8% for wheat. The use of the *lacZ* reporter gene, in the present study, has facilitated detection of bacteria at points of lateral root emergence and suggests that previous reports of interactions between rhizobia and the non-legumes rice and wheat, where rhizobia were observed at the bases of emerging short, thick lateral roots, underestimated the proportion of plants that were colonized (Cocking et al., 1992, 1994).

A certain specificity was found in the colonization ability of *A. caulinodans* when the colonization of LRCs of wheat by other rhizobia and plant-associated bacteria was investigated. Among these, *Rhizobium meliloti* was unable to colonize LRCs of wheat. However, *Azospirillum brasilense*, like *Azorhizobium caulinodans*, was able to colonize these sites.

The flavonoid naringenin stimulates intercellular colonization of rice and wheat roots

Rhizobia have adapted to use flavonoids, a class of phenolics released by plant roots, as part of a regulatory system to initiate the transcription of their nodulation (*nod*) genes (Rolfe, 1988). Flavonoids are produced by both legumes and non-legumes. Indeed, wheat was one of the plants in which traces of flavonoids were first detected in root exudates (Lundegårdh and Stenlid, 1944). The *nodD*1 gene product of the *Rhizobium* strain NGR234 responds to activation by the simple phenolic compounds vanillin and isovanillin present in wheat seedling extracts (Le Strange et al., 1990). More generally, phenolic compounds secreted into the rhizosphere by plants induce expression of a variety of genes in plant-associated bacteria. Flavonoids also enhance the growth rate of certain rhizobia (Hartwig et al., 1991) and phenolic compounds can also act as chemo-attractants (Peters and Verma, 1990).

Four families of flavonoids (flavones, flavonols, flavanones and isoflavones) have been shown to have *nod* gene-inducing activity (Dénarié et al., 1992). The

Table 1. Effect of flavonoids and succinate on colonization of lateral root cracks (LRCs) of wheat by *Azorhizobium caulinodans* strain ORS571 (pXL-GD4)

Treatment	% of LRCs colonized per plant $\pm$ SD
ORS571	10.8 ± 6.6^a
ORS571 + 10^{-4} *M* naringenin	33.2 ± 9.1^b
ORS571 + 10^{-5} *M* naringenin	38.0 ± 12.0^b
ORS571 + 10^{-4} *M* daidzein	7.3 ± 3.2^a
ORS571 + 10^{-4} *M* luteolin	11.0 ± 6.7^a
ORS571 + 10^{-4} *M* succinate	9.0 ± 3.7^a

Treatments with different letters differ significantly at the $p = 0.01$ level. Analysis of variance with Fisher's test (Snedecor and Cochran, 1989). Ten replicates per treatment.

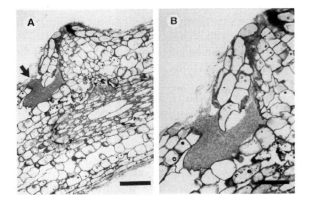

Figure 1. (**A**) Light micrograph of a longitudinal section of a rice root showing colonization (arrowed) of a LRC by *A. caulinodans* strain ORS571 in the presence of 10^{-4} *M* naringenin (*Bar* = 100 μm). (**B**) Higher magnification of (A) showing the large intercellular pocket of azorhizobia at the site of the LRC (*Bar* = 50 μm).

flavanone, naringenin, at both 10^{-4} and 10^{-5} *M*, significantly stimulated (at the $p = 0.01$ level) the colonization of LRCs of wheat by ORS571 (pXLGD4), but both daidzein (an isoflavone) and luteolin (a flavone) had no effect (Table 1). Succinate, a suitable carbon source for ORS571, at 10^{-4} *M*, also had no significant effect on colonization of wheat roots, indicating that naringenin does not act as a simple carbon source. Interestingly, naringenin at 10^{-4} *M* did not stimulate *R. meliloti* to colonize wheat roots (data not shown), suggesting some specificity towards *A. caulinodans* strain ORS571. Preliminary results on the colonization of rice by *A. caulinodans* also suggest that naringenin at 10^{-4} *M* is stimulatory. Colonization of rice LRCs was increased from 7.1% to 19.9% in the presence of naringenin.

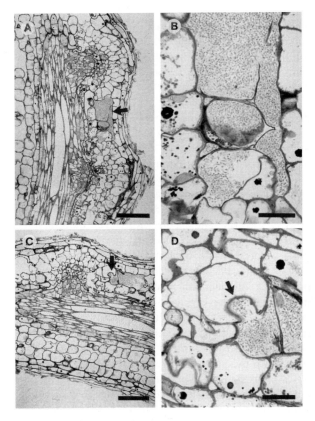

Figure 2. (**A**) Spread of azorhizobia from LRCs into the main root cortex of rice; bacteria (arrowed) appear to be 3 to 4 cell layers beneath the root surface (*Bar* = 100 μm). (**B**) Higher magnification of (A) showing large pockets of bacteria within the cortex (*Bar* = 20 μm). (**C**) Light micrograph showing the spread of azorhizobia into neighbouring cortical cells by invagination (arrowed) of an adjacent plant cell wall (Bar = 100 μm). (**D**) Higher magnification of the arrowed region shown in (C). (*Bar* = 20 μm).

In rice inoculated with ORS571 in the presence of 10^{-4} *M* naringenin, toluidine blue staining of sections showed that bacteria had entered at LRCs, resulting in the presence of large intercellular pockets of bacteria in the region of emerging lateral roots (Figure 1A, B). These large intercellular pockets closely resemble the intercellular infection pockets filled with azorhizobia observed in sections of roots of *S. rostrata* a few days after inoculation (Ndoye et al., 1994). The further colonization of rice roots by azorhizobia (Figure 1A, B) resulted from their spread inwards over a distance of three to four plant cell layers and their forming larger pockets of azorhizobia within the main root cortex (Figure 2A, B). Further spread of azorhizobia from these pockets appeared to be by invagination of adjacent plant cell walls (Figure 2C, D), similar to the spread of rhizobia during the early stages of nodule development on *Aeschynomene afraspera* (Alazard and Duhoux, 1990).

Nod *genes are not involved in intercellular colonization of wheat by* Azorhizobium caulinodans

Naringenin is one of the most efficient inducers of the expression of *nod* genes in ORS571, and *nod* genes are required for nodule formation in *S. rostrata* (Goethals et al., 1989). Naringenin has not been detected in *Sesbania rostrata* seedling exudates, but the major *Azorhizobium nod* gene-inducing factor present, 7,4′dihydroxyflavanone (liquiritigenin; Messens et al., 1991), is closely related structurally to naringenin. It was therefore important to determine whether naringenin was stimulating colonization of these non-legumes via *nod* genes. In the presence of specific flavonoids, NodD proteins activate the transcription of *nod* operons, which, in turn, control the synthesis of specific lipo-chitooligosaccharides, the so-called Nod factors (Dénarié et al., 1996).

We first assessed the possible role of Nod factors in the colonization of LRCs of wheat by *A. caulinodans* strain ORS571 (pXLGD4) and, currently, rice is being assessed in a similar way. In this study with wheat, a *nodC* mutant of ORS571, which does not produce Nod factors (Geelen et al., 1993), was tested. No difference was found between the levels of colonization by the *nodC* mutant and the wild-type strain. Colonization of wheat roots by the *nodC* mutant was still stimulated by naringenin (Figure 3). An ORS571 *nodD* mutant (Goethals et al., 1990) was also evaluated, as it was thought possible that NodD could activate genes other than *nod* genes in the presence of naringenin. However,

the *nodD* mutant was able to colonize LRCs of wheat as well as the wild-type strain (Figure 3). Colonization by the *nodD* mutant was also stimulated by naringenin, indicating that the mechanism by which naringenin stimulates LRC colonization of this non-legume is not mediated by Nod factors or the NodD protein.

Are flavonoids acting as signal molecules in root colonization?

In the present study, we have now demonstrated that *A. caulinodans* can colonize cracks at the points of emergence of lateral roots of both rice and wheat and can subsequently colonize intercellular spaces of the root cortex. In wheat, we have demonstrated that this colonization is *nod* gene-independent. The demonstration of this LRC colonization in these cereals and in *Arabidopsis thaliana* (Gough et al., 1997) implies that this is likely to be a general occurrence in non-leguminous plants.

The mechanism(s) of stimulation of colonization of wheat and rice LRCs and subsequent intercellular colonization adjacent to LRCs by naringenin are unknown. However, it does not appear to be due to a carbon source effect, since naringenin is active at low concentrations, while succinate at a similar molarity is not stimulatory. Therefore, naringenin may act as a signal molecule and not as a substrate. Recently, it has been proposed that plant signals, including flavonoids, are involved in the selective stimulation of specific microbial communities in the rhizosphere, explaining why plant roots are only colonized by specific soil bacteria (Phillips and Streit, 1996). The present results may support this hypothesis. The precise colonization site that we have described for *A. caulinodans*, at LRCs in wheat and rice, is one of the principal sites of entry into roots of soil-borne fungal and bacterial pathogens (Forster, 1986; Vasse et al., 1995). The discovery and understanding of why some plant signals stimulate colonization of this site may therefore have important potential applications other than those in the area of nitrogen fixation. These include colonization by plant growth promoting bacteria and by bacteria used for biocontrol of plant root pathogens. Genetic engineering may lead to the generation of plants which are able to over-produce such signal molecules and, consequently, influence their own rhizosphere towards colonization by beneficial micro-organisms (O'Connell et al., 1996).

It may be relevant that the induction by flavonoids of rhizobial genes, which are not *nod* genes, has already

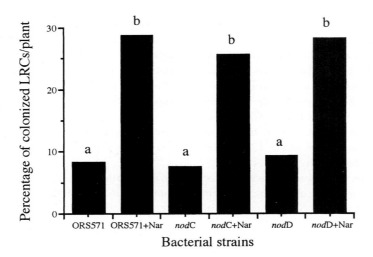

Figure 3. Effect of 10^{-4} M naringenin on colonization of LRCs of wheat by *A. caulinodans* strain ORS571 (pXLGD4) and its *nod* mutants. Treatments with different letters differ significantly at the $p = 0.01$ level. Analysis of variance with Fisher's test (Snedecor and Cochran, 1989). Ten replicates per treatment.

been reported (Sadowsky et al., 1988). This control by host-specific flavonoids of rhizobial genes, other than *nod* genes, may be important for early steps in the establishment of symbiosis in legumes and also in colonization of non-legumes. Therefore, it will be important to determine whether naringenin, and any other flavonoids that may be found to behave similarly, induce a new class of genes in *A. caulinodans*, which could be involved both in the symbiotic interaction between *A. caulinodans* and its natural host *Sesbania rostrata* and in interactions between *A. caulinodans* and non-legume crops, such as rice and wheat. It would be interesting to determine whether naringenin (or other flavonoids) influence the production of cell wall degrading enzymes, such as cellulase and pectinase, which may improve the colonization ability of *A. caulinodans*. It has been suggested that the intercellular colonization of roots by diazotrophic bacteria might be improved by the use of diazotrophs possessing genes for the synthesis of pectin-degrading enzymes that could facilitate root invasion (Okon, 1985).

Possibilities for endophytic nitrogen fixation in rice and wheat

The ultimate aim of establishing endophytic interactions between diazotrophs and non-legumes is that the diazotrophs should fix nitrogen and transfer this fixed nitrogen to the plant. The establishment of endosymbiotic nitrogen fixation in cereals would be one of the

most significant contributions that biotechnology could make to agriculture (Cocking et al., 1994). Previously, low levels of nitrogenase activity, up to 112 nmoles of ethylene accumulated over 24 h, have been reported in similar tube experiments using wheat cv. Wembley inoculated with *A. caulinodans* strain ORS571 (Cocking et al., 1995). In the present study, using wheat cv. Canon, we could detect nitrogenase activity up to 41.3 nmoles of ethylene accumulated over 24 h by the acetylene reduction assay, only when succinate was added to the plant growth medium. This suggests, that under our experimental growth conditions, carbon may be limiting for nitrogen fixation. However, in both experiments, what is being measured is nitrogen fixation from bacteria within intercellular spaces and from bacteria colonizing the root surface. The dependence on externally added succinate for nitrogen fixation by wheat cv. Canon indicates that it is more likely to be the nitrogenase activity from surface bacteria that is being measured. The results may also suggest that some cultivars of wheat exude more carbon containing compounds from their root systems than others. In the present study on the interaction of *A. caulinodans* with wheat cv. Canon, we were able to demonstrate that conditions are appropriate for nitrogen fixation at *Azorhizobium* colonization sites by detecting low level expression of *nif* promoter *lacZ* fusion.

Recent findings have shown that when wheat plants are grown in pots in growth chambers, with aseptic precautions taken as fully as possible, high levels of acetylene reduction activity can be detected in plants inoculated with *A. caulinodans*. Nitrogenase activity was not detected in uninoculated plants or in plants inoculated with a *nif⁻* strain of *A. caulinodans* (Sabry et al., 1997). In these experiments, plants were removed from their pots for acetylene reduction activity measurements, thus exposing all surface colonizing bacteria to oxygen concentrations inhibitory to nitrogen fixation. The latter work also demonstrated that wheat plants inoculated with *A. caulinodans* significantly increased in dry weight and total nitrogen content, as compared to uninoculated controls. These results suggest that an adequate supply of carbon is being provided from photosynthesis in pot-grown plants. Light and electron microscopy of sections cut from the short lateral roots observed by Sabry et al. (1997) on wheat inoculated with *A. caulinodans*, showed that the short lateral roots were invaded by bacteria presumed to be azorhizobia. The invading bacteria were observed within cells of the lateral root meristem, between cells of the cortex and within some of the xylem elements. No bacteria were observed in uninoculated wheat plants similarly grown in growth chambers and similarly analysed by light and electron microscopy.

The fact that we have defined conditions which give reproducible invasion of lateral root cracks and subsequent intercellular colonization of the roots of rice and wheat by rhizobia at high frequency, provides important information to researchers attempting to extend rhizobial colonization and endophytic nitrogen fixation to non-legume crops.

Acknowledgements

GW and CG were supported by the grant "Assessing opportunities for biological nitrogen fixation in rice" from the Danish International Development Agency, GW also by the University of Nottingham and CG also by a fellowship from INRA. CAB acknowledges support by the ODA (Plant Sciences Programme), KJO'C by MAFF and ECC by the Leverhulme Trust and the Rockefeller Foundation. SLK was a Rockefeller Foundation Biotechnology Career Fellow.

References

Alazard D and Duhoux E 1990 Development of stem nodules in a tropical forage legume, *Aeschynomene afraspera*. J. Exp. Bot. 41, 1199–1206.

Al-Mallah M K, Davey M R and Cocking E C 1987 Enzymatic treatment of clover root hairs removes a barrier to *Rhizobium*-host specificity. Bio/Technol. 5, 1319–1322.

Al-Mallah M K, Davey M R and Cocking E C 1989 Formation of nodular structures on rice seedlings by rhizobia. J. Exp. Bot. 40, 473–478.

Al-Mallah M K, Davey M R and Cocking E C 1990 Nodulation of oilseed rape (*Brassica napus*) by rhizobia. J. Exp. Bot. 41, 1567–1572.

Boddey R M, de Oliveira O C, Urquiga S, Reis V M, de Oliveira F L, Baldani V L D and Döbereiner J 1995 Biological nitrogen fixation associated with sugar cane and rice: Contributions and prospects for improvement. Plant Soil 174, 195–209.

Boivin C, Camut S, Malpica C A, Truchet G and Rosenberg C 1990 *Rhizobium meliloti* genes encoding catabolism of trigonelline are induced under symbiotic conditions. Plant Cell 2, 1157–1170.

Cocking E C, Al-Mallah M K, Benson E and Davey M R 1990 Nodulation of non-legumes by rhizobia. *In* Nitrogen Fixation: Achievements and Objectives. Eds. P M Gresshoff, E C Roth, G Stacey and W E Newton. pp 813–823. Chapman and Hall, New York, USA.

Cocking E C, Davey M R, Kothari S L, Srivastava J S, Jing Y, Ridge R W and Rolfe B G 1992 Altering the specificity control of the interaction between rhizobia and plants. Symbiosis 14, 123–130.

Cocking E C, Webster G, Batchelor C A and Davey M R 1994 Nodulation of non-legume crops. A new look. Agro-Food-Industry Hi-Tech. January/February, 21–24.

Cocking E C, Kothari S L, Batchelor C A, Jain S, Webster G, Jones J, Jotham J and Davey M R 1995 Interaction of *Rhizobium* with non-legume crops for symbiotic nitrogen fixation. *In* Azospirillum VI and Related Microorganisms, NATO ASI Series, Vol. G37. Eds. I Fendrik, M del Gallo, J Vanderleyden and M de Zamaroczy. pp 197–205. Springer Verlag, Berlin, Germany.

Davey M R, Webster G, Manders G, Ringrose F L, Power J B and Cocking E C 1993 Effective nodulation of micro-propagated shoots of the non-legume *Parasponia andersonii* by *Bradyrhizobium*. J. Exp. Bot. 44, 863–867.

Dénarié J, Debellé F and Rosenberg C 1992 Signaling and host range variation in nodulation. Ann. Rev. Microbiol. 46, 497–531.

Dénarié J, Debellé F and Promé J C 1996 *Rhizobium* lipochitooligosaccharide nodulation factors: Signaling molecules mediating recognition and morphogenesis. Ann. Rev. Biochem. 65, 503–535.

Fåhraeus G 1957 The infection of clover root hairs by nodule bacteria studied by a simple glass slide technique. J. Gen. Microbiol. 16, 374–381.

Forster R C 1986 The ultrastructure of the rhizoplane and rhizosphere. Ann. Rev. Phytopathol. 24, 211–234.

Geelen D, Mergaert P, Gememia R A, Goormachtig S, Van Montagu M and Holsters M 1993 Identification of *nodSUIJ* genes in Nod locus 1 of *Azorhizobium caulinodans*: evidence that *nodS* encodes a methyltransferase involved in Nod factor modification. Mol. Microbiol. 9, 145–154.

Goethals K, Gao M, Tomekpe K, Van Montagu M and Holsters M 1989 Common *nodABC* genes in Nod locus 1 of *Azorhizobium caulinodans*: Nucleotide sequence and plant-inducible expression. Mol. Gen. Genet. 219, 289–298.

Goethals K, Van Den Eede G, Van Montagu M and Holsters M 1990 Identification and characterization of a functional *nodD* gene of *Azorhizobium caulinodans* ORS571. J. Bacteriol. 172, 2658–2666.

Gough C, Webster G, Vasse J, Galera C, Batchelor C, O'Callaghan K, Davey M, Kothari S, Dénarié J and Cocking E 1996 Specific flavonoids stimulate intercellular colonization of non-legumes by

Azorhizobium caulinodans. *In* Biology of Plant-Microbe Interactions. Eds. G Stacey, B Mullin and P M Gresshoff. pp 409–415. International Society for Molecular Plant-Microbe Interactions, Minnesota.

Gough C, Vasse J, Galera C, Webster G, Cocking E and Dénarié J 1997 Interactions between bacterial diazotrophs and non-legume dicots: *Arabidopsis thaliana* as a model plant. Plant Soil 194, 123–130.

Hartwig U A, Joseph C M and Phillips D A 1991 Flavonoids released naturally from alfalfa seeds enhance growth rate of *Rhizobium meliloti*. Plant Physiol. 95, 797–803.

Kitts C L and Ludwig R A 1994 *Azorhizobium caulinodans* respires with at least 4 terminal oxidases. J. Bacteriol. 176, 886–895.

Ladha J K and Reddy P M 1995 Extension of nitrogen fixation to rice - necessity and possibilities. Geo. J. 35, 363–372.

Leong S A, Williams P H and Ditta G S 1985 Analysis of the $5'$ regulatory region of the gene for δ-aminolevulinic acid synthetase of *Rhizobium meliloti*. Nucl. Acids Res. 13, 5965–5976.

Le Strange K, Bender G L, Djordjevic M A, Rolfe B G and Redmond J W 1990 The *Rhizobium* strain NGR234 *nod*D1 gene product responds to activation by the simple phenolic compounds vanillin and isovanillin present in wheat seedling extracts. Mol. Plant-Microbe Interact. 3, 214–220.

Lundegårdh H and Stenlid G 1944 The exudation of nucleotides and flavanones from living roots. Ark. Bot. 31A, 1–27.

Messens E, Geelen D, Van Montagu M and Holsters M 1991 $7,4'$-Dihydroxyflavanone is the major *Azorhizobium nod* gene-inducing factor present in *Sesbania rostrata* seedling exudate. Mol. Plant-Microbe Interact. 4, 262–267.

Ndoye I, De Billy F, Vasse J, Dreyfus B and Truchet G 1994 Root nodulation of *Sesbania rostrata*. J. Bacteriol. 176, 1060–1068.

O'Connell K P, Goodman R M and Handelsman J 1996 Engineering the rhizosphere: Expressing a bias. TIBTech 14, 83–88.

Okon Y P G 1985 *Azospirillum* as a potential inoculant for agriculture. TIBTech 3, 223–228.

Pawlowski K, Ratet P, Schell J and De Bruijn F J 1987 Cloning and characterization of *nifA* and *ntrC* genes of the stem nodulating bacterium ORS571, the nitrogen fixing symbiont of *Sesbania rostrata*: Regulation of nitrogen fixation (*nif*) genes in the free living versus symbiotic state. Mol. Gen. Genet. 206, 207–219.

Peters N K and Verma D P S 1990 Phenolic compounds as regulators of gene expression in plant-microbe interactions. Mol. Plant-Microbe Interact. 3, 4–8.

Phillips D A and Streit W 1996 Legume signals to rhizobial symbionts: A new approach for defining rhizosphere colonization. *In* Plant-Microbe Interactions. Eds. G Stacey and N T Keen. pp 236–271. Chapman and Hall, New York, USA.

Quispel A 1991 A critical evaluation of the prospects for nitrogen fixation with non-legumes. Plant Soil 137, 1–11.

Rolfe B G 1988 Flavones and isoflavones as inducing substances of legume nodulation. BioFactors 1, 3–10.

Sabry R S, Saleh S A, Batchelor C A, Jones J, Jotham J, Webster G, Kothari S L, Davey M R and Cocking E C 1997 Endophytic establishment of *Azorhizobium caulinodans* in wheat. Proc. Roy. Soc. Lond.: Biol. Sci. 264, 341–346.

Sadowsky M J, Olson E R, Foster V E, Kosslak R M and Verma D P S 1988 Two host-inducible genes of *Rhizobium fredii* and characterization of the inducing compound. J. Bacteriol. 170, 171–178.

Sloger C and van Berkum P 1992 Approaches for enhancing nitrogen fixation in cereal crops. *In* Biological Nitrogen Fixation Associated with Rice Production. Eds. S K Dutta and C Sloger. pp 229–234. Oxford and IBH Publishing, New Delhi, India.

Snedecor G W and Cochran W E 1989 Statistical Methods, 8th Edn. Iowa State University Press, Ames, USA. 503 p.

Somasegaran P and Hoben H J 1994 Handbook for Rhizobia. Springer-Verlag, New York, USA. 450 p.

Sprent J I and Raven J A 1992 Evolution of nitrogen fixing symbioses. *In* Biological Nitrogen Fixation. Eds. G Stacey, R H Burris and H J Evans. pp 461–496. Chapman and Hall, New York, USA.

Tsien H C, Dreyfus B L and Schmidt E L 1983 Initial stages in the morphogenesis of nitrogen-fixing stem nodules of *Sesbania rostrata*. J. Bacteriol. 156, 888–897.

Van Berkum P and Bohlool B B 1980 Evaluation of nitrogen fixation by bacteria in association with roots of tropical grasses. Microbiol. Rev. 44, 491–517.

Vasse J, Frey P and Trigalet A 1995 Microscopic studies of intercellular infection and protoxylem invasion of tomato roots by *Pseudomonas solanacearum*. Mol. Plant-Microbe Interact. 8, 241–251.

Webster G, Davey M R and Cocking E C 1995a *Parasponia* with rhizobia: A neglected non-legume nitrogen-fixing symbiosis. AgBiotech News Info 7, 119N–124N.

Webster G, Poulton P R, Cocking E C and Davey M R 1995b The nodulation of micro-propagated plants of *Parasponia andersonii* by tropical legume rhizobia. J. Exp. Bot. 46, 1131–1137.

Yanni Y G, Rizk R Y, Corich V, Squartini A, Ninke K, Philip-Hollingsworth S, Orgambide G, De Bruijn F, Stoltzfus J, Buckley D, Schmidt T M, Mateos P F, Ladha J K and Dazzo F B 1997 Natural endophytic association between *Rhizobium leguminosarum* bv. *trifolii* and rice roots and assessment of its potential to promote rice growth. Plant Soil 194, 99–114.

Guest editors: J K Ladha, F J de Bruijn and K A Malik

Plant and Soil **194**: 123–130, 1997.
© 1997 *Kluwer Academic Publishers. Printed in the Netherlands.*

123

Interactions between bacterial diazotrophs and non-legume dicots: *Arabidopsis thaliana* as a model plant

Clare Gough[1], Jacques Vasse[1], Christine Galera[1], Gordon Webster[1], Edward Cocking[2] and Jean Dénarié[1]

[1]*Laboratoire de Biologie Moléculaire des Relations Plantes-Microorganismes, INRA-CNRS, BP27, 31326 Castanet-Tolosan, France* and [2]*Plant Genetic Manipulation Group, Department of Life Science, University of Nottingham, Nottingham NG7 2RD, U.K.*

Key words: Arabidopsis thaliana, Azorhizobium caulinodans, flavonoids, lateral root crack, *nod* genes, rhizobia

Abstract

When interactions between diazotrophic bacteria and non-legume plants are studied within the context of trying to extend biological nitrogen fixation to non-legume crops, an important first step is to establish reproducible internal colonization at high frequency of these plants. Using *Azorhizobium caulinodans* ORS571 (which induces stem and root nodules on the tropical legume *Sesbania rostrata*), tagged with a constitutively expressed *lacZ* reporter gene, we have studied the possibilities of internal colonization of the root system of the model dicot *Arabidopsis thaliana*. ORS571 was found to be able to enter *A. thaliana* roots after first colonizing lateral root cracks (LRCs), at the points of emergence of lateral roots. Cytological studies showed that after LRC colonization, bacteria moved into the intercellular space between the cortical and endodermal cell layers of roots. In our experimental conditions, this LRC and intercellular colonization are reproducible and occur at high frequency, although the level of colonization at each site is low. The flavonoids naringenin and daidzein, at low concentrations, were found to significantly stimulate (at the $p=0.01$ level) the frequency of LRC and intercellular colonization of *A. thaliana* roots by *A. caulinodans*. The role in colonization of the structural *nodABC* genes, as well as the regulatory gene *nodD*, was studied and it was found that both colonization and flavonoid stimulation of colonization are *nod* gene-independent. These systems should now enable the various genetic and physiological factors which are limiting both for rhizobial colonization and for endophytic nitrogen fixation in non-legumes, to be investigated. In particular, the use of *A. thaliana*, which has many advantages over other plants for molecular genetic studies, to study interactions between diazotrophic bacteria and non-legume dicots, should provide the means of identifying and understanding the mechanisms by which plant genes are involved in these interactions.

Introduction

Due to the ever-increasing amounts of nitrogenous fertilizer which are used for crop production and the limited supply of fossil fuel energy available to manufacture this fertilizer, it is becoming increasingly important to try to extend biological nitrogen fixation to non-legume crops. Several strategies have already been envisaged to achieve this goal, including the transfer of bacterial *nif* genes from a free-living diazotroph into non-legume crops and the induction on the roots of non-legume crops, of structures resembling root nodules

induced by rhizobia on legume plants. The rationale for this latter approach was that if rhizobia have to be inside nodules to fix nitrogen when in symbiotic association with legumes, then such structures would have to be induced on non-legumes. The discovery that bacteria of the genera *Rhizobium* and *Bradyrhizobium* can form effective nitrogen-fixing nodules on the non-legume *Parasponia* (Trinick, 1973), provided added impetus to this approach. However, while many attempts were made to induce what were called "nodule-like" structures or "*para*-nodules" on non-legumes such as wheat, maize and rice, few, inconsistent and often controversial results were obtained (de Bruijn et al., 1995). Moreover, many of these struc-

* FAX No: +3361285061. E-mail: gough@toulouse.inra.fr

tures had to be artificially induced by adding 2,4-D or cell wall degrading enzymes.

More recently it has been discovered that the endophytic diazotrophs *Acetobacter diazotrophicus* and *Herbaspirillum* spp. can contribute substantially to the growth of sugarcane via nitrogen fixation and without the formation of any nodule-like structures (Boddey et al., 1995). Opinion has consequently changed and the emphasis is now to try to establish stable endophytic associations between diazotrophic bacteria and non-legume crops. The same conclusion has also come from studies on associative nitrogen fixing systems between soil diazotrophs and non-legume crops. In these systems, bacteria only colonize the surface of roots and remain vulnerable to competition from other micro-organisms of the rhizosphere. Compared to symbiotic associations between rhizobia and legume plants, where bacteria are protected from competition with rhizospheric micro-organisms and have the possibility of intimate metabolic exchange with host plants, the yield of fixed nitrogen from associative systems is low.

Two major approaches are currently being undertaken to try to extend biological nitrogen fixation to non-legume crops. The first is to look for natural endophytic diazotrophs of non-legume crops and to assess such endophytic nitrogen fixation. This is being exploited with sugarcane (Boddey et al., 1995) and other crops such as rice (Ladha and Reddy, 1995). The second approach is to determine whether rhizobia can internally colonize and ultimately fix nitrogen in non-legumes. Both approaches obviously have their own advantages and drawbacks, the main advantage of the second approach being the extensive knowledge of the mechanism of rhizobial nitrogen fixation. However, if the second approach is to be at all feasible, then internal rhizobial colonization needs to occur at high frequency and at a high level. The first step is therefore to study the possibility of intercellular rhizobial colonization of non-legume roots, including entry, spread and the mechanisms involved. A sensitive and quantifiable system is needed for this. Reports of intercellular rhizobial colonization of non-legumes have been made, but these have used sectioning as a method of detection (Cocking et al., 1992, 1994), which is difficult to quantify. Microscopical observation of colonization is greatly facilitated by using a bacterial strain tagged with a reporter gene.

The primary aim of this work was to assess and quantify intercellular colonization of *Arabidopsis thaliana* roots by *Azorhizobium caulinodans* strain ORS571 carrying a constitutively expressed *lacZ* reporter gene. The choice of the rhizobial strain ORS571, which forms stem and root nodules on the tropical legume *Sesbania rostrata*, was based on three characteristics (de Bruijn, 1989). Firstly, as it is unlikely that there will be bacteroid differentiation in non-legume plants, ORS571 has the unusual ability among rhizobia to fix nitrogen in free-living conditions without differentiation into bacteroids. Secondly, as it is also unlikely that non-legumes will be able to reproduce the sophisticated oxygen barriers which exist in legume nodules to protect the nitrogenase enzyme complex from inhibitory oxygen levels, ORS571 can tolerate up to 12 μM dissolved oxygen while fixing nitrogen in culture. Thirdly, ORS571 forms nodules on its host plant following crack entry infection *ie* intercellularly, between adjacent cells. This type of rhizobial infection does not involve the formation of infection threads at the tips of root hairs, which is considered to be an advanced character, confined to certain legume tribes (Sprent et al., 1987), and the only non-legume, *Parasponia*, to form effective nitrogen-fixing nodules with rhizobia is infected by crack entry (Webster et al., 1995).

For the choice of the non-legume plant, *A. thaliana* could at first sight seem inappropriate, as this plant obviously has no agronomic importance. However, largely because of its small genome size and rapid generation time, *A. thaliana* has become over the last few decades a model plant for molecular genetic studies. *A. thaliana* is currently by far the best plant system for identifying and studying the role of genes involved in a given plant response. For example, many biochemical and morphological mutants of *A. thaliana* have been mapped and the corresponding genes cloned. From rhizobia-legume symbioses, it is known that certain plant genes (coding for so-called nodulins), are specifically expressed in response to rhizobial infection. The function of many of the early nodulins (those detected during infection and morphogenesis of the nodule), is still unknown, but late nodulins (whose production coincides with the nitrogen-fixing state of the mature nodule) include several proteins of known function (leghaemoglobin, glutamine synthetase, glutamate synthetase, aspartate transferase, sucrose synthetase). Whether any of these proteins will be involved in associations between rhizobia and non-legume crops is difficult to predict, but *A. thaliana* offers the best possibility among non-legume plants of addressing this question. The identification of any plant genes specif-

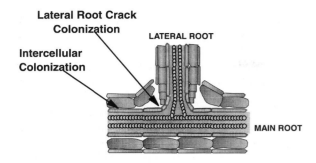

Lateral Root Crack Colonization

Intercellular Colonization

LATERAL ROOT

MAIN ROOT

Figure 1. Schematic representation of the point of emergence of a lateral root of *A. thaliana*, showing the localization of the observed lateral root crack colonization and intercellular colonization by *A. caulinodans*.

ically involved in rhizobia-non-legume interactions, should also be feasible using *A. thaliana*.

Having described and quantified intercellular colonization of *A. thaliana* by *A. caulinodans* ORS571, the role in colonization of flavonoids was investigated and specific flavonoids, at low concentrations, were found to significantly stimulate colonization. The use of bacterial mutants showed that neither colonization nor the stimulation of colonization by flavonoids were mediated by *nod* genes or Nod factors. Lastly, the use of *A. thaliana* is discussed as a means of defining factors limiting biological nitrogen fixation in non-legume dicots and of understanding the molecular genetic mechanisms involved in interactions between diazotrophic bacteria and non-legume dicots.

Lateral Root Cracks (LRCs) of *A. thaliana* are colonized by *A. caulinodans* and this leads to intercellular colonization

Arabidopis thaliana plants were grown in sterile conditions in tubes containing soft Fahraeus agar medium, and inoculated with *Azorhizobium caulinodans* ORS571(pXLGD4). pXLGD4 is a broad host-range plasmid carrying a constitutively expressed fusion between the promoter of the *hemA* gene of *Rhizobium meliloti* and the *lacZ* gene of *Escherichia coli* (Leong et al., 1985). Bacterial colonization was visualised on whole roots by the blue pigment resulting from the degradation of X-gal, according to Boivin et al. (1990) and sections for light and electron microscopy were prepared as described by Vasse et al. (1995).

A. thaliana plants which had been inoculated with ORS571(pXLGD4) did not show any obvious differ-ences in root morphology compared to uninoculated control plants. Inoculated roots which were histochemically stained were found to be colonized on the surface of root tips and inside lateral root cracks, LRCs, (the natural fissures at the bases of lateral roots resulting from lateral root emergence). LRCs, being more heavily colonized than root tips, were studied and were found to provide an entry point into the root system for *A. caulinodans*. The colonization of LRCs started at an early stage in lateral root formation and as these laterals developed, bacteria entered deeper into the cracks and from here into intercellular spaces between the cortex and the endodermis, in the main root. These two stages of colonization, LRC colonization and intercellular colonization, are represented in Figure 1. Occasionally, whole epidermal or endodermal cells were entirely filled with bacteria, indicating that such cells were no longer viable. No other evidence of intracellular colonization was found. No evidence was found either for the presence of bacteria in the vascular system, although this was only studied one and two weeks after inoculation.

LRC colonization (including both colonization of the LRC and intercellular colonization), was reproducible and between 63 and 100% of *A. thaliana* plants were found to have LRCs colonized by ORS571(pXLGD4). This unexpected finding of such high colonization frequencies, is certainly a reflection of the sensitivity of the *lacZ* reporter gene for detecting the presence of colonizing bacteria. The relatively low level of colonization at each site would have been much more difficult to detect by previous techniques such as sectioning. A scoring system was developed which consisted in counting the number of colonized LRCs and expressing this as a percentage of the total number of LRCs per plant. The mean result for four independent experiments was that 10.8% of LRCs were colonized per plant.

A. caulinodans ORS571 can therefore intercellularly colonize LRCs of *A. thaliana* reproducibly and at high frequency. This work was performed in parallel with a similar study on an important non-legume crop, wheat, for which LRC and intercellular colonization by *A. caulinodans* ORS571 were also found to occur and at a similar frequency to that reported here for *A. thaliana* (Gough et al., 1996; Webster et al., this volume). The demonstration of this quantifiable intercellular colonization in two very different non-legume plants, represents a major breakthrough in studies of interactions between diazotrophic bacteria and non-legume plants, as the principal bottlenecks which existed previously in this field - the rarity of internal colonization and the difficulty of obtaining reproducible and quantitative data - have now been overcome. The use of a reporter gene to visualise bacteria has been a key element in this work, enabling a relatively easy and very sensitive method of detecting LRC and intercellular colonization.

LRCs of plants are often preferentially colonized by soil bacteria and are one of the principal entry sites into plants of soil-borne fungal and bacterial plant pathogens. Moreover, LRCs of *A. thaliana* have been shown by scanning electron microscopy to be fissures in the otherwise smooth root surface (Dolan et al., 1993), and as such they would seem to be suitable niches for bacterial colonization. An interesting aspect of LRC colonization of *A. thaliana* by *A. caulinodans*, however, is its apparently dynamic nature. That is, *A. caulinodans* starts by colonizing the LRCs of young emerging lateral roots, then seems to migrate away from the superficial part of the LRC, down along the fissure into the root, to occupy the intercellular space between the cortex and the endodermis, within the main root adjacent to the base of the lateral root. Possible explanations for this include aerotactic or chemotactic behaviour. Alternatively, the disappearance of bacteria from LRCs could be explained by the activation of plant defence reactions. No obvious signs of defence were seen, but this does not exclude the existence of more subtle responses to the presence of bacteria.

Whatever the reason for the progression from LRC to intercellular colonization, this phenomenon has the advantage of moving bacteria deeper into the root system, but the disadvantage that it seems to be accompanied by a decrease in bacterial numbers. The actual number of bacteria which are intercellular at LRCs is low and almost certainly too low to allow the detection of any nitrogenase activity. This low level of colonization therefore constitutes a first important limiting factor for biological nitrogen fixation.

The cortical layer of *A. thaliana* is only one or sometimes two cell layers thick, while plants such as rice and wheat have greater numbers of cortical cell layers and consequently the possibility of greater intercellular colonization in the cortex. However, nutrient conditions are probably more favourable for bacterial multiplication and nitrogen fixation in the vascular system of plants than in cortical intercellular spaces. In two of the best known examples of endophytic biological nitrogen fixation in non-nodulating non-legumes, in sugarcane by *Herbaspirillum* spp. and *Acetobacter*

diazotrophicus and in Kaller grass by *Azoarcus*, the diazotrophic bacteria colonize the vascular systems of these plants and xylem vessels are considered to act as a niche for nitrogen fixation and the exchange of metabolites between *A. diazotrophicus* and sugarcane (James et al., 1994). Entry into the xylem of sugarcane by *A. diazotrophicus* takes place intercellularly at the root tips and at cracks at lateral root junctions. LRC colonization of *A. thaliana* by *A. caulinodans* may therefore represent an important preliminary stage necessary for entry of bacteria into the plant, but ultimately the means will probably have to be found to get *A. caulinodans* into the vascular system of *A. thaliana* and other non-legumes. One possibility for this could be to identify and transfer bacterial genes responsible for entry into the xylem, into *A. caulinodans*.

LRC and intercellular colonization by *A. caulinodans* is stimulated by specific flavonoids

As the first stage of infection by *A. caulinodans* of its legume host plant, *Sesbania rostrata*, is also intercellular and at the points of emergence of lateral roots (Ndoye et al., 1994), this suggested that some factors controlling the symbiotic association may also be involved in interactions between *A. caulinodans* and *A. thaliana*. As in all other rhizobia-legume symbioses, specific flavonoids, which are a class of plant phenolics secreted in the rhizosphere by plants, induce nodulation (*nod*) genes of *A. caulinodans* that are required for nodulation of *S. rostrata*. Flavonoids also enhance the growth rate of certain rhizobia (Hartwig et al., 1991) and phenolic compounds in general can induce expression of a variety of genes in plant-associated bacteria (Peters and Verma, 1990). Four families of flavonoids (flavones, flavonols, flavanones and isoflavones), have been shown to have *nod* gene inducing activity (Dénarié et al., 1992). One molecule of each family was tested for its effect on LRC colonization of *A. thaliana* by ORS571(pXLGD4). Acetosyringone, a plant phenolic which induces expression of virulence genes in *Agrobacterium tumefaciens*, and succinate, one of the best carbon sources for *A. caulinodans*, were also tested.

The flavanone naringenin and the isoflavone daidzein (both at 5×10^{-5} M) significantly stimulated (at the $p=0.01$ level), colonization of LRCs of *A. thaliana* (Figure 2). Daidzein, but not naringenin, at 10^{-5} M also significantly stimulated (at the $p=0.01$ level) colonization, while succinate, at both 10^{-5} M

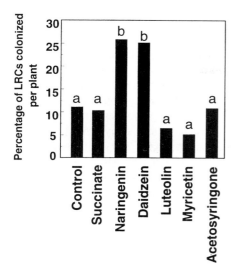

Figure 2. Effect of flavonoids, acetosyringone and succinate on colonization of LRCs of *A. thaliana* by *A. caulinodans* ORS571(pXLGD4). Compounds were added to the plant growth medium at 5×10^{-5} M. Statistical analysis: treatments with different letters differ significantly at the p=0.01 level. Analysis of variance with Fisher's test (Snedecor and Cochran, 1980). 16 plants were scored per treatment.

and 5×10^{-5} M, had no significant effect on colonization. Naringenin and daidzein contain nearly four times as much carbon as succinate, but the fact that 10^{-5} M daidzein significantly stimulated colonization and 5×10^{-5} M succinate did not, indicates that these flavonoids are not being used as carbon sources. This, together with the fact that stimulation of LRC colonization was restricted to specific flavonoids at low concentrations, suggests that flavonoids might be acting as signals and not as substrates in these *A. caulinodans-A. thaliana* interactions.

A better understanding of the mechanism by which flavonoids stimulate LRC colonization of *A. thaliana* is needed, because this could suggest ways of increasing colonization levels. This information could, in turn, be extended to other non-legumes such as wheat, for which naringenin has also been shown to significantly stimulate LRC colonization by *A. caulinodans* (Gough et al., 1996; Webster et al., this volume). Obvious candidates for bacterial genes being induced by flavonoids in the interaction *A. caulinodans-A. thaliana* are rhizobial *nod* genes, as naringenin is one of the most efficient inducers of the expression of *nod* genes in *A. caulinodans* (Goethals et al., 1989). The possibility that naringenin was stimulating colonization of *A. thaliana* via *nod* genes was therefore investigated.

nod genes are not involved in LRC and intercellular colonization by *A. caulinodans*

The induction of *nod* genes by specific flavonoids is mediated by NodD proteins. Once activated by flavonoids these proteins induce the transcription of *nod* operons, which in turn control the synthesis of specific lipo-chitooligosaccharides, the so-called Nod factors, required for infection, nodulation and the control of host specificity (Dénarié and Cullimore 1993). A *nodC* mutant of *A. caulinodans* ORS571 (Geelen et al., 1993), was used to assess the possible role of Nod factors in colonization of LRCs of *A. thaliana*. *nodC* is a structural gene involved in the synthesis of the core chitooligosacharidic backbone of Nod factors (Geremia et al., 1994; Dénarié et al., 1996). No difference was found in the levels of colonization between the *nodC* mutant and the wild-type strain and naringenin stimulated colonization by the two strains to the same degree.

As it is possible that NodD could be activated by naringenin to induce genes other than *nod* genes, an *A. caulinodans* ORS571 *nodD* mutant was also tested, *A. caulinodans* ORS571 only containing one copy of *nodD* (Goethals et al., 1990). Again, the *nodD* mutant colonized *A. thaliana* as well as the wild-type strain and naringenin stimulated colonization by the *nodD* mutant. The mechanism by which naringenin stimulates LRC colonization of *A. thaliana* is therefore not mediated by Nod factors or the NodD protein. The fact that daidzein also stimulates LRC colonization of *A. thaliana*, but does not induce *nod* gene expression in ORS571 (Goethals et al., 1989), supports this conclusion.

nod genes are therefore not involved, either in LRC colonization or in the stimulation of LRC colonization by flavonoids. If *nod* genes had been shown to be required for colonization of non-legumes, this would have had considerable implications for future lines of research. For example, the extensive knowledge of rhizobial infection into legumes could have been applied and this would have greatly facilitated the genetic analysis of this system.

Now the question that needs to be addressed concerning the mechanism of stimulation of LRC colonization by flavonoids, is whether it is the plant or the bacterium which is sensitive and responding to these molecules. Mutants of *A. thaliana* could be studied to determine whether flavonoids are acting via an effect on the plant. Despite the fact that rhizobial *nod* genes were shown not to be involved, however, there is a lot of precedent for flavonoids inducing bacterial genes to suggest that it is *A. caulinodans* and not *A. thaliana* that is responding to flavonoids. In particular, host-specific flavonoids have been reported to induce rhizobial genes which are not *nod* genes (Sadowsky et al., 1988). It is also possible that naringenin and daidzein increase the growth rate of *A. caulinodans*. It has been reported, for example, that luteolin at 10^{-5} M can reduce the doubling time of *Rhizobium meliloti* by a factor of five and that this effect is independent of *nodD* genes (Hartwig et al., 1991). Future experiments are planned to investigate whether flavonoids can increase the growth rate of *A. caulinodans* and whether flavonoids can stimulate LRC colonization of *A. thaliana* by bacteria other than *A. caulinodans*. If flavonoid stimulation of LRC colonization was confined to a limited number of bacteria, then this would indicate that it is indeed the bacterial partner that is responding to flavonoids.

Factors limiting endophytic biological nitrogen fixation in *A. thaliana*

The plasmid pPR3408, carrying a translational fusion between *nifD* of ORS571 and the *lacZ* reporter gene, (Pawlowski et al., 1987), was used to determine whether expression of *nif* genes of *A. caulinodans* ORS571 could be detected at colonized LRCs and in intercellular spaces of *A. thaliana*. The gene *nifD* is one of three genes in the *nifHDK* operon, coding for proteins of the nitrogenase complex. Transcription of this operon is activated by the central regulatory protein NifA in both the free-living and the symbiotic state and expression of *nifA* is, in turn, controlled by the cellular nitrogen and oxygen status (de Bruijn, 1989).

Repression of the expression of the *nifD-lacZ* fusion by oxygen was verified by growing the strain ORS571(pPR3408) in liquid culture at 21% O_2. Under these conditions, no β-galactosidase activity could be detected. When this strain was inoculated onto *A. thaliana* plants, however, it was possible to detect some β-galactosidase activity histochemically on the root surface, especially at root tips, and intercellularly at LRCs. Both the level and the frequency of activity were low, with fewer than 10% of colonized LRCs showing *nifD* expression. Nevertheless, the fact that a certain level of expression can be detected, indicates that intercellular conditions of nitrogen and oxygen in roots of *A. thaliana* can be conducive for *nif* gene expression.

Concerning the expression at the root surface, roots are growing in soft agar medium and in such conditions, at a certain distance from the surface, the oxygen concentration is likely to be permissible for *nifD* expression. Nitrogen fixation by *A. caulinodans* on the root surface of agar-grown plants is therefore not surprising and complicates the interpretation of any acetylene reduction assays (ARA) which are performed on such plants. For example, nitrogen fixation could be detected by ARA from *A. caulinodans* ORS571-inoculated *A. thaliana* plants grown in agar, but only when carbon sources such as succinate were added. As the addition of succinate greatly increases colonization of the root surface of *A. thaliana* by ORS571(pXLGD4), it is likely that this surface colonization is responsible for the majority of the detected ARA. The use of *nif* promoter-*lacZ* fusions is therefore better than ARA measurements to study intercellular nitrogen fixation by *A. caulinodans*.

As already mentioned, the level of intercellular colonization at LRCs of *A. thaliana* is certainly a limiting factor in the interaction between *A. caulinodans* and *A. thaliana*, there being too few bacteria to detect any nitrogenase activity. Other potential limiting factors include the partial pressure of oxygen, the availability of a carbon source and the level of combined nitrogen, all of which will affect bacterial nitrogenase activity. The fact that only a low level of the *nifD-lacZ* fusion could be detected, indicates that at least one of these factors is limiting for *nif* gene expression in intercellular spaces of *A. thaliana*, the exact nature of which could be studied with such fusions. Another factor which could potentially limit biological nitrogen fixation in *A. thaliana*, as well as in other non-legume plants, is the transfer of the fixed nitrogen to the plant. In free-living conditions, without differentiation into bacteroids, *A. caulinodans* does not liberate the nitrogen it fixes (Donald and Ludwig, 1984) and no evidence was found that *A. caulinodans* differentiates into bacteroids in *A. thaliana*. A way around this could be to use bacterial mutants which uncouple nitrogen fixation and assimilation.

It is obvious that the prospect of an efficient biological nitrogen fixing system with high levels of nitrogen fixation and significant benefits to the plant, is a long-term goal. This work confirms the prediction that the particular characteristics of nitrogen fixation of *A. caulinodans* (that it can fix nitrogen without differentiation into bacteroids and at relatively high oxygen concentrations), will represent genuine advantages over other rhizobia which could have been chosen and will probably outweigh any colonization advantages offered by other rhizobial species as it should be possible to increase the level of colonization by *A. caulinodans*. In addition, now that conditions have been defined for reproducible LRC colonization of non-legume plants, it should be possible to identify and study the various genetic and physiological factors which are limiting in these associations for nitrogen fixation and bacterial multiplication.

A. thaliana as a model plant for studying interactions between bacterial diazotrophs and non-legume dicots

Finally, the use of *A. thaliana* as a model plant for studying interactions between rhizobia, as well as other diazotrophic bacteria, and non-legume dicots should be discussed in the light of this work. The main advantages encountered using *A. thaliana* have been the ability to grow large numbers of plants in a restricted space, the almost transparent nature of the roots which has meant that that it has been easy to detect the slightest blue coloration, and the fact that the cellular organization of the *A. thaliana* root has been described (Dolan et al., 1993). This detailed knowledge of the cellular organization of *A. thaliana* roots, including knowledge of root development and the precise number, position and morphology of root cells, could be a particular advantage in describing intercellular colonization of roots of non-legume dicots, as well as in describing any effects that intercellular colonization could have on roots.

The main reason for using the model plant *A. thaliana*, however, has not yet been exploited and future studies should make use of *A. thaliana* mutants to identify and understand the mechanisms by which plant genes are involved in this interaction. In the case of rhizobia-legume symbioses, substantial progress has been made in understanding the mechanisms involved, but this understanding has been almost entirely made on the side of the bacterial partner, due to bacterial genetics. Progress has recently slowed down, because of the difficulty of genetic studies on the plant partners of these associations. Comparatively little is known, for example, concerning which plant genes are necessary for symbiosis and the mechanisms involved in their activation and co-ordinated regulation. *A. thaliana* offers the possibility of genetic and molecular genetic studies, including mutant analysis and the molecular cloning of genes. These features of *A. thaliana* are currently being exploited in the analysis of plant-pathogen interactions, where fast progress is being made to understand the response of the plant to pathogen infection at a molecular genetic level. The application of *A. thaliana* as a model plant in studying interactions between diazotrophic bacteria and non-legume dicots should also enable rapid progress to be made in understanding the mechanisms involved in these interactions.

Acknowledgements

C G was supported by a fellowship from the Institut National de la Recherche Agronomique, and G W by the grant "Assessing opportunities for biological nitrogen fixation in rice" from the Danish International Development Agency and E C by the Leverhulme Trust and the Rockefeller Foundation.

References

Boddey R M, de Oliveira O C, Urquiaga S, Reis V M, de Oliveira F L, Baldani V L D and Döbereiner J 1995 Biological nitrogen fixation associated with sugarcane and rice: contributions and prospects for improvement. Plant Soil 174, 195–209.

Boivin C, Camut S, Malpica C A, Truchet G and Rosenberg C 1990 *Rhizobium meliloti* genes encoding catabolism of trigonellin are induced under symbiotic conditions. Plant Cell 2, 1157–1170

Cocking E C, Webster G, Batchelor C A and Davey M R 1994 Nodulation of non-legume crops. A new look. Agro-Food-Industry Hi-Tech. January/Febuary, 21-24.

Cocking E C, Davey M R, Kothari S L, Srivastava J S, Jing Y, Ridge R W and Rolfe B G 1992 Altering the specificity control of the interaction between rhizobia and plants. Symbiosis 14, 123–130.

de Bruijn F J, Ying Y and Dazzo F B 1995 Potential and pitfalls of trying to extend symbiotic interactions of nitrogen-fixing organisms to presently non-nodulated plants, such as rice. Plant Soil 174, 225–240.

de Bruijn F 1989 The unusual symbiosis between the diazotrophic stem-nodulating bacterium *Azorhizobium caulinodans* ORS571 and its host, the tropical legume *Sesbania rostrata*. *In* Plant-Microbe Interactions: Molecular and Genetic Perspectives. Vol. 3. Ed. T Kosuge and E W Nester pp 457–504 McGraw-Hill Publishing Company.

Dénarié J, Debellé F and Promé J C 1996 *Rhizobium* lipo-chitooligosaccharide nodulation factors: signaling molecules mediating recognition and morphogenesis. Annu. Rev. Biochem. 65, 503–535.

Dénarié J and Cullimore J 1993 Lipo-chitooligosaccharide nodulation factors: a new class of signaling molecules mediating recognition and morphogenesis. Cell 74, 951–954.

Dénarié J, Debellé F and Rosenberg C 1992 Signaling and host range variation in nodulation. Annu. Rev. Microbiol. 46, 497–531.

Dolan L, Janmaat K, Linstead P, Poethig S, Roberts K and Scheres B 1993 Cellular organisation of the *Arabidopsis thaliana* root. Development 119, 71–84.

130

Donald R G K and Ludwig R A 1984 *Rhizobium* sp. strain ORS571 ammonium assimilation and nitrogen fixation. J. Bacteriol. 158, 1144–1151.

Geelen D, Mergaert P, Gememia R A, Goormachtig S, Van Montagu M and Holsters M 1993 Identification of *nodSUIJ* genes in Nod locus 1 of *Azorhizobium caulinodans*: evidence that *nodS* encodes a methyltransferase involved in Nod factor modification. Mol. Microbiol. 9, 145–154.

Geremia R A, Mergaert P, Geelen D, Van Montagu M and Holsters M 1994 The NodC protein of *Azorhizobium caulinodans* is an N-acetylglucosaminyltransferase. Proc. Natl. Acad. Sci. USA. 91, 2669–2673.

Goethals K, Van Den Eede G, Van Montagu M and Holsters M 1990 Identification and characterization of a functional *nodD* gene in *Azorhizobium caulinodans* ORS571. J. Bacteriol. 172, 2658–2666.

Goethals K, Gao M, Tomekpe K, Van Montagu M. and Holsters M 1989 Common *nodABC* genes in Nod locus 1 of *Azorhizobium caulinodans*: nucleotide sequence and plant-inducible expression. Mol. Gen. Genet. 219, 289–298.

Gough C, Webster G, Vasse J, Galera C, Batchelor C, O'Callaghan K, Davey M, Kothari S, Dénarié J and Cocking E C 1996 Specific flavonoids stimulate intercellular colonization of non-legumes by *Azorhizobium caulinodans*. *In* Biology of Plant-Microbe Interactions. Eds. G Stacey, B Mullin and P M Gresshoff. pp 409–415. Int. Soc. for Molec Plant-Microbe Interact., Minnesota.

Hartwig U A, Joseph C M and Phillips D A 1991 Flavonoids released naturally from alfalfa seeds enhance growth rate of *Rhizobium meliloti*. Plant Physiol. 95, 797–803.

James E K, Reis V M, Olivares F L, Baldani J I and Döbereiner J 1994 Infection of sugarcane by the nitrogen-fixing bacterium *Acetobacter diazotrophicus*. J. Exp. Bot. 45, 757–766.

Ladha J K and Reddy P M 1995 Extension of nitrogen fixation to rice - necessity and possibilities. GeoJournal 35, 363–372.

Leong S A Williams P H and Ditta G S 1985 Analysis of the $5'$ regulatory region of the gene for δ-aminolevulinic acid synthetase of *Rhizobium meliloti*. Nucl. Acids Res. 13, 5965–5976.

Ndoye I, De Billy F, Vasse J, Dreyfus B and Truchet G 1994 Root nodulation of *Sesbania rostrata*. J. Bacteriol. 176, 1060–1068.

Peters N K and Verma D P S 1990 Phenolic compounds as regulators of gene expression in plant-microbe interactions. Mol. Plant-Microbe Interact. 3, 4–8.

Pawlowski K, Ratet P, Schell J and de Bruijn F J 1987 Cloning and characterization of *nifA* and *ntrC* genes of the stem nodulating bacterium ORS571, the nitrogen fixing symbiont of *Sesbania rostrata*: regulation of nitrogen fixation (*nif*) genes in the free living versus symbiotic state. Mol. Gen. Genet. 206, 207–219.

Sadowsky M J, Olson E R, Foster V E, Kosslak R M and Verma D P S 1988 Two host-inducible genes of *Rhizobium fredii* and characterization of the inducing compound. J. Bacteriol. 170, 171–178.

Snedecor G and Cochran W 1980 Statistical Methods. (Ames, IA, Iowa State University Press).

Sprent J I, Sutherland J M and de Faria S M 1987 Some aspects of the biology of nitrogen-fixing organisms. Phil. Trans. R. Soc. Lond. B 317, 111–129.

Trinick M J 1973 Symbiosis between *Rhizobium* and the non-legume, *Trema aspera*. Nature 244, 459–460.

Vasse J, Frey P and Trigalet A 1995 Microscopic studies of intercellular infection and protoxylem invasion of tomato roots by *Pseudomonas solanacearum*. Mol. Plant-Microbe Interact. 8, 241–251.

Webster G, Davey M R and Cocking E C 1995 Parasponia with rhizobia: a neglected non-legume nitrogen-fixing symbiosis. AgBiotech News Info. 7, 119N–124N.

Guest editors: J K Ladha, F J de Bruijn and K A Malik

Plant and Soil **194**: 131–144, 1997.

Root morphogenesis in legumes and cereals and the effect of bacterial inoculation on root development

B.G. Rolfe[1], M.A. Djordjevic[1], J.J. Weinman[1], U. Mathesius[1], C. Pittock[1], E. Gärtner[1], K.M. Ride[1], Zhongmin Dong[2], Margaret McCully[2] and J. McIver[1]
[1]*PMI Group, RSBS, ANU, Canberra, ACT 0200, Australia* and* [2]*Biology Department, Carlton University, Ottawa, Canada*

Key words: Acetobacter, auxin, 2,4-D, clover, nitrogen fixation, *Rhizobium*, rice, transgenic plants

Abstract

Root morphology is both genetically programmed and environmentally determined. We have begun an analysis into the components of root development by: (a) constructing a range of transgenic clover plants to assess some of the genetic programs involved as both roots and nodules are initiated and develop. These transgenic plants report on auxin activity, flavonoid synthesis and chitinase expression and suggest a role for flavonoids as regulators of auxin levels; and (b) determining in cereals the effect of both added auxin and specific microorganisms on the initiation and development of modified root outgrowths and lateral roots. Appropriate combinations of auxin, the nitrogen fixing *Acetobacter diazotrophicus,* and rice variety did give rise to some plants which grew slowly for over 12 months in a nitrogen-free medium.

Abbreviations: CHS – chalcone synthase, 2,4-D – 2,4-Dichlorophenoxyacetic acid, GUS – β-glucuronidase, MRO – modified root outgrowth, NFM – nitrogen free medium, RCA – rooted cotyledon assay, RLA – rooted leaf assay, X-gluc – 5-bromo-4-chloro-3-indolyl β-D-glucuronic acid.

Introduction

Roots in terrestrial plants serve a multitude of functions: conduits to supply both nutrients and water to the plant from the soil, a location for the synthesis and the exchange of various plant hormones, storage organs of plant resources and the anchorage of the plant (Fitter, 1991; Schiefelbein and Benfey, 1991). Root systems vary widely, both within and between species, and many attempts have been made to produce classifications of the different types of root systems (Fitter, 1991). Plant roots grow in a highly heterogeneous environment - the soil, and possess an ability to react to this heterogeneity and modify the form of their root system as a consequence. The root morphology is directed by a genetic program but environmental factors largely determine the ultimate configuration of the root system. Thus, root morphology has an enormous "phenotypic plasticity" and this is a major reason for the difficulty in developing a comprehensive root classification system. However, the different root systems observed illustrate the genetic diversity that is possible in whole plant systems and that some differences are characteristic of a species (Zobel, 1991).

Root development is clearly influenced by the phytohormone auxin. The most common and widely distributed auxin in higher plants is indole-3-acetic acid (IAA). Auxins are known to influence a number of plant functions such as promotion of cell elongation and cell division, apical dominance, root initiation, differentiation of vascular tissue, ethylene biosynthesis, mediation of tropistic responses, and the alteration of the expression of specific genes (Chasan, 1993; Key, 1989; Sachs, 1993; Warren-Wilson and Warren-Wilson, 1993). Auxins move in the vascular systems and enter into cells but also are unique in that they undergo polarised transport.

In normal plant growth, it has been proposed that the initiation of a lateral root occurs during the movement of auxin from the shoots to the root tip (Tsvi

* FAX: +6162490754. E-mail: rolfe@rsbs-central.anu.edu.au

Sachs, personal communication). New meristems are initiated along this pathway of auxin flow. Auxin acts as a "correlative signal", coordinating the development and organization of tissues within the plant. Because auxin can act over great distances through the plant, carrying information about the presence and state of a specific organ and its location, it can indirectly influence many aspects of cell differentiation and specification of cell orientation. The exogenous application of auxin promotes root initiation in many plants, and it has been suggested by Sachs (1984) that the accumulation of endogenous auxin promotes root initiation. Accumulation of auxin could be due to two major factors: a localised formation of auxin or the blockage of auxin transport. One way to interfere with auxin fluxes in the root is to interrupt the auxin transport. Certain flavonoids have been mooted to be endogenous polar auxin transport regulators (Jacobs and Rubery, 1988).

The exogenous application of the synthetic auxin, 2,4-D, to either wheat or rice plants induces modified root outgrowths (MROs) which result from the induction of meristems (Rolfe and McIver, 1994). The initiation of multiple meristems in condensed zones on the root may be the result of the plant initiating sinks of auxin in order to reduce auxin levels. However, at certain specific concentrations of externally applied 2,4-D, the initiated meristems were found to be so close together that they often fused. Within the 2,4-D-induced MROs there was a regionally within the structures, and at least three cell types were present (Ridge et al., 1993). It was concluded that these induced MROs were modified lateral roots containing large amounts of starch carbon reserves rather than specialised structures. Furthermore, adding microorganisms can modulate or interfere with the development of these outgrowths (Ridge et al., 1992, 1993). Although 2,4-D may modify the frequency of lateral root formation, initiation still occurs in the pericycle and the distal limit of the primordium still terminates in a cap-like process. It was also concluded that the added synthetic auxin disturbs the orientation of cell division, causing abnormal root growth. Different strains of *Rhizobium, Azospirillum, Agrobacterium* and *Escherichia coli* were added separately with 2,4-D to wheat seedling roots, and showed different effects on seedling health and on the growth and internal structure of these modified root outgrowths. Generally, bacteria caused a less-organised internal structure to the induced-MROs and an earlier 'senescence'. Some bacteria also caused stunting and death of seedlings at concentrations of 2,4-D addition that would have no such effect alone (Ridge et al., 1992).

Bacterial association with root growth is an important new field of agricultural research. Recently, a very interesting group of nitrogen-fixing, gram negative bacteria have been isolated from the fluid of intercellular spaces of sugarcane stems and roots (Dong et al., 1994). Techniques were developed to aseptically remove the solution that filled the intercellular spaces of sugar cane. These solutions, containing about 12% sucrose, had up to 10^4 bacteria/mL of extracted fluid (Dong et al., 1994). The bacterium was shown to be a nitrogen-fixing *Acetobacter diazotrophicus*. This bacterium is a chief candidate as the producer of fixed nitrogen for the sugarcane crops where it occurs, although the details of the "symbiotic" association are still unknown (Dong et al., 1994). Moreover, these bacteria have unusual growth requirements and might form a valuable group of future endophytic bacteria, either in their natural form or in a specifically modified form for the inoculation of rice or wheat.

One of the goals of our laboratory is to study plant root morphogenesis and whether an appropriate structure in root tissues may be a prerequisite for highly successful bacterial colonization and endophytic bacterial growth. Comparative studies have been initiated on both legumes and rice plants to examine the formation of lateral roots, their position along the root axis, the factors that cause their initiation or modify their growth, and the regulation of their spacing relative to one another. The long term goal is to apply the findings on the ontogeny of legume root morphogenesis to the study of root growth on the cereal rice. To approach the study of auxin accumulation and their role in the initiation of lateral root growth we have developed, (a) a number of plant bioassays using clovers (Rolfe and McIver, 1994, 1996), and (b) specific transgenic plants carrying reporter genes, to serve as tools with which to examine the systemic physiological controls of root morphogenesis and nodule formation (Mathesius et al., 1996).

Materials and methods

Media

All plant and bacterial growth media used in this work have been previously described (Rolfe et al., 1980; Ridge et al., 1992, 1993).

Inoculation of transgenic clover plants with Rhizobium *cells*

Strains were grown overnight on BMM plates and a suspension of cells made up in sterile water to a final optical density equivalent to 1 to 3×10^8 cells per mL. A single drop of this suspension of between 1 to 2 μL was placed on the root tip of a plantlet being tested. The final bacterial titre was about 2×10^5 cells per drop. The position of the root tip at the time of inoculation was marked on the bottom of the petri dish.

Generation of rooted leaves in clovers

Rooted leaves (RLA) were produced from clovers by a modified version of the method described by Lie (1971). The two youngest leaves with several centimetres of petiole were cut from young plants, washed in water, treated with 1.5% (w/v) sodium hypochlorite for 1 min with constant shaking, and then washed thoroughly 4 times with sterile water to remove all traces of hypochlorite. Petioles were then cut to approximately 2 cm in length and the ends pushed under the agar on freshly poured Fahraeus medium (1.2% agar). After sealing the lower half of the plate with Nescofilm, plates were placed on edge in a growth chamber with high humidity and low light conditions. The photon flux density used was approximately 35 μmol m^{-2}s^{-1}, and the plate temperature 22 °C. These detached leaves were prepared and incubated without hormone treatment. Usually 5 or 6 leaves were used per Fahraeus plate to minimise potential cross contamination. Roots appeared within 6 to 8 days after cutting for white clovers, and 7 to 14 days for subterranean clovers. When the roots started to form, the rooted leaves were then transferred to new Fahraeus agar plates (1.5% agar), inoculated with a chosen *Rhizobium* strain, or treated with a particular phytohormone, and the plates were incubated vertically under a photon flux density of 300 μmol m^{-2}s^{-1} and a plate temperature of 24 °C. Leaves from young clover plants readily formed roots which were nodulated within a few days of inoculation. It was not possible to obtain plants which were sterile but contamination was minimised by using healthy young sink leaves from vigorously growing plants. Thus, it was found best to take only the two most recently emerged sink leaves as they were the least contaminated and could form roots more readily than older source leaves.

Generation and testing of transgenic clover plants

An *Agrobacterium*-mediated transformation system was used to introduce promoter:*GUS* constructs into white (Larkin et al., 1996) and subterranean clovers (Khan et al., 1994). The drug resistance of the transformants was confirmed and integration checked by Southern blotting, and the GUS activity was demonstrated using *in vivo* staining with X-gluc (Jefferson et al., 1987). Promoter:*GUS* constructs used in this study are listed in Table 1. For the study of expression of the *GH3* and *CHS1* promoters in transgenic white clover plants, rooted leaf material was used. Plant roots were fixed in 0.5% paraformaldehyde in phosphate buffer, stained with X-Gluc (G Truchet, personal comun.), embedded in 3% agarose and sectioned on a Lancer vibratome (Series 1000) in sections of 80 μm to 100 μm. For microscopic examination, a NIKON Optiphot light microscope was used.

Bacterial inoculation of rice

We have used different *Rhizobium* bacteria to inoculate rice seedlings (with or without 2,4-D in the agar): *Azorhizobium* strain ORS 571 (Dreyfus et al., 1988), IHP100 (Bender, 1988) because of their capacity for *ex planta* nitrogen fixation, and strain NGR234 (Trinick, 1980) because it has the broadest host range of all tested rhizobia. Furthermore, strain NGR234 was shown to respond to chemical signals isolated from the roots of both wheat (le Strange et al., 1990) and rice seedlings (Rolfe, unpublished) and grow in the rhizosphere of the roots of rice plants (Ridge et al., 1993).

Several plant assay procedures were used; large test tubes containing Fahraeus F agar or perlite plus 10 mls of liquid F medium; large agar plates containing nitrogen-free medium (NFM) were also used. All plants were grown in growth cabinets with a photon flux density of 500 μmol m^{-2}s^{-1} using a 16 hour day cycle. Where seedlings were inoculated with tagged rhizobia they were checked after two weeks to examine the success of their root association and the level of invasion into root tissues. The general overall appearance, vigour and size of the rice plants was assessed and three replicates were done for each test strain. Controls consisted of seedlings inoculated with half strength F medium either alone or containing 5 mM KNO$_3$.

In one set of experiments studying root growth and lateral root formation on rice cultivars, the bacterial cultures were grown in liquid BMM in flasks on a shaker for 2 days at 28 °C. Seedlings were added to

Table 1. Promoter: *GUS* constructs used in this study.

promoter	plant source	T-DNA vector	3'-terminator	comment/features	reference
CHS1	subterranean clover (Arioli et al., 1994)	pTAB10 (Khan et al., 1994)	pea vicilin 3'-sequence	1.2 kb of *CHS1* sequence upstream of an engineered *Nco*I site spanning the *CHS1* initiator fused to a *Nco*I site spanning the *GUS* initiator methionine.	This work
GH3	soybean (Hagen et al., 1991)	pTAB10	pea vicilin 3'-sequence	749 bp of *GH3* sequence upstream of a *Nco*I site spanning the *GH3* initiator methionine fused to a *Nco*I site spanning the *GUS* initiator methionine. Auxin responsive.	Larkin et al., 1996
basic chitinase	*tabacum* (Neale et al., 1990)	pTAB10	*tumefaciens NOS* 3'-sequence	1764 bp of promoter sequence PCR engineered and cloned in front of the *GUS* gene of pBI101.3 (Clontech) was a gift of Alan Neale. The T-DNA cassette from this was then sub-cloned into pTAB10.	This work

the flasks (which were returned to the shaker for 2-4 h) before being planted as described above.

Inoculation of rice roots with genetically marked strains

Several of the *Rhizobium* strains used to inoculate rice seedlings show some degree of differential association with some rice cultivars. Each of these strains has been labelled with a reporter gene so that it can be used to follow the location of these bacteria on the surface of the roots of rice seedlings. The labelling was also used to monitor the degree of bacterial penetration into the internal tissues of the roots. The reporter gene system used is the *GUS* (the *E. coli uidA*) gene encoding the enzyme β-glucuronidase which

hydrolyses a wide variety of synthetic glucuronides in a colour forming reaction. The different strains have been labelled with the transposon mTn5SS*gusA*$_{20}$ which has the *uidA* gene driven from the promoter that drives the kanamycin resistance gene in Tn5 (Wilson, 1995). These tagged "screening" rhizobia were tested in experiments using the short grain rice cultivar Calrose.

Analysis of the root development of Australian short and long grain rice varieties

Plant tests

Seeds of the different rice cultivars were obtained from the Australian Ricegrowers' Co-operative Limited, Leeton. Short grain (japonica) rice varieties Calrose, Echuca, Bahia, Bogan and Amaroo and long grain (indica) rice varieties Pelde and Doongara were dehusked by grinding lightly with a pestle in a mortar. Seeds were washed in 95% ethanol, rinsed three times with sterile distilled water (SDW), soaked in 6% (w/v available chlorine) sodium hypochlorite (commercial grade) with a drop of Tween 20 in flasks for 1 h in a shaker at a speed that maintained swirling in the flasks. The seeds were then washed in SDW and left to imbibe for 1-2 h on the shaker. They were again washed in SDW and placed on BMM medium (Rolfe et al., 1980) plates at 28 °C in the dark for 2 days to germinate. BMM was used to reveal any residual microbial contamination.

Germinated seedlings were transferred to large Petri dishes containing 1.5% agar (made with a half-strength Fahraeus solution) plus or minus 2,4-D. Each dish contained 4 seedlings fixed to the surface of the agar with a drop of molten sterile 2% water agar. The Petri dishes were sealed with Nescofilm, small holes were added for ventilation and placed vertically in a growth chamber (16 h day, 27 °C, 500 μmol m^{-2} s^{-1} photon flux density).

Measurements and statistical analysis

All rice plant tests were done at the same time under the same conditions. Petri dishes in the growth cabinet were placed in racks in a random order according to a table of random numbers. Cultivar and treatment was randomly allocated to 35 dishes within three replicate blocks, 4 plants grown in each plate. The root of each seedling was measured before and after the experimental period. After five days of growth, the plants were removed from the Petri dishes and photocopied with a black background at 200% enlargement. The photocopies were used to record and measure root length and lateral root incidence. Lateral root numbers and root length were examined with or without 2,4-D or ORS 571 addition. Analysis of variance (Genstat 5 Release 1.3 on a Vax/VMS4) was used to calculate the means of the square root of the number of roots, and analysis of co variance was used to calculate the means of the log of the final length adjusted for initial length.

Testing of a putative rice endophyte

Different rice cultivars were inoculated with the nitrogen-fixing bacterium *Acetobacter diazotrophicus* strain PAL 5 (Dong et al., 1994), with and without the addition of 2,4-D, and measurements were taken of plant growth stimulation. Root association of the chosen endophytic strain was examined by microscopy and plating out the root washings for colony formation. The influence of added auxin (2,4-D) to the enhancement of this association was also tested.

Results

Studies on root morphogenesis in legumes: Bioassay for the analysis of transformed clover plants

Transformed white clovers can be rapidly grown and clonal material amplified by a variety of vegetative propagation methods. A simple bioassay can be generated from plant runners for the analysis of these plants. The rooted leaf assay (RLA) (Rolfe and McIver, 1996) was developed so that the symbiotic and physiological properties of large numbers of transgenic plants could be studied. This is a rapid technique that can use either non-transformed or transformed plant material, and avoids the potential problem of a disturbed hormone balance that is inherent in the use of "hairy roots" transformed with *A. rhizogenes*. Many rooted leaves can be made from each plant and grown on agar plates to give access to genetically-identical biological material. The rooted leaves have no shoot meristem or cotyledons and therefore enable an analysis to be made of root morphogenesis without either of these plant tissues. In spite of this, the rapidly produced rooted leaves still form lateral roots and nodules and can be examined for their root responses under a variety of treatments. Three types of leaves were examined for rooting and nodulation: young unopened folded, young opened and developing, and mature leaves. The mature leaves usually failed to root, while the young unopened folded and the young opened and developing leaves readily rooted and could be nodulated with various *Rhizobium* strains.

The use of transgenic clover plants to investigate endogenous auxin activity in root tissues

A study of the cellular and molecular events occurring during the formation of spontaneous or auxin-induced lateral roots in radish and *Arabidopsis* showed that it was a two stage process (Sussex et al., 1995). First, a primordium was formed and subsequently a subset of primordial cells took over as the lateral root apical meristem. We examined this process using a set of transgenic clover plants carrying different reporter genes. An indication of the activity of auxin in clover plants during lateral root formation was obtained using transgenic clovers containing a *GH3:GUS* gene fusion. The *GH3* region of the construct is an auxin responsive promoter (Hagen et al., 1991) isolated from soybean. *GH3* expression correlates with the proposed internal auxin concentrations and cell susceptibility to auxin. In white clover, *GH3* expression occurs in vascular bundles in the root, where auxin is transported, rapidly shifts to the bottom side of gravistimulated roots following lateral auxin transport and is specifically up regulated from endogenous levels by auxin addition (Larkin et al., 1996).

As auxin has been found to be involved in lateral root initiation (Whiteman et al., 1980), we investigated the inferred cellular distribution of auxin during this process using plants containing the *GH3:GUS* transgene. Transverse sections of roots during different stages of lateral root development showed that *GH3:GUS* expression occurs in pericycle cells in front of xylem elements before cell division (Figure 1). Subsequently, dividing pericycle cells are also stained, accompanied by increased staining around the respective xylem element. After the initial primordium formation in the pericycle, the dividing cells no longer show expression of the *GH3:GUS* gene construct. Instead, cells in the outer cortex in front of the growing lateral primordium express *GH3:GUS*. The *GH3:GUS* reporter gene expresses consistently in these outer cortical cells of the root and effectively marks the site where a lateral root has initiated beneath. A developing lateral root invariably grows towards, and through, this area of *GH3:GUS* expression (Figure 2). However, there is no expression in the outer cortical cells directly opposite the pericycle cells where an initial expression of the GH3:GUS reporter gene has occurred. During lateral root elongation, *GH3:GUS* expression occurs in the differentiating vascular bundles.

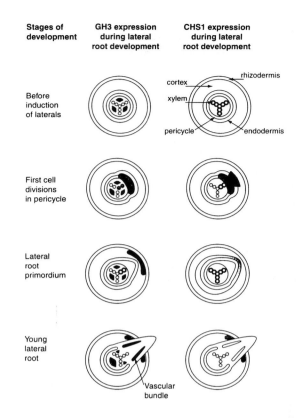

Figure 1. Transgenic clover plants. Comparison of *GH3* and *CHS1* expression during different stages of lateral root development. Cells expressing the promoter, as determined by the activity of the fused *GUS* reporter gene, are shaded. The results are discussed in the text.

Transgenic clover plants to investigate flavonoid gene expression during lateral root formation

Our earlier studies had shown that physical wounding and *Rhizobium* inoculation of clovers results in a rapid induction of the chalcone synthase gene(s) which codes for the first enzyme of the flavonoid pathway (Lawson et al., 1994). Using the *CHS1:GUS* gene construct we found that *CHS1* expression occurs in the cortex and vascular bundle, including phloem, pericycle and xylem parenchyma in the roots of transgenic white clovers. Expression also occurs in front of the xylem poles in endodermis and inner cortex cells. Dividing pericycle cells and very early lateral primordia also express the *CHS1:GUS* gene fusion, whereas expression stops in primordia reaching a size of 0.3 - 0.5 of the tap root diameter. The *CHS1:GUS* gene expression is increased around the xylem elements of the pole in front of the developing lateral root primordium (Figure 1). Later, this expression occurs in front of

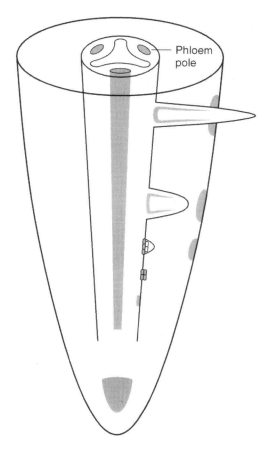

Figure 2. Transgenic clover plants containing the GH₃: GUS reporter. Summary of the expression of the GH3:GUS gene fusion in the different tissues during root morphogenesis. The reporter gene was used to examine endogenous auxin presence. Cells expressing the promoter, as determined by the activity of the fused *GUS* reporter gene, are shaded. Elevated levels of auxin were detected first in the pericycle cells before division and in the early dividing cells of the primordium. Only after the lateral root apical meristem has formed can the elevated levels of auxin be detected in the outer cortical cells which now serves as a marker for where lateral roots will form on the main root.

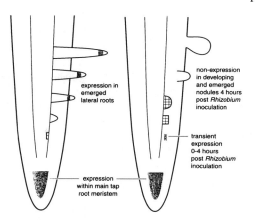

Figure 3. Transgenic clover plants containing the chitinase-reporter. Expression of the basic-chitinase:*GUS* construct during different stages of lateral root development and nodule formation. Cells expressing the promoter, as determined by the activity of the fused *GUS* reporter gene, are shaded. The results are discussed in the text.

the growing lateral primordium in the cortex forming a clear blue patch of stained material and remains in the cortex cells after the lateral root emerges from the tap root. The lateral root invariably grows towards and through the same cell zone in the outer cortex which also showed *GH3:GUS* activity (Figures 1 and 2).

Transgenic clover plants to investigate chitinase gene expression during lateral root and nodule formation

A tobacco basic-chitinase promoter:*GUS* gene (Neale et al., 1990) construct was transformed into clovers to test its expression pattern and the effect of various stimuli on its expression. Expression of this construct was found to show comparable developmental expression patterns in tobacco and clovers: high expression in root tips and very low expression in healthy aerial tissues. Wounding in all tissues tested induced expression to high levels. The similar expression of this construct in a heterologous system provides a basis for its use as a tool for the investigation of molecular events in clovers during infection, symbiosis and pathogenesis. Expression of the reporter construct was observed during the early stages of the emerging lateral root and the subsequent root but not in the lateral root primordia before it emerges from the root (Figure 3). In contrast, experiments using the same transgenic clovers to study *Rhizobium*-clover interactions found no significant expression in nodules over a period of 1 day to 14 days post-infection. However, *Rhizobium* infection induces a temporal expression in the cortical cells in the zone of infection between 1 and 4 hours post-infection which was either absent or could not be detected thereafter.

Studies on root morphogenesis in rice: Development of root systems on rice cultivars

There are three major groups of rice: indica, japonica and javanica and there is only limited literature available on the differences between the roots of the various rice cultivars (Tsunoda and Takahashi, 1984; Oka, 1988). Thus, we have chosen representatives of the two

AVERAGE ROOT LENGTHS FOR RICE VARIETIES (5d Growth)

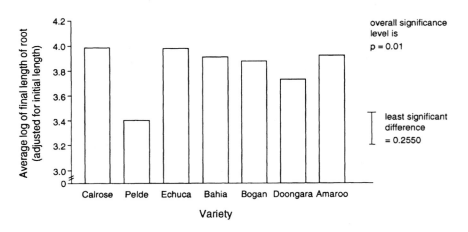

Figure 4. Average root lengths measured for seven Australian rice varieties after five days of growth.

AVERAGE NUMBER OF LATERAL ROOTS FOR RICE VARIETIES (5d Growth)

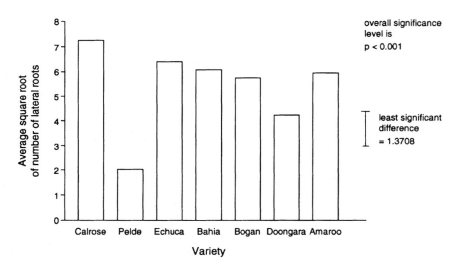

Figure 5. Average number of lateral roots formed by seven rice varieties after five days of growth.

most disparate rice groups, indica (long grained rice) and japonica (short grained rice), to initiate an investigation of rice cultivar root systems. Of the two long grain varieties, only Pelde shows statistically significant differences to the short grain varieties (Figures 4 and 5) for both root length and average number of lateral roots. Even though the length of the root is shorter for Pelde, it also shows a reduced number of lateral roots per unit of root length. Pelde also shows a significant difference to the other long grain tested, Doongara, for

lateral root number. Doongara shows significant differences to the short grains for the number of lateral roots (Figure 5) but not root length (Figure 4).

Within the short grain group, Bogan was found to be significantly different to Calrose on the average number of lateral roots (Figure 4). The other varieties show slight differences of means, but none are statistically significant. All varieties tested show a significant response to 10^{-5} M 2,4-D for root length (Figure 6) and a significant response to 10^{-5} M and 10^{-6} M 2,4-D for

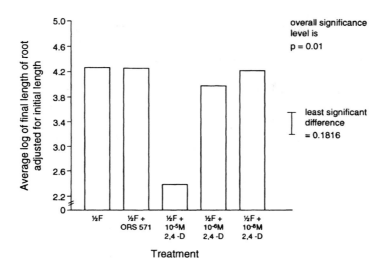

AVERAGE ROOT LENGTHS AFTER TREATMENT WITH
2,4 -D AND ORS 571 FOR ALL RICE VARIETIES (5d Growth)

Figure 6. Average root lengths measured after five days of growth for seven rice varieties treated with auxin 2,4-D or *Rhizobium* strain ORS 571. 1/2 F is half concentration of F-medium.

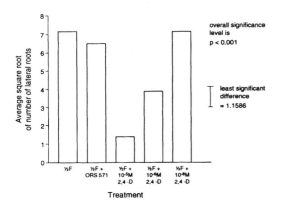

AVERAGE NUMBER OF LATERAL ROOTS AFTER TREATMENT WITH
2,4 -D AND ORS 571 FOR ALL RICE VARIETIES (5d Growth)

Figure 7. Number of lateral roots measured after five days of growth for seven rice varieties treated with auxin 2,4-D or *Rhizobium* strain ORS 571.

average lateral root number (Figure 7) when compared to the control.

Modified root outgrowths are induced on rice varieties Calrose and Pelde using the synthetic phytohormone 2,4-D in a concentration range of 2.5×10^{-6} M to 1×10^{-5} M (Ridge et al., 1993). These MROs appear to result from the fusion of multiple meristems induced in very close proximity to each other. The 2,4-D also induces a range of other phenomena, including various levels of fusion of primordia and MRO structures with a callus-like surface instead of an epidermis, but with a differentiated internal anatomy. Although these spheroid MROs are similar in external morphology to determinate legume nodules they show no internal infection or colonisation by *Rhizobium* strain NGR234 or the *Azorhizobium* strain ORS 571. However, on the rice variety Calrose, the strain ORS 571 sometimes extensively colonises the surface of the MROs in comparison to levels found on emerging lateral roots and to the root surface of the same plants. The strain NGR234 does not colonise the surface of the MROs.

Inoculation of rice cultivars with Rhizobium *strains*

The initial experiments to examine the colonization profiles of marked bacterial strains along the roots of cultivar Calrose were done with plants also treated with 2,4-D. The results showed that plants treated with 2,4-D developed root structures which outwardly resemble nodules or tumour outgrowths. The upper part of the main roots still have emerging lateral roots but these often develop into short stunted structures. However, further examination showed them to be MROs. The most prolific formation of MRO structures occurs at the root tip region of the auxin-treated plants. The genetically marked strains colonize the roots of these auxin-treated rice seedlings. However, this colonization is not evenly distributed along the roots. When the rice roots were stained for GUS activity (indication of bacterial colonization), some regions of the root showed intense blue staining while other areas had no activity at all. The greatest colonization is associated with the 2,4-D induced MROs, particularly at the root tip region.

Cytological studies showed that these root outgrowths have an epidermis, an outer cortex-layer and a single central vascular bundle which is connected to the central vascular system of the main root. The OR571::*GUS* marked strain colonises the surface epidermal tissues of the 2,4-D induced structures. This generally appears to be an intercellular association but occasionally some intracellular occupancy seems to occur but these could be invaded dying epidermal cells. However, the results are encouraging and suggest that some strains might succeed in colonizing these modified root outgrowths.

Inoculation of rice cultivars with Acetobacter diazotrophicus

The first series of experiments involved using germinated seedlings of the rice varieties Calrose and Pelde inoculated with broth cultures of *Acetobacter diazotrophicus* strain PAL 5 and their subsequent incubation for two weeks in large test tubes containing an agar slope of nitrogen-free-medium (NFM). Both the roots and agar surfaces were examined by microscopy. Root washings were plated for bacterial growth. All the roots of cv. Calrose were covered with bacterial growth while the majority of the roots of cv. Pelde had either no apparent bacterial growth or a limited association. Washings from cv. Calrose contained very high numbers of bacteria (later identified as *Acetobacter*) while those from cv. Pelde had far fewer bacterial cells present.

In another series of inoculation experiments the synthetic auxin 2,4-D was added at 2.5×10^{-6} M to the agar and the plants examined. After two weeks, it was found that the roots of the cv. Pelde had enhanced numbers of *Acetobacter* in the presence of 2,4-D. In contrast, 2,4-D had no effect upon the numbers of *Acetobacter* colonising the roots of cv. Calrose. Thus, it appeared that the added auxin could enhanced some bacterial-rice associations but not others over a relatively short incubation time (2 weeks).

In the next series of experiments the same level of synthetic auxin was added to induce MROs on rice roots (Ridge et al., 1993) and the rice plants were also inoculated with *Acetobacter*. The experiments were run for longer than the two week period used above to assess the long term effects. The results showed that some of the roots of the rice varieties. Calrose and Pelde, seemed to become closely associated with the inoculum *Acetobacter* strain, since these rice plants continued to grow slowly on a N-free-medium when compared with the uninoculated controls and plants inoculated with *Acetobacter* but without the added auxin. About every 6 weeks the inoculated plants were carefully transferred to fresh sterile plates containing N-free medium. Excess dead or dying leaves were removed and plates carefully resealed and placed in the growth chamber. Of the original inoculated plants about 12% continued to grow slowly and survive on NFM for over 12 months, with some plants even setting seed. As expected, uninoculated controls died off between 4 and 6 weeks after the start of the experiment. These initial experiments of inoculating rice seedlings with *Acetobacter diazotrophicus* have indicated that the cv. Pelde might form the most successful long term plant-microbe association with this bacterium. Recently, the roots of these 12 month old plants were microscopically examined after they were either stained with toluidine blue or treated with periodic-acid Schiff's reaction and then stained with DAPI (Gochnauer et al., 1989) (Figure 8). Interestingly, there were many tiny branched roots, of about 125 μm or less in diameter, originating from the larger nodal roots (Figure 8b). These tiny roots had one layer of epidermis, one layer of cortical cells and an endodermis and stele all lignified. The bacteria which colonised the tiny roots on their outer surfaces were largely in the groves between adjacent epidermal cells. After numerous sections and examinations it was concluded that there were no or few bacteria within the roots. On the older nodal roots the colonising bacteria appear to clump more readily and to be covered with a coating possibly of polysaccharide origin (Figure 8e and f). These experiments are an interesting start but whether the basis of the plant growth was due to nitrogen-fixation or some kind of plant growth promoting association or a combination of the both is still unknown.

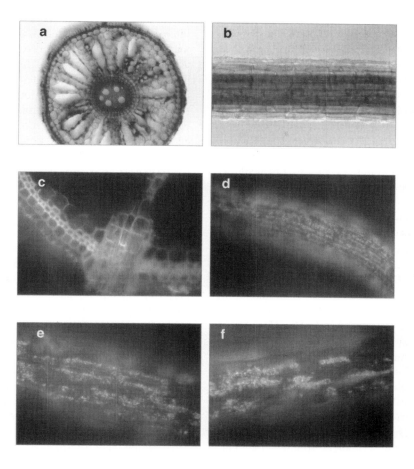

Figure 8. Staining and sections of rice roots from plants previously treated with 2,4-D and inoculated with *Acetobacter diazotrophicus* and grown continuously on N-free medium for over 12 months. Fig. **a** is a transverse section of a nodal root stained with toluidine blue and viewed with bright field; **b** a tiny branched root stained with toluidine blue and colonised with bacteria; **c** a transverse section of one of the nodal roots with its epidermis, hypodermis and two underlying suberised layers and the aerenchyma space, showing the exit point of one of the tiny branched roots and showing the tight junction. The bacteria show up on the epidermal surface because of the DAPI staining used; **d** whole mount of a branched root, DAPI stained, showing the bacteria lined up along the groves between adjacent epidermal cells; **e** higher magnification of a branched root bacteria still mainly in the groves; **f** bacterial distribution on the surface of older branched roots where some of the layers have been lost and the bacteria appear to be more clumped.

Discussion

Roots provide excellent material to serve as a model for examining the molecular events required for organ development in plants (Dolan et al., 1993; Benfey and Schiefelbein, 1994). Because of the difficulties of interpreting the effects of exogenously added phytohormones on plant responses, Schiefelbein and Benfey (1991) proposed the use of transgenic reporter plants with modified hormone ratios as an alternative approach to analyse root development. We have developed the rooted leaf bioassay (RLA) so that the symbiotic and plant physiological properties of large numbers of transgenic plants could be studied. Many genetically-identical rooted leaves can be made from each transgenic plant thus enabling large numbers of assays to be done rapidly. The RLA amplifies a large amount of transgenic plant material and thus provides a statistically ideal system for studies of root morphogenesis.

In this study we have used a series of transgenic clovers to examine aspects of auxin activity, expression of phenylpropanoid pathway genes and a basic-chitinase gene during root morphogenesis. The transgenic clovers used contained the *GUS* reporter gene fused to either the soybean *GH3* (auxin responsive) promoter (Larkin et al., 1996), to a clover *CHS* (chalcone synthase) promoter (Arioli et al., 1993; Howles et al., 1994), or to a tobacco basic-chitinase promoter (Neale et al., 1990). Recently we have shown that there are three characteristics of the expression of the *GH3:GUS* construct in clovers which indicate that the use of this genetic tool is both a reliable and rapid assay of auxin mediated responses in the transgenic plants (Larkin et al., 1996). The *GH3:GUS* expression in untreated transgenic white clover is consistently induced in the outer cortex of the root effectively marking the site of an initiated lateral root beneath. A developing lateral root invariably grows towards, and through, this area of *GH3:GUS* expression. In addition, the *GH3:GUS* is expressed within 60-90 minutes on the non-elongating side of gravistimulated roots concurrently with root curvature, in accordance with the proposed role of auxin in gravistimulation. Finally, expression of the *GH3:GUS* constructs is significantly enhanced over endogenous levels by the addition of auxin but not other phytohormones (Larkin et al., 1996). The influence of auxin on lateral root formation is firmly established (Whiteman et al., 1980; Hirsch, 1992), and if we assume that the *GH3:GUS* expression reflects locations of auxin action, we conclude that auxin levels have to be elevated in pericycle cells triggered for division for lateral root primordia as well as during the initial cell divisions of this primordium (Whiteman et al., 1980; Duditz et al., 1993; Sussex et al., 1995). During later stages of lateral root development the requirement for auxin seems to be reduced. The *GH3:GUS* expression in the developing vascular bundle cells underlines the action of auxin during vascular bundle formation (Sachs et al., 1993). The characteristic staining in the cortex in front of the growing lateral root might indicate cell division in the outer cortex cells to compensate for damage during the passage of the lateral root though the tap root (Bell and McCully, 1970).

A comparison of *GH3:GUS* with *CHS1:GUS* expression during different stages of lateral root formation shows that the expression patterns overlap in many cell types, especially around the xylem poles and in the first dividing cells of the primordium. This observation indicates a possible interaction of flavonoids, resulting from chalcone synthase activity, with auxin transport or uptake in those tissues. In particular, increased *CHS* expression around the xylem poles could result in flavonoid accumulation that may interfere with the auxin transport in the vascular bundle (Jacobs and Rubery, 1988).

The studies with transgenic plants containing the basic-chitinase promoter:*GUS* construct show a very interesting and unexpected result with the indication that there is a transitory expression of this gene before any indication of cortical cell division can be detected. This could be indicating a role of hydrolytic enzymes in preparing cortical cells before their division (Ashford and McCully, 1970; Bell and McCully, 1970). The gene is expressed in the tap root meristem and detected in the meristematic cells of emerged lateral roots. However, as we were unable to observe the expression of this gene in young pre- emerging lateral root primordia, it provides us with another genetic marker in the early differences between the developmental ontogeny of lateral roots and nodules.

The exogenous application of the synthetic auxin, 2,4-D, to rice plants induces modified root outgrowths (MROs). These are probably modified lateral roots, rather than specialised structures, with carbon reserves (as starch in amyloplasts) similar to those found in the cortex or roots. They result from the induction of meristems initiated at or close to the root pericycle. The initiation of multiple meristems in condensed zones on the root may be the result of the plant initiating sinks of auxin in order to reduce auxin levels. However, at certain specific concentrations of externally applied 2,4-D, the initiated meristems are found to be so close together that they often fuse and may result from the inter-lateral root spacing being reduced to zero. Within the 2,4-D-induced MROs there is a well-defined internal structure and at least three cell types are present: a cell division zone, amyloplast (starch-containing) cells, and highly vacuolated cells (Ridge et al., 1993). While vascular tissue enters the MROs, there is no organised vascular system as is seen in *Rhizobium*-induced nodules on legumes. If new meristems form where auxin accumulates and xylem tissue develops along auxin flow pathways, a lateral root would presumably have only a polarised and single auxin flow path, and as a result have a single vascular stele. Some types of legume nodules are thus likely to be the result of the stimulation of multiple meristems that contribute both to the body of the nodule by separate zones of peripheral division, and to the development of a number of vascular bundles arranged around the nodule. Thus, the formation of a site of multiple meristems enables nodules to have multiple vascular bundles at an early stage in development.

Scant literature is available about the differences of root formation in the various cultivars of rice, we have studied representatives of the two most disparate groups, indica (long grain) and japonica (short grain) rice to investigate root and lateral root growth. Our initial experiments on rice root growth show that there are statistically significant differences in the timing and pattern of the development of tap and lateral roots between differing Australian rice varieties. Some of these differences cross the major group line between indica and japonica types, but we have also shown that there are significant differences within the groups. To examine the potential plasticity of these root systems, we have started to look at how different environmental conditions such as growth temperatures and growth media effect the initiation and emergence of lateral roots.

A number of attempts have been made to use *Acetobacter diazotrophicus* as a single inoculum of sugarcane, sorgum and sugar beet but these were reported as failures (Dobereiner et al., 1993). Thus, we decided to use a different approach with this group of bacteria using rice cultivars inoculated with both *Acetobacter* and the auxin 2,4-D at the same time or separately. The initial experiments have suggested that *Acetobacter* may associate beneficially with the synthetic auxin-treated rice seedlings which form MROs. The results raise many questions that we can not answer at the moment but are encouraging and suggest that *Acetobacter* might succeed in colonizing these modified root outgrowths or plants possessing these outgrowths. *Acetobacter diazotrophicus* was shown to be a nitrogen-fixing bacterium which could grow on medium containing 10% or more sucrose and that a tolerance of its nitrogenase activity to oxygen exists in aerobically grown cells (Dong et al., 1994). This bacterium is the only member of the *Acetobacteriaceae* able to fix nitrogen and is a prime suspect as the producer of fixed nitrogen for sugar cane crops where it occurs in Cuba, Brazil and Australia and the estimated levels of nitrogen fixed are greater than 110 kg per ha per year (Dong et al., 1994; J.K. Ladha, personal comun.). Moreover, it has a wide range of desirable characteristics necessary for a possible inoculum as nitrogen-fixing endophyte of rice. The bacterium possesses no nitrate reductase, and N-fixation is not affected by high levels (30 mM) of nitrate. Also, ammonium causes only partial inhibition of nitrogenase (Cavalcante and Dobereiner, 1988). This incomplete inhibition of N-fixation by ammonium in the bacterium, as well as the lack of nitrate reductase, may make it particularly effective in fixing nitrogen in rice fields that have already received nitrogen fertiliser. Thus, this bacterium might form part of a valuable group of endophytic bacteria, either in their natural form or in a specifically modified form, to be used as rice inoculum strains.

Our studies involving the exogenous addition of 2,4-D to promote altered root morphogenesis and ultimately the potential association of endophytic bacteria is revealing some of the innate developmental programs of plant roots. The formation of the tiny branched roots was totally unexpected. This type of information will be necessary if these root developmental programs are to be experimentally manipulated by transgenic means in the future.

Acknowledgments

The authors thank the Rockefeller Foundation for support, Ms D Barnard for excellent technical assistance, Dr RW Ridge for useful discussions, and Alan Neale for his gift of the basic chitinase:*GUS* reporter construct.

References

Arioli T, Howles P A Weinman J J and Rolfe B G 1994 In *Trifolium subterraneum*, chalcone synthase is encoded by a multigene family. Gene 138, 79–86.

Ashford A E and McCully M E 1970 Locatization of Naphthol AS-B1 phosphatase activity in lateral and main root meristems of pea and corn. Protoplasma 70, 441–456.

Bell J K and McCully M E 1970 A Histological Study of lateral root initiation and development in *Zea mays*. Protoplasma 70, 179–205.

Bender G L 1988 PhD Thesis, The genetic basis of *Rhizobium* host range extension to the non-legume *Parasponia*. Australian National University.

Benfey P N and Schiefelbein J W 1994 Getting to the root of plant development: the genetics of *Arabidopsis* root formation. TIG 10, 84–88.

Cavalcante V A and Dobereiner J 1988 A new acid-tolerant nitrogen fixing bacterium associated with sugarcane. Plant Soil 108, 23–31.

Chasan R 1993 Embryogenesis: new molecular insights. The Plant Cell 5, 597–599.

Dobereiner J, Reis V M, Paula M A and Olivares F 1993 Endophytic diazotrophs in sugar cane, cereals and tuber plants. *In* New Horizons in Nitrogen Fixation. Eds. R Palacios, Mora, W Newton. pp. 671–676. Kluwer Academic Publishers, Dordrecht, The Netherlands.

Dolan L, Janmaat K, Willemsen V, Linstead P, Poethig S, Roberts K and Scheres B 1993 Cellular organisation of the *Arabidopsis thaliana* root. Development 119, 71–84.

Dong Z, Canny M J, McCully M E, Roboredo M R, Cabadilla C F, Ortega E and Rodes R 1994 A nitrogen-fixing endophyte of sugarcane stems. Plant Physiol. 105, 1139–1147.

Dreyfus B, Garcia J L and Gillis M 1988 Characterization of *Azorhizobium caulinodans*. Gen. Nov. sp. Nov, a stem-nodulating nitrogen-fixing bacterium isolated from *Sesbania rostrata*. Int. Systemic Bacteriol. 38, 89–98.

144

Duditz D, Börge L, Bakó L, Dedeoglu D, Magyar Z, Kapros T, Felföldi F and Györgyey J 1993 Key components in cell cycle control during auxin-induced cell division. In: Molecular and cell biology of the cell cycle. Eds. J C Ormrod and D Francis. pp 111–131. Kluwer Academic Publishers, Dordrecht, The Netherlands.

Fitter A H 1991 Characteristics and functions of root systems. *In* Plant Roots: The Hidden Half. Eds. Y Waisel, A Eshel and U Kafkafi. pp. 3–25. Marcel Dekker, Inc., New York.

Gallagher S R 1992 GUS protocols: Using the GUS gene as a reporter of gene expression. Academic Press.

Gochnauer M B, McCully M E and Labbe H 1989 Different populations of bacteria associated with sheathed and bare regions of roots of field-grown maize. Plant Soil 114, 107–120.

Hagen G, Martin G, Li Y and Guilfoyle T J 1991 Auxin-induced expression of the soybean GH3 promoter in transgenic tobacco plants. Plant Molec. Biol. 17, 567–569.

Hirsch A M 1992 Developmental biology of legume nodulation. New Phytol. 122, 211–237.

Howles P A, Arioli T and Weinman J J 1994 Characterization of a phenylalanine ammonia-lyase multigene family in *Trifolium subteraneum.* Gene 138, 87–92.

Jacobs M and Rubery P H 1988 Naturally occurring auxin transport regulators. Science 241, 346–349.

Jefferson R A, Kavanagh T A and Bevan M W 1987 GUS fusions: beta-glucuronidase as a sensitive and versatile gene fusion marker in higher plants. EMBO J. 6, 3901–3907.

Key J L 1989 Modulation of gene expression by auxin. BioEssays 11, 52–57.

Khan M R I, Tabe L M, Heath L C, Spencer D and Higgins T J V 1994 *Agrobacterium*-mediated transformation of subterranean clover (*Trifolium subterraneum* L.). Plant Physiol. 105, 81–88.

Larkin P J, Gibson J M, Mathesius U, Weinman J J, Gärtner E, Hall E, Tanner G J, Rolfe B G and Djordjevic M A 1996 Transgenic white clover. Studies with the auxin responsive promoter, GH3, in root gravitropism and secondary root development. Transgenic Research 5, 325–335.

Lawson C G R, Djordjevic M A, Weinman J J and Rolfe B G 1994 *Rhizobium* inoculation and physical wounding results in the rapid induction of the same chalcone synthase copy in *Trifolium subterraneum.* Molec. Plant-Microbe Interact. 7, 498–507.

le Strange K K, Bender G L, Djordjevic M A, Rolfe B G and Redmond J W 1990 The *Rhizobium* strain NGR234 *nodD*1 gene product responds to activation by the simple phenolic compounds vanillin and isovanillin present in wheat seedling extracts. Molec. Plant-Microbe Interact. 3, 214–220.

Lie T A 1971 Nodulation of rooted leaves in leguminous plants. Plant and Soil 34, 663–673.

Mathesius U, Schlaman H R M, Meijer D, Lugtenberg B J J, Spaink H P, Weinman J J, Roddam L F, Sautter C, Rolfe B G and Djordjevic M A 1996 New tools for investigating nodule initiation and ontogeny: Spot inoculation and microtargeting of transgenic white clover roots shows auxin involvement and suggests a role for flavonoids. 8th International Congress Molecular Plant-Microbe Interactions, Knoxville, TN.

Neale A D, Wahleithner J A, Lund M, Bonnett H T, Kelly A, Meeks-Wagner D R, Peacock W J and Dennis E S 1990 Chitinase, β-1,3-glucanase, osmotin, and extensin are expressed in tobacco explants during flower formation. The Plant Cell 2, 673–684.

Oka H I 1988 Origin of cultivated rice, Japan Scientific Societies Press/Elsevier, Tokyo.

Ridge R W, Bender G L and Rolfe B G 1992 Nodule-like structures induced on the roots of wheat seedlings by the addition of the synthetic auxin 2,4-Dichlorophenoxyacetic acid and the effects of microorganisms. Aust. J. Plant Physiol, 19, 481–492.

Ridge R W, Ride K M and Rolfe B G 1993 Nodule-like structures induced on the roots of rice seedlings by addition of the synthetic auxin 2,4-Dichlorophenoxyacetic acid. Aust. J. Plant Physiol., 20, 705–717.

Rolfe B G and McIver J M 1994 The rooted cotyledon and rooted leaf bioassay. Proc. 1st European Nitrogen Fixation Conference, Szeged, Hungary, 122–126.

Rolfe B G and McIver J M 1996 Single-leaf plant bioassays for the study of root morphogenesis and *Rhizobium*-legume nodulation. Aust. J. Plant Physiol., 23, 271–283.

Sachs T 1993 The role of auxin in the polar organization of apical meristems. Aust. J. Plant Physiol. 20, 541–553.

Sachs T 1984 Positional Controls in Plant Development. Eds. P W Barlow and D J Carr. pp. 193–224. Cambridge University Press, Cambridge, UK .

Schiefelbein J W and Benfey P N 1991 The development of plant roots: New approaches to underground problems. The Plant Cell 3, 1147–1154.

Sussex I M, Godoy J A, Kerk N M, Laskowski M J, Nusbaum H C, Welsch J A and Williams M E 1995 Cellular and molecular events in a newly organizing lateral root meristem. Phil. Trans. R. Soc. Lond. B 350, 39–43.

Trinick M J 1980 Relations among the fast-growing rhizobia of *Lablab purpureus, Leucaena leucocephala, Mimosa* spp., *Acacia farnesiana* and *Sesbania grandiflora* and their affinities with other rhizobial groups. J. Appl. Bacteriol. 49, 39–53.

Tsunoda S and Takahashi N 1984 Biology of rice, Japan Scientific Societies Press, Elsevier, Tokyo.

Warren-Wilson J and Warren-Wilson P M 1993 Mechanisms of auxin regulation of structural and physiologic polarity in plants, tissues, cells and embryos. Aust. J. Plant Physiol. 20, 555–571.

Whiteman F, Schneider E and Thimann K V 1980 Hormonal factors controlling the initiation and development of lateral roots. Physiol. Plant. 49, 304–314.

Wilson K J 1995 Molecular techniques for the study of rhizobial ecology in the field. Soil Biology and Biochemistry 27, 501–514.

Zobel R W 1991 Genetic control of root systems. *In* Plant Roots: The Hidden Half. Eds. Y Waisel, A Eshel and U Kafkafi. pp. 27–38. Marcel Dekker, Inc., New York.

Guest editors: J K Ladha, F J de Bruijn and K A Malik

Plant and Soil **194**: 145–154, 1997.
© 1997 *Kluwer Academic Publishers. Printed in the Netherlands.*

Strategies for increased ammonium production in free-living or plant associated nitrogen fixing bacteria

Rita Colnaghi[1], Andrew Green, Luhong He[2], Paul Rudnick and Christina Kennedy[3]
Department of Plant Pathology, College of Agriculture, PO Box 210036, The University of Arizona, Tucson, AZ 85721, USA. Current addresses: [1]*Departimento di Scienze Molecolari Agroalimentari, University of Milano, Italy and* [2]*111 Koshland Hall, Department of Plant Biology, University of California at Berkeley, Berkeley, CA 94720, USA.* [3]*Corresponding author**

Key words: ammonium, *Azotobacter*, diazotroph, glutamine synthetase, NifA, nitrogen fixation

Abstract

Strategies considered and studied for achieving ammonium excretion in nitrogen fixing bacteria include 1) inhibition of ammonium assimilation and 2) interference with the mechanisms by which ammonium inhibits either nitrogenase synthesis or activity. These aspects of nitrogen fixation have been best studied in diazotrophic Proteobacteria and Cyanobacteria and those of the former are reviewed in this paper. Ammonium assimilation by glutamine synthetase (GS) can be diminished or prevented by treatment of bacteria with chemicals that inhibit GS activity and in some diazotrophs, such treatment results in excretion of up to 15mM ammonium into liquid growth medium. Also, mutants with altered GS activity, isolated by selection for resistance to GS inhibitors, often excrete ammonium. In Proteobacteria, ammonium inhibits nitrogenase activity and/or synthesis, the latter by preventing activity or expression of NifA, a transcriptional activator required for expression of other *nif* genes. In *Azotobacter vinelandii*, ammonium inhibits NifA activity but not its synthesis; NifL mediates this effect by interacting directly with NifA causing its inactivation. In *nifL* insertion mutants, NifA is constitutively active and up to 10 mM ammonium is excreted during nitrogen fixation. GlnD insertion/deletion mutations are unable to be stably maintained in *A. vinelandii* wild type but are stable and viable in a mutant that produces constitutively active GS (cannot be adenylylated). This confirms the hypothesis that GlnD is required for activity of GS, an essential enzyme in *A. vinelandii*. In addition, the stable *glnD* mutants are Nif⁻, supporting also the previous conclusion that GlnD is involved in mediating NifL/NifA interaction. Mechanisms of inhibition of synthesis or activity of NifA by ammonium in other diazotrophs are discussed and compared.

Introduction

Nitrogen fixing bacteria, diazotrophs, convert N_2 to NH_3 by electron reduction and protonation of gaseous dinitrogen. Nitrogenase is the enzyme complex in diazotrophs responsible for N_2 fixation; its biosynthesis is determined by 15 to 20 different nitrogen fixation (*nif*) gene products (Dean and Jacobson, 1992). The *nif* genes may be carried on plasmids as in most *Rhizobium* species or, apparently more commonly, are located on the major chromosome in free-living and associative nitrogen-fixing bacteria. These genes and their genomic arrangement in various diazotrophs is

reviewed elsewhere (Dean and Jacobson, 1992; see also Lee et al., 1997).

The activity and/or synthesis of nitrogenase enzyme in diazotrophs is inhibited by an abundance of fixed nitrogen compounds in the environment. Thus, in plentiful supplies of ammonium, nitrogenase enzyme activity may be inhibited, as occurs in several photosynthetic nitrogen-fixing bacteria such as *Cyanobacteria* or in species of *Rhodobacter*, *Rhodospirillum* and *Azospirillum* (Roberts and Ludden, 1992; Zhang et al., 1996). In the latter three organisms, nitrogenase Fe protein is modified by attachment of an ADP-ribose moiety by the DraT enzyme in cells grown with plentiful ammonium, causing the protein to be inactive. The ADP-ribose is removed by the DraG enzyme when

* FAX No: 15206219290. E-mail: kennedy@biosci.arizona.edu

fixed nitrogen becomes limiting (Roberts and Ludden, 1992). In addition, in all free-living or associative diazotrophs examined, excess ammonium prevents *nif* gene expression, the details of which are discussed in this paper.

The product of nitrogen fixation, ammonia, is assimilated in diazotrophs by the enzyme glutamine synthetase (GS). In this reaction, ammonia (ammonium) is incorporated into glutamate to yield glutamine:

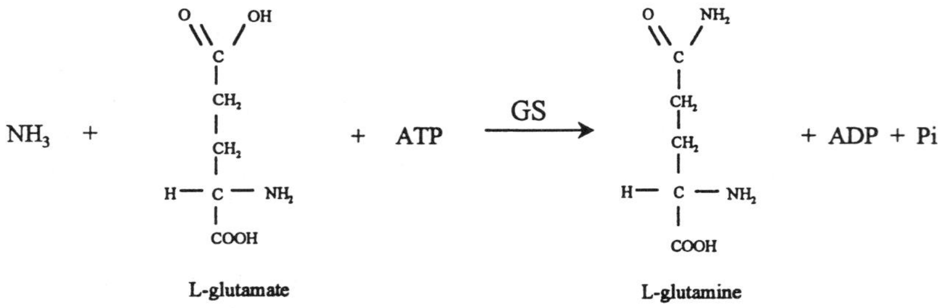

down of other amino acids. These latter two pathways provide less energy-requiring mechanisms for ammonium assimilation in many bacteria. GS activity can be prevented either by mutation in the *glnA* gene which encodes this enzyme or by treatment of cells with a specific chemical inhibitor of GS activity such as methionine sulfoximine (MSX).

Overcoming the inhibitory effects of excess environmental ammonium (or other forms of fixed nitro-

Glutamate is regenerated by the activity of GOGAT (glutamate synthase):

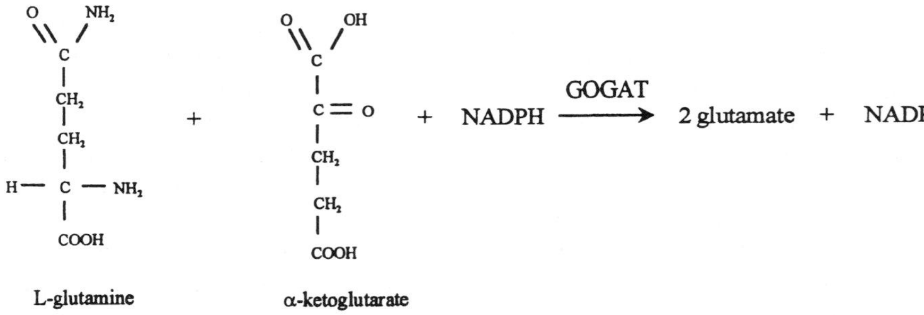

gen) on nitrogenase synthesis or activity so that nitrogen fixation is constitutive, or preventing ammoni-

By the concerted activities of these two enzymes, GS and GOGAT, cells are supplied with the key intermediates of N metabolism, glutamate and glutamine. Glutamate is the source of α amino groups of all amino acids, half the N in pyrimidine, purine and imidazole rings, and the amino group of adenine: these represent about 90% of all N-containing metabolites (Reitzer, 1996). Glutamine provides N in amino–sugars, nicotinamide adenine dinucleotide (NAD), paraminobenzoic acid (PABA) and the other N's in purines, pyrimidines, histidine and tryptophan: these represent about 10% of all N-containing metabolites (Reitzer, 1996). If GS is not present, glutamine is not synthesized and ammonium assimilation is impaired, or is completely prevented in organisms such as *A. vinelandii* (see below). In GOGAT-deficient mutants, glutamate can also be formed by amination of α-ketoglutarate (αKT) via glutamate dehydrogenase (GDH) or from the break-

um assimilation so that the ammonium formed during nitrogen fixation is excreted, have both been considered as strategies to increase the amount of fixed nitrogen transferred from bacterial to plant partner in associative or symbiotic plant-diazotroph relationships. This paper will review the relevant aspects of ammonium assimilation and the regulation of *nif* gene expression in several diazotrophs and how manipulation of genes or biochemical interference of enzymes involved in these processes has resulted in diazotrophs which excrete ammonium. Our recent results on ammonium assimilation and the regulation of *nif* gene expression in *A. vinelandii* imply future strategies for achieving the goal of increasing ammonium production in agriculturally important plant-associating diazotrophs.

The central role of glutamine synthetase in diazotrophs: interference with ammonium assimilation can lead to ammonium excretion

The first report that a nitrogen fixing bacterium might excrete ammonium came from work on mutant strains of *Klebsiella pneumoniae* impaired in their ability to assimilate ammonium (Shanmugam and Valentine, 1975). These strains had lesions not in *glnA* encoding GS but in one or both of the other pathways for ammonium assimilation, viz. GOGAT and GDH. While GS allowed biosynthesis of glutamine, the absence of GOGAT and/or GDH blocked formation of glutamate, which as described earlier provides the α amino groups to all the other amino acids. In these mutant strains grown under N_2 fixing conditions (anaerobic, absence of ammonium), the ammonium formed by nitrogenase was unable to be assimilated further than into glutamine and the excess was excreted into the medium, leading to an extracellular concentration of up to 5 mm in the culture medium. In subsequent work, GS-deficient mutants were also shown to excrete ammonium during nitrogen fixation, accumulating up to 3mM in the extracellular medium (Andersen and Shanmugam, 1977). Another case in which GOGAT deficiency led to ammonium excretion ($\sim$1 mM) was that of *Rhodospirillum rubrum* where glutamate auxotrophs were isolated and examined (Weare, 1978).

Work with other diazotrophs aimed at achieving ammonium excretion focussed on genetic or biochemical interference with GS activity. Wall and Gest (1979) isolated glutamine auxotrophs of the photosynthetic bacterium *Rhodobacter capsulatus* by ethylmetane sulfonate (EMS) mutagenesis and classic indirect selection of mutants using penicillin. These mutants absolutely required the addition of glutamine for growth and nitrogenase activity was not repressed by the addition of ammonium. Whereas when N_2 was the sole source of N, the mutants fixed nitrogen and excreted ammonium leading to a concentration of 15 mM in the culture medium, the highest level ever reported for ammonium excretion by a diazotroph.

Methylammonium, $CH_3NH_3^+$, an analog of ammonium used to study ammonium uptake rates in bacterial cells (because it can be radiolabelled with the stable isotope ^{14}C), inhibits the growth of *A. vinelandii*. Methylammonium as well as methylalanine resistant mutants of this organism were reported to excrete up to 0.5 mM ammonium into the culture medium during nitrogen fixation (Gordon and Jacobson, 1983). While the target for these inhibitors was not identified, it is possible or likely that they are inhibitors of GS activity, since methylammonium resistant mutants of *E. coli* mapped within the *glnA* locus (La Rossa, 1996).

During the years 1975-1990, a number of papers were published describing the effects of the GS inhibitors on nitrogen fixation and ammonium excretion in various diazotrophs including, in particular, various species of Cyanobacteria. Inhibition of GS activity by L-methionine-D,L-sulfoximine (MSX), an irreversible inhibitor of GS activity, led to excretion of about 1 mM ammonium in *Anabaena cylindrica* (Stewart and Rowell, 1975), and to excretion of 8 mM ammonium in *Anabaena* sp. 33047 (Ramos et al., 1984). Addition of 5-hydroxylysine, a reversible GS inhibitor, to cultures of N_2-fixing *A. cylindrica*, had similar effects and also alleviated NH_4^+ inhibition of heterocyst formation (Ladha et al., 1978). In addition, mutants of *Nostoc muscorum* resistant to MSX, presumably arising due to alterations in GS protein, excreted up to about 1mM ammonium (Singh et al., 1983).

Of particular relevance to the theme of the workshop and symposium on nitrogen fixation in rice and other non-legume plants are reports of research on Cyanobacteria able to colonize *Azolla anabaena*, a tropical fern used in rice cultivation, or to colonize rice or wheat plants directly. In the former group of organisms, including *Anabaena* strain 2B and *Anabaena azollae*, treatment of free-living cultures with MSX led to excretion of up to 3mM ammonium (Newton and Cavins, 1985; Zimmerman and Boussiba, 1987). An MSX-resistant mutant of *A. siamensis* isolated from rice fields excreted low amounts of ammonium (about 0.05 mm) (Thomas et al., 1990). Among several MSX-resistant mutants of *A. variabilis* ATCC29413, one with very low GS activity, named SA1, is of particular interest (Spiller et al., 1986). SA1 excretes up to 3mM of ammonium during diazotroph growth in liquid medium. Inoculation of rice seedlings with the wild type strain SA0 and the mutant SA1 resulted in dramatic differences in growth under N-limiting growth conditions in the laboratory (Latorre et al., 1986). Plants inoculated with ammonium-excreting strain SA1 weighed 5-fold more then those inoculated with wild-type SA0 and as much as uninoculated plants supplied with 5 mM ammonium sulfate. In similar experiments with wheat plants, SA1- inoculated plants weighed twice as much as uninoculated plants or those inoculated with the wild type strain SA0 (Spiller and Gunasekaran, 1990). Plants supplied with ammonium had three times the weight as the N-starved uninoculat-

148

ed plants. Thus the potential for ammonium-excreting mutants of diazotrophic bacteria to significantly stimulate plant growth is evident from these results. However, SA1 failed to survive when it was introduced into the rhizosphere of field-grown plants, presumably because its slow growth made it unable to compete with other rhizosphere organisms (K T Shanmugam, personal communication). This emphasizes the need for ammonium- excreting mutants to not be disadvantaged with respect to growth and survival if they are to be used in agriculture.

Ethylenediamine (EDA) inhibits growth of both *Anabaena* and the plant-associating *Azospirillum brasilense* (Polukhina et al., 1982; Machado et al., 1991). Mutants of these organisms resistant to EDA were found to excrete ammonium (1.5 and 5 mM, respectively). The EDA-resistant mutants of *A. variabilis* were unable to grow with ammonium as N source but could grow with added glutamine, had three-times higher nitrogenase activity than the wild type parental strain, and formed more heterocysts than normal. Also cells of *A. variabilis* EDA resistant mutants immobilized on alginate beads showed light-dependent ammonium production from N_2 (Kerby et al., 1986; Musgrave et al., 1982).

Control of GS activity by reversible adenylylation

Another important aspect of GS activity, crucial to understanding how the fixed N status in bacterial cells influences its role in ammonium assimilation, is its reversible adenylylation/deadenylylation in response to changing fluxes of fixed N supply. The following 'model' was deduced from the growth and biochemical phenotypes of mutant strains of a number of enteric bacteria, including *Escherichia coli*, *Salmonella typhimurium*, *Klebsiella aerogenes*, and diazotrophic strains of the enteric *Klebsiella pneumoniae* (most recently reviewed by Merrick and Edwards, 1995b). In high levels of ammonium, most of the subunits of GS (12 identical subunits of about 60 kDa in size are contained in the holoenzyme) are adenylylated at a Tyr residue at about amino acid 400. Adenylylation results in a dramatic decrease in GS activity, and while adenylylated GS in enteric bacteria is sufficiently active to provide glutamine as an amino acid for protein biosynthesis, its activity with respect to ammonium assimilation is probably negligible.

The product of the *glnE* gene is responsible for both adenylylation and deadenylylation of GS. GlnE activity is dependent on the state of another protein, PII,

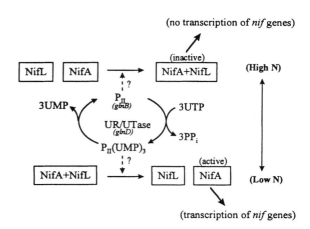

Figure 1. Model illustrating the possible involvement of PII protein in A. vinelandii in mediating the influence of High N (+ammonium) and LowN (nitrogen-fixing conditions) on the ability of NifL to inactivate Nif

which is reversibly uridylylated and deuridylylated in response to fixed N conditions by the *glnD*-encoded GlnD protein. GlnD is thought to sense the intracellular ratio of glutamine / α-ketoglutarate (αKG): if the ratio is high, then PII is not uridylylated, if the ratio is low, then PII is uridylylated by attachment of a uridine monophosphate group (UMP) to a conserved tyrosine residue (Tyr51). Uridylylation of PII is greater if it is bound to αKG, which occurs at a high affinity (Kamberov et al., 1995). Deuridylylated PII stimulates the adenylyl transferase activity of GlnE and leads to the formation of relatively inactive GS (non-assimilatory but able to provide glutamine).

Another activity of PII is its influence on the NtrB/C two component regulatory system in which NtrB is the sensor component of ammonium supply via PII and NtrC the response regulator (Keener and Kustu, 1988). PII-UMP stimulates the phosphorylase activity of NtrB; deuridylylated PII stimulates its phosphatase activity. In low ammonium, NtrB phosphorylates NtrC; NtrC-P activates the expression of a number of genes and operons involved in nitrogen assimilation, including *glnA* (see below) in enteric and other bacteria, and *nifLA* in *K. pneumoniae*.

Regulation of *nif* gene expression: synthesis and/or activity of NifA in proteobacterial diazotrophs can be inhibited by ammonium

Expression of the 15 to 20 genes involved in nitrogenase synthesis and activity in all the Proteobacterial diazotrophs examined requires an activator of transcription, NifA. Each *nif* gene-containing operon or transcriptional unit is preceded by a promoter region that is recognized by an RNA polymerase holoenzyme that carries the specialized sigma factor, σ^{54} also known as σ^{N}; this promoter region contains the consensus base pair sequences tGGc-N8-tGCa , where *C* is located 12 bp upstream of the base pair at which transcription is initiated (reviewed in Merrick and Edwards, 1995b). σ^{54} -recognized promoters precede not only *nif* genes in Proteobacteria, but also genes involved in diverse cellular activities such as dicarboxylic acid transport, chemotaxis, degradation of aromatic hydrocarbons, and also growth on poor nitrogen sources other than N_2, such as histidine or arginine in diverse Proteobacterial organisms. Transcription of these genes requires not only σ^{54} for recognition but also a transcriptional activator protein that binds at a site approximately 100 bases upstream of the σ^{54} binding site. In the case of *nif* genes, this transcriptional activator is NifA (for review of NifA and other transcriptional activators of σ^{54}-dependent genes see Merrick and Edwards, 1995b). NifA is now clearly recognized to be a key target for ammonium inhibition of nitrogen fixation. In some organisms, transcription of the *nifA* gene itself is inhibited by ammonium. In others, NifA activity is prevented in ammonium. In a few organisms, both synthesis and activity of NifA are prevented.

In *K. pneumoniae*, *R. capsulatus*, and *Herbaspirillum seropedicae*, expression of the *nifA* gene (two identical *nifA* genes in *R. capsulatus*) is prevented in cells grown in high ammonium (Merrick et al., 1982; Fosterhartnett and Kranz, 1992; Souza et al., 1995a). The *nifA* gene is contained in an operon, the *nifLA* operon, in *K. pneumoniae* or is a singularly transcribed gene in the latter two organisms. In all three, transcription of *nifA* requires the transcriptional activator protein NtrC in its phosphorylated form, but interestingly, the *nifA* genes of *R. capsulatus* do not require σ^{54} for expression as do all other NtrC-activated genes previously characterized (Fosterhartnett and Kranz, 1992). Since NtrC-PO$_4$ is only available in cells grown under limiting fixed N conditions, *nifA* transcription is prevented in high levels of ammonium in these organisms.

In *K. pneumoniae*, *A. vinelandii*, *A. brasilense*, *H. seropedicae* and *R. capsulatus*, the activity of the NifA protein is inhibited in cells grown in ammonium. In the former two but not the latter three organisms, the *nifA* gene lies downstream from and is cotranscribed with *nifL* (Blanco et al., 1993). While in *K. pneumoniae*,

the *nifLA* operon requires NtrC-PO$_4$ for transcription and is therefore itself regulated by ammonium supply, in *A. vinelandii* the *nifLA* operon is transcribed constitutively and independently of NtrC. In *A. vinelandii* and *K. pneumoniae*, the NifL protein is a sensor for the environmental status of both ammonium and oxygen supply and if either is plentiful or excessive, NifL inactivates NifA so that transcription of the nitrogenase (*nif*) genes does not occur. NifL was recently shown to be a flavoprotein, indicating that its conversion to the form in which it inactivates NifA, at least in response to oxygen, apparently involves a redox reaction (Hill et al., 1996). How NifL is converted from its inert form to an active NifA-inhibitory form in cells provided with ammonium (or other repressive fixed N sources), and conversely, how active NifL is converted to its inert form in ammonium or oxygen, are key questions in understanding *nif* gene regulation in these two nitrogen fixing members of the gamma group of Proteobacteria (see below).

In *A. brasilense*, NifA activity is inhibited in ammonium-grown cells by a process that requires PII protein (Liang et al., 1992). How this occurs is not known, but recent results indicate that the N-terminal region of NifA is involved . N-terminal deletion mutations in this region of NifA result in synthesis of a constitutively active NifA which is not inhibited by ammonium suggesting that the intact N-terminal region can block activity of the other domains by a PII-dependent conformational change in the presence of ammonium (Arsene et al., 1996). Similar results were reported for NifA in *H. seropedicae* (Souza et al., 1995 b). In *R. capsulatus*, ammonium inactivation of the two NifA proteins occurs by a mechanism specific for these proteins since introduction of *nifA* genes from either *K. pneumoniae* or *Rhizobium meliloti* resulted in ammonium-insensitive nif gene activation (Masepohl and Klipp, 1996). Tn5 insertion mutations in a gene identified as *hvrA*, known to be involved in regulation of photosynthesis, apparently relieve ammonium inhibition of NifA activity in *R. capsulatus* by an as yet uncharacterized mechanism.

One goal of our research is to understand how in *A. vinelandii* the signal of ammonium sufficiency is transduced to the NifL protein, converting it from an apparently inert form into its active form that binds to and prevents NifA from transcribing the other *nif* genes.

Ammonium excretion in *nifL* mutants of *A. vinelandii*

Analysis of the phenotype of mutants carrying gene cassette insertion mutations in the *nifLA* region led to the discovery that a *nifL*:KIXX mutant expressed nitrogenase constitutively, even when ammonium is added to the growth medium (Bali et al., 1992). In addition, this mutant strain growing on N_2 excreted ammonium into the culture medium at the end of exponential phase growth, leading to the accumulation of up to 10 mM ammonium. The pH in these cultures increased to about 9 which apparently inhibited further nitrogenase activity. These results indicate that NifL is the single target of ammonium inhibition of nitrogen fixation in *A. vinelandii*. The next question became: how is the activity of NifL influenced by high ammonium levels in wild type *A. vinelandii* cells? Are the GlnD or PII proteins involved in determining the activity of NifL?

A gene originally identified as *nfrX* in *A. vinelandii* was later determined to be a *glnD* homolgue on the basis of amino acid sequence similarity deduced from the nucleotide sequence of *nfrX* compared to a partial sequence of the *glnD* gene of *E. coli* (Contreras et al., 1991), then later to the full sequence of the *E. coli* gene (van Heeswijk et al., 1993). Significantly, the Nif⁻ phenotype of *nfrX(glnD)*::Tn5 mutants was suppressed (i.e. the strain became Nif⁺) by introduction of the *nifL*:KIXX mutation described above (Santero et al., 1988; Contreras et al., 1991). The simplest model to explain this result is that in the *glnD* mutants, NifL is in its 'active' conformation and prevents NifA activity even under conditions of N-limitation. An alternative explanation is that in the *glnD* mutants, NifA is in a state or conformation that is constitutively receptive to inactivation by NifL.

Interestingly, all of the 5 original *glnD* (*nfrX*) mutations had Tn5 inserted at the 3' end of this gene within the 3' terminal approximately 200 of its 2700 base pairs (Contreras et al., 1991). Because of the possibility of a partially active protein being present, we sought to isolate true *glnD* null mutants by deletion of central DNA and insertion of an antibiotic-resistance gene. While a number of such mutations were created and after transformation into *A. vinelandii* replaced the wild type gene by homologous recombination, such mutant strains were unstable and never established the mutation in a homogenous state: upon Southern analysis, the mutated and wild type *glnD* regions were both always present and upon removal of the antibiotic encoded by the gene cassette inserted and used for selection, antibiotic resistance was quickly lost and only wild type copies of the chromosome were present.

In another set of experiments to determine whether the PII protein encoded by *glnB* was also involved in NifL-mediated regulation of NifA activity, the *glnB* gene was isolated, sequenced and deletion/insertion mutations created. Once again, the mutated genes were unable to be stably maintained in *A. vinelandii* transformant strains and the results reported above for the *glnD* mutants were similarly observed (Meletzus et al., 1997).

The only other type of mutation known to behave in such a manner in *A. vinelandii*, i.e. appears to be lethal, is that of *glnA*::Tn5 or *glnA*::KIXX insertions (Toukdarian et al., 1990). Since *A. vinelandii* is unable to transport glutamine, auxotrophs cannot be isolated. Also, as described previously, GS is essential for ammonium assimilation in *A. vinelandii*, there being no GDH present as an alternative pathway for assimilation. The simplest model to explain the apparent lethality of the *glnD* and *glnB* insertion mutations is that, following the enteric model for the ways in which the state of the PII protein dictates the alternative adenylating/deadenylylating activities of GlnE, PII-UMP is essential for the deadenylylation of GS in *A. vinelandii*. Since *glnB* mutants would be devoid of PII, and *glnD* mutants unable to uridylylate PII, then in neither mutant would there be PII-UMP available and GS should be constitutively adenylylated, inactive, and unavailable for ammonium assimilation, a completely disabling state for *A. vinelandii* survival.

In order to test this model, *glnA* mutations were constructed in which the codon for Tyr407 TAC, the site of adenylylation by GlnE, was changed by site-directed mutagenesis to the Phe codon TTC. Insertion of this mutation into the *A. vinelandii* chromosome resulted in a mutant strain (named GSPhe) that was unable to adenylylate GS. The constitutively (hyper-) active GS enzyme in this GSPhe mutant resulted in its inability to grow well on ammonium-containing medium. With substrate present in excess, the highly active GS presumably consumed so much ATP during constitutive conversion of ammonium to glutamine that growth was impaired. Consistent with this idea is that the GSPhe mutant grew well on N-free medium or on ammonium medium to which MSX was added. The GSPhe mutant was not impaired in nitrogen fixation and nitrogen fixation was repressed by added ammonium.

The GSPhe mutant was able to be stably transformed with the *glnD*:omega deletion/insertion mutants, in contrast to the wild type strain. After several generations of growth on BS+Str medium, the mutation appeared to be stably integrated. The GSPhe *glnD*:omega transformants were Nif⁻, consistent with the observation that GlnD is required for the conversion of NifL^A to NifL^I (where NifL^I does not inactivate NifA).

In low ammonia, GlnD uridylylates PII and PII-UMP might mediate the dissociation of NifL from NifA and *nif* gene transcription can therefore occur. In cells growing in or exposed to high ammonium, GlnD deuridylylates PII-UMP; PII might therefore stimulate NifL binding to NifA (or NifA becoming receptive to NifL binding), resulting in prevention of *nif* gene expression. Experiments are in progress to determine whether the *glnB* mutations can be stabilized in the GSPhe mutant. If so, then we can answer the important question of whether the state of the PII protein in *A. vinelandii* is involved in the conversion of 'active' to 'inactive' NifL, or alternatively, if PII is involved in the conversion of NifA to a form which is sensitive to NifL inactivation to one which is insensitive. A model presenting this hypothesis is shown below (Figure 1). If the state of PII does not influence nif gene transcription, then GlnD must mediate the NifL/NifA interaction by some other mechanism, possibly involving some other protein. One candidate is a 40 kDa protein which, like PII, was observed to be uridylylated in *A. vinelandii* cells grown in the absence of ammonium (Luhong He, unpublished experiments).

Prospects for significant ammonium excretion in other diazotrophs

The advances made in recent years in understanding the mechanisms by which environmental fixed nitrogen influences nitrogenase synthesis and activity as well as ammonium assimilation by GS are significant. It should now be theoretically possible to create ammonium excreting mutants of any particular diazotroph by adopting appropriate mutational strategies to disarm the mechanisms by which ammonium inhibits their ability to fix nitrogen and assimilate ammonium. Of practical importance is that the 'disarming' mutations should not result in mutant strains unable to grow in their natural environment or have a disdvantage in competing with other microbes for survival.

Our results with *A. vinelandii* have defined a strategy for construction of *nif*-derepressed and ammonium-excreting mutants via manipulation of the *nifL* gene

which negatively regulates *nif* gene expression. However, since NifA activity is regulated by NifL in species of *Azotobacter* and *Klebsiella*, but not in other Proteobacterial diazotrophs, different strategies for construction of ammonium excreting mutants in other diazotrophs must be adopted. In each of the diazotrophic genera characterized, the mechanisms of ammonium inhibition of nitrogen fixation are somewhat different, and usually involve more than a single 'target' of inhibition. As examples, we will consider how ammonium excretion might be achieved in three relatively well-studied diazotrophic genera by manipulation of genes affecting ammonium control of nitrogenase synthesis and activity.

a. Azospirillum

While *nifA* expression is constitutive in this organism, NifA is inactivated in ammonium-grown cells. Also, NifA activity requires the PII protein by a mechanism not understood, but probably involving PII-UMP in the absence of fixed N. Since GlnD is required for uridylylation of PII, *glnD* mutants of *A. brasilense* might be expected to be Nif⁻ due to inactivity of NifA. Mutants of NifA lacking the N-terminal region do not require PII for activity (Arsene et al., 1996); they are also expected not to require GlnD for activity although characterization of *glnD* in species of *Azospirillum* has not so far been reported. Are the the NifA N-terminal deletion mutants constitutive for the production of active nitrogenase? Since *A. brasilense* also contains a DRAT/DRAG system for inactivation of nitrogenase enzyme in response to ammonium, mutations should also be made in the DraT gene to preclude ADP-ribosylation of nitrogenase in cells exposed to ammonium. Analysis of a *nifA* N-terminal deletion strain in a *draT⁻* background indicated that the strain was only partially constitutive for nitrogenase activity, showing about 25% of nitrogenase activity in ammonium-containing medium as compared to N$_2$-grown cells (Arsene et al., 1996). Whether this mutant strain excretes ammonium was not reported, but the fact that ammonium is still at least partially repressive in terms of nitrogen fixation suggested that factors other than NifA and DRAT/DRAG, not dependent on PII activity, are responsive to ammonium. A second mechanism of post-translational regulation of nitrogenase by ammonium was recently reported (Zhang et al., 1996) accounting at least in part for the less than fully constitutive activity in the truncated *nifA draT* double mutant. In addition, the recent report of a second PII-encoding gene in *A. brasilense*, *glnZ*, is interesting; the GlnZ gene product cannot apparently substitute for GlnB in its role in being required for active NifA in N_2-grown cells (De Zamaroczy et al., 1996). Whether GlnZ is involved in some other aspect of regulation of nitrogen fixation in *A. brasilense*, possibly in regulating nitrogenase activity independently of the DRAT/DRAG system, is not yet known. Further elucidation of the 'second mechanism' for post-translational control of nitrogenase activity by ammonium in *A. brasilense*, as well as of involvement of GlnZ in *nif* gene expression, will add necessary knowledge required in order to achieve a strategy for construction of ammonium-excreting mutants in *Azospirillum*.

B. Klebsiella

In *Klebsiella* species, there are two targets of ammonium control of nitrogen fixation: 1) ammonium prevents expression of the *nifLA* operon (via a requirement of NtrC-P for activation of transcription of *nifLA*); and 2) ammonium in the cells environment causes NifL to be in its 'active' form that inhibits NifA activity. While both *glnB* (PII) and GlnD are involved in expression of the *nifLA* operon, neither appear to be required for or are involved in NifL activity (Holtel and Merrick, 1989; Merrick and Edwards, 1995a). Is there a second *glnB* (*glnK*) in *K. pneumoniae* as in *E. coli*? In their report of the second *glnB*-like gene, *glnK*, in *E. coli*, van Heeyswijk et al. (1996) mentioned unpublished evidence for there being a second *glnB*-like gene in *K. pneumoniae*. If this is so then the involvement of PII in responses of NifL to fixed nitrogen must be reevaluated. While mutant constructs of *Klebsiella oxytoca* containing an insertion mutation in *nifL* and a plasmid expressing NifA from a strong promoter were reported to synthesize nitrogenase constitutively even under high levels of ammonium supply, these authors did not report whether these mutants excreted ammonium during nitrogen fixation (Kim et al., 1989). It seems likely that these mutants might excrete ammonium since both targets of ammonium control were eliminated.

C. Rhizobium

While rhizobial species have not so far been considered in this discussion of free-living and associative diazotrophs, one recent publication should be mentioned in the context of ammonium excretion by diazotrophs. Fully-differentiated nitrogen-fixing bacteroids within root nodules excrete ammonium that is assimilated by plant nodule cell GS enzymes. While many rhizobial species have been shown to contain two or three different GS enzymes (De Bruijn et al., 1989), these appear to be inactivated, possibly by adenylylation, or not synthesized in bacteroids, which almost certainly accounts for the ability of bacteroids to excrete ammonium produced by nitrogenase. A recent study of R. meliloti *glnA* mutants unable to be adenylylated due to a Tyr/Phe mutation (constructed similarly to the *A. vinelandii* GSPhe mutant discussed above) showed that adenylylation of GSI is not particularly important in the change of *R. meliloti* from an ammonium assimilating to an ammonium excreting organism within alfalfa root nodules (Arcondeguy et al., 1996). Total amount of GSI in bacteroids was much less, as shown by Western blot experiments, indicating that expression of *glnA1* and not adenylylation of GSI is probably more important in the conversion of bacteroids to ammonium-excreting organisms.

D. Endophytic diazotrophs

A few endophytic diazotrophs able to colonize monocots such as sugarcane and rice have recently been identified (e.g. for *Acetobacter diazotrophicus*, see Lee et al., 1997 and for *Azoarcus* sp., see Hurek et al., this volume). If such nitrogen-fixing bacteria prove to be significant and important in providing fixed N for plant growth, then understanding the molecular mechanisms by which ammonium inhibits nitrogen fixation and applying mutational strategies for overcoming its effects in these organisms, possibly combined with mutations which decrease the activity of bacterial GS, could have significant benefit for the yield of monocot crops without nitrogenous fertilizer application. A study of the roles of *nifA, glnA, glnB, glnD, ntrBC*, and other regulatory genes is already underway in *A. diazotrophicus* (Lee et al., 1997) and could lead to improved strains for sugarcane production. Applying the strategies discussed in this review to other associative, endophytic, and symbiotic diazotrophs is similarly of potential value to other crops.

Acknowledgements

The work on *Azotobacter vinelandii* was supported by a grant from the United States Department of Agriculture, National Research Initiative Competitive Grants Program, award no. 95-37305-2067.

References

Andersen K and Shanmugam K T 1977 Energetics of biological nitrogen fixation: determination of the ratio of formation of H_2 to NH_4^+ catalysed by nitrogenase of *Klebsiella pneumoniae in vivo*. J. Gen. Microbiol. 103, 107–122.

Arcondeguy T, Huez I, Fourment J and Kahn D 1996 Symbiotic nitrogen fixation does not require adenylylation of glutamine synthetase I in *Rhizobium meliloti*. FEMS Microbiol. Letts. 145, 33–40.

Arsene F, Kaminski P A and Elmerich C 1996 Modulation of NifA activitiy by PII in *Azospirillum brasilense*: Evidence for a regulatory role of the NifA N-terminal domain. J. Bacteriol. 178, 4830–4838.

Bali A, Blanco G, Hill S and Kennedy C 1992 Excretion of ammonium by a *nifL* mutant of nitrogen fixing *Azotobacter vinelandii*. Appl. Environ. Microbiol. 58, 1711–1718.

Blanco G, Drummond M D, Kennedy C and Woodley P 1993 Sequence and molecular analysis of the *nifL* gene of *Azotobacter vinelandii*. Mol. Microbiol. 9, 869–879.

Contreras C, Drummond M, Bali A, Blanco G, Garcia E, Bush G, Kennedy C and Merrick M 1991 The product of the nitrogen fixation regulatory gene *nfrX* of *Azotobacter vinelandii* is functionally and structurally homologous to the uridylyltransferase encoded by *glnD* in enteric bacteria. J. Bacteriol. 173, 7741–7749.

Dean D and Jacobson M R 1992 Biochemical genetics of nitrogenase. *In* Biological Nitrogen Fixation Eds. G Stacey, H J Evans and R Burris. pp 763–834. Chapman and Hall, New York.

De Bruijn F, Rossbach S, Schneider M, Ratet P, Messmer S, Szeto W, Ausubel F and Schell J 1989 *Rhizobium meliloti* 1021 has three differentially regulated loci involved in glutamine biosynthesis, none of which is essential for symbiotic nitrogen fixation. J. Bacteriol. 171, 1673–1682.

De Zamaroczy M, Paquelin A, Peltre G, Forchhammer K and Elmerich 1996 Coexistence of two structurally similar but functionally different PII proteins in *Azospirillium brasilense*. J. Bacteriol. 178, 4143–4149.

Fosterhartnett D and Kranz R G 1992 Analysis of the promoters and upstream sequences of *nifA*1 and *nifA*2 in *Rhodobacter capsulatus* - activation requires NtrC but not RpoN. Mol. Microbiol. 6, 1049–1060.

Gordon J K and Jacobson M R 1983 Isolation and characterization of *Azotobacter vinelandii* mutant strains with potential as bacterial fertilizer. Can. J. Microbiol. 29, 973–978.

Hill S, Austin S, Eydmann T, Jones T and Dixon R 1996 *Azotobacter vinelandii* NIFL is a flavoprotein that modulates transcriptional activation of nitrogen-fixation genes via a redox-sensitive switch. Proc. Natl. Acad. Sci. USA 93, 2143–2148.

Holtel A and Merrick M J 1989 The *Klebsiella pneumoniae* PII protein (*glnB* gene product) is not absolutely required for nitrogen regulation and is not involved in NifL-mediated *nif* gene regulation. Mol. Gen. Genet. 217, 474–480.

Kamberov E S, Atkinson M R and Ninfa A J 1995 The *Escherichia coli* PII signal transduction protein is activated upon binding 2-ketoglutarate and ATP. J. Biol. Chem. 270, 17797–17807.

Keener J and Kustu S 1988 Protein kinase and phosphoprotein phosphatase activities of nitrogen regulatory proteins NTRB and NTRC of enteric bacteria: Roles of the conserved amino-terminal domain of NTRC. Proc. Natl. Acad. Sci. USA 85, 4976–4980.

Kerby N W, Musgrave S C, Rowell P, Shestakov S V and Stewart W D 1986 Photoproduction of ammonium by immobilized mutant strains of *Anabaena variabilis*. Appl. Microbiol. Biotechnol. 24, 42–46.

Kim Y M, Hidaka M, Masaki H, Beppu T and Uozumi T 1989 Constitutive expression of nitrogenase system in *Klebsiella oxytoca* by gene targeting mutation to the chromosomal *nifLA* operon. J. Biotechnol. 10, 293–301.

La Rossa R A 1996 Mutant selections linking physiology, inhibitors, and genotypes. *In Escherichia coli* and *Salmonella* Cellular and Molecular Biology. Ed. F C Neidhardt. pp 2527–2587. ASM Press, Washington D.C.

Ladha J K, Rowell P and Stewart W D 1978 Effects of 5-hydroxyllysine on acetylene reduction and ammonium assimilation in the cyanobacterium Anabaena cylindricum. Biochem. Biophys. Res. Com. 83, 688–696.

Latorre C, Lee J H, Spiller H and Shanmugam KT 1986 Ammonium ion-excreting cyanobacterial mutant as a source of nitrogen for growth of rice: a feasibility study. Biotech. Letts. 8, 507–512.

Lee S, Sevilla M, Meletzus D, Teixeira K, de Oliveira A L, Baldani I and Kennedy C 1997 Analysis of nitrogen fixation and regulatory genes in the sugarcane endophyte *Acetobacter diazotrophicus*. Plant and Soil (*submitted*).

Liang Y Y, De Zamaroczy M, Arsene F, Paquelin A and Elmerich C 1992 Regulation of nitrogen fixation in *Azospirillum brasilense* Sp7: Involvement of *nifA*, *glnA* and *glnB* gene products. FEMS 100, 113–120.

Machado H B, Funayama S, Rigo L U and Pedrosa F O 1991 Excretion of ammonia by *Azospirillum brasilense* resistant to ethylenediamine. Can. J. Microbiol. 37, 549–553.

Masepohl B and Klipp W 1996 Organization and regulation of genes encoding the molybdenum nitrogenase and the alternative nitrogenase in *Rhodobacter capsulatus*. Arch. Microbiol. 165, 80–90.

Meletzus D, Rudnick P, Doetsch N, Green A, and Kennedy C 1997 Characterization of the *glnB amtB* operon of *Azotobacter vinelandii* (submitted, J. Bacteriol).

Merrick M, Hill S, Hennecke H, Hahn M, Dixon R and Kennedy C 1982 Repressor properties of the *nifL* gene product of *Klebsiella pneumoniae*. Mol. Gen. Genet. 185, 75–81.

Merrick M and Edwards R 1995a The role of uridylyltransferase in the control of *Klebsiella pneumoniae nif* gene regulation. Mol. Gen. Genet. 247, 189–198.

Merrick M J and Edwards R A 1995b Nitrogen Control in Bacteria. Microbiol. Revs. 59, 604–622.

Musgrave S C, Kerby N W, Codd G A and Stewart W D P 1982 Sustained ammonia production by immobilized filaments of the nitrogen-fixing cyanobacterium *Anabaena* 27893. Biotech. Letts. 4, 647–652.

Newton J W and Cavins J F 1985 Liberation of ammonia during nitrogen fixation by a facultatively heterotrophic cyanobacterium. Biochim. Biophys. Acta 809, 44–50.

Polukhina L E, Sakhurieva G N and Shestakov S V 1982 Ethylenediamine-resistant *Anabaena variabilis* mutants with derepressed nitrogen-fixing system. Microbiology 51, 90–95.

Ramos J L, Guerrero M G and Losada M 1984 Sustained photoproduction of ammonia from dinitrogen and water by the nitrogen-fixing cyanobacterium *Anabaena* sp. strain ATCC 33047. Appl. Environ. Microbiol. 48, 114–118.

Reitzer L J 1996 Ammonia assimilation and the biosynthesis of glutamine, glutamate, aspartate, asparagine, L-alanine, and D-alanine. *In Escherichia coli* and *Salmonella*. Ed. F C Neidhardt. pp 391–407. ASM Press, Washington D.C.

Roberts G P and Ludden P W 1992 Control of nitrogen fixation in photosynthetic bacteria. *In* Biological Nitrogen Fixation. Eds.

154

G Stacey, R H Burris and H J Evans. pp 135–165. Chapman and Hall, New York.

Santero E, Toukdarian A, Humphrey R and Kennedy C 1988 Identification and characterisation of two nitrogen fixation regulatory regions *nifA* and *nfrX* in *Azotobacter vinelandii* and *Azotobacter chroococcum*. Mol. Microbiol. 2, 303–314.

Shanmugam K T and Valentine R C 1975 Microbial production of ammonium ion from nitrogen. Proc. Natl. Acad. Sci. USA 72, 136–139.

Singh H N, Singh R K and Sharma R 1983 An L-methionine-D,L-sulfoximine-resistant mutant of the cyanobacterium *Nostoc muscorum* showing inhibitor resistant g-glutamyl-transferase, defective glutamine synthetase and producing extracellular ammonia during N_2 fixation. FEBS Lett. 153, 10–14.

Souza E M, Machado H B and Yates M G 1995a Deletion analysis of the promoter region of the *nifA* gene from *Herbaspirillum seropedicae*. *In* Nitrogen Fixation: Fundamentals and Applications. Eds. I A Tikhonovich, N A Provorov, V I Romanov and W E Newton. pp 259. Kluwer Academic, Dordrecht.

Souza E M, Pedrosa F O, Machado H B, Drummond M and Yates M G 1995b The N-terminus of the NifA protein of *Herbaspirillum seropedicae* is probably involved in sensing of ammonia. *In* Nitrogen Fixation: Fundamentals and Applications. Eds. I A Tikhonovich, N A Provorov, V I Romanov and W E Newton. p 260. Kluwer Academic, Dordrecht.

Spiller H, Latore C, Hassan M E and Shanmugam K T 1986 Isolation and characterization of nitrogenase-derepressed mutant strains of Cyanobacterium *Anabaena variabilis*. J. Bacteriol. 165, 412–419.

Spiller H and Gunasekaran M 1990 Ammonium-excreting mutant strain of the cyanobacterium *Anabaena variabilis* supports growth of wheat. Applied Microbiology and Biotechnology 33, 477–480.

Stewart W D P and Rowell P 1975 Effects of L-methionine-D,L-sulfoximine on the assimilation of newly fixed NH_3, acetylene reduction and heterocyst production in *Anabaena cylindrica*. Biochem. Biophys. Res. Commun. 65, 846–856.

Thomas S P, Zaritsky A and Boussiba S 1990 Ammonium excretion by an L-methionine-DL-sulfoximine-resistant mutant of the rice field cyanobacterium *Anabaena siamensis*. Appl. Environ. Microbiol. 56, 3499–3504.

Toukdarian A, Saunders G, Selman-Sosa G, Santero E, Woodley P and Kennedy C 1990 Molecular analysis of the *Azotobacter vinelandii glnA* gene encoding glutamine synthetase. J. Bacteriol. 172, 6529–6539.

van Heeswijk W C, Rabenberg M, Westerhoff H V and Kahn D 1993 Genes of the glutamine synthetase adenylylation cascade are not regulated by nitrogen in *Escherichia coli*. Mol. Microbiol. 9, 443–457.

van Heeswijk W, Hoving S, Molenaar D, Stegeman B, Kahn D and Westerhoff H V 1996 An alternative PII protein in the regulation of glutamine synthetase in *Escherichia coli*. Mol. Microbiol. 21, 133–146.

Wall J D and Gest H 1979 Derepression of nitrogenase activity in glutamine auxotrophs of *Rhodopseudomonas capsulata*. J. Bacteriol. 137, 1459–1463.

Weare N M 1978 The photoproduction of H_2 and NH_4^+ fixed from N_2 by a derepressed mutant of *Rhodospirillum rubrum*. Biochim. Biophys. Acta 502, 486–494.

Zhang Y P, Burris R H, Ludden P and Roberts G 1996 Presence of a second mechanism for the posttranslational regulation of nitrogenase activity in *Azospirillum brasilense* in response to ammonium. J. Bacteriol. 178, 2948–2953.

Zimmerman W J and Boussiba S 1987 Ammonia assimilation and excretion in an asymbiotic strain of *Anabaena azollae* from *Azolla filiculoides* Lam. J. Plant Physiol. 127, 443–450.

Guest editors: J K Ladha, F J de Bruijn and K A Malik

Plant and Soil **194**: 155–160, 1997.
© 1997 *Kluwer Academic Publishers. Printed in the Netherlands.*

Genetics of *Azospirillum brasilense* with respect to ammonium transport, sugar uptake, and chemotaxis

A. Van Dommelen, E. Van Bastelaere, V. Keijers and J. Vanderleyden
*F.A. Janssens Laboratory of Genetics, Willem de Croylaan 42, B-3001 Heverlee, Belgium**

Key words: ammonium, *Azospirillum*, chemotaxis, glutamine synthetase, transport

Abstract

This paper describes molecular aspects of *Azospirillum*-plant root association with respect to nitrogen flux and carbon utilization. In the first part, biochemical and genetic data are reported on the transport of ammonium and methylammonium in *A. brasilense* cells. Ammonium excreting *A. brasilense* mutants reported so far appear to result from alterations in genes encoding for enzymes involved in ammonium assimilation. Solid genetic evidence is given on the occurrence of a postulated ammonium transporter in *A. brasilense*. In the second part, biochemical and genetic evidence is likewise given for the occurrence of a high-affinity uptake system for D-galactose in *A. brasilense*. A sugar-binding protein that is part of this uptake system is required for chemotaxis of *A. brasilense* towards particular sugars, including D-galactose.

Introduction

The use of biological nitrogen fixation as a nitrogen source for plants has been increasingly studied. In this context, the release of ammonium by nitrogen-fixing bacteria associated with economically important crops is of great interest. *Azospirillum brasilense* belongs to a group of plant growth-promoting bacteria, which is capable of fixing nitrogen under micro-aerobic conditions. Enhancement of plant growth, nitrogen content, and plant yield have often been observed following *Azospirillum* inoculation (Okon, 1994). This paper deals with two aspects of the *Azospirillum*-plant interaction: (1) ammonium excretion and uptake by *A. brasilense*; and (2) occurrence of signalling between plant roots and *A. brasilense*.

Excretion of ammonium by *A. brasilense*

Various ammonium excreting *A. brasilense* mutants have been obtained by selection on ethylenediamine (Machado et al., 1990; Christiansen-Weniger and van Veen, 1991). These mutants are not genetically characterised. We have observed that the mutants isolated by Christiansen-Weniger and van Veen (1991) can

be complemented with the plasmid pAB463 containing the *glnA* gene of *A. brasilense* (Bozouklian et al., 1986). *glnA* codes for the glutamine synthetase enzyme, which is responsible for the major ammonium assimilation pathway in *A. brasilense*. It catalyses the synthesis of glutamine: GLUTAMATE + NH_3 + ATP $\rightarrow$ GLUTAMINE + ADP + P_i. Glutamine is then converted into glutamate by the glutamate synthetase: GLUTAMINE + α-KETOGLUTARATE + NADPH $\rightarrow$ 2 GLUTAMATE + $NADP^+$.

Hartmann et al. (1984) reported that strains selected as glutamine synthetase and glutamate synthase mutants excrete ammonium. The ammonium-excreting, ethylenediamine resistant mutants described by Machado et al. (1990) are also deficient in NH_4^+ assimilation. These observations confirm the crucial role of the glutamine synthetase in retaining fixed nitrogen. Reducing the ammonium assimilation capacity in other bacteria has also been reported to cause ammonium excretion (Boussiba and Gibson, 1990 and references therein; Singh et al., 1983; Shanmugam and Valentine, 1975). Often mutants with reduced ammonium assimilation capacity are obtained by selecting L-methionine-DL-sulfoximine resistant strains that can be used in agriculture as a source of nitrogen because of their ammonium excretion capacity (Thomas et al., 1990). L-methionine-DL-sulfoximine is a glutamate analogue that inhibits the glutamine synthetase activity.

Besides lower ammonium assimilation, as is the case in glutamine synthetase and glutamate synthase mutants, the absence of a postulated ammonium carrier can theoretically also provoke ammonium excretion. After comparing intra- and extra-cellular ammonium concentrations, Kleiner (1985) proposed an active ammonium uptake mechanism to maintain the higher intracellular ammonium concentration despite the diffusion of the non-protonated NH_3 through the cell membrane.

Most prokaryotic NH_4^+ carriers are repressed by ammonium (Kleiner, 1985). The *glnKamtB* operon of *Rhizobium etli* and *Escherichia coli* is transcribed from a single nitrogen-regulated promoter which requires the RpoN (NtrA) protein as a sigma factor and the phosphorylated NtrC protein as a transcription activator (Patriarca et al., 1996; Van Heeswijk et al., 1996). In *Klebsiella pneumoniae*, a nitrogen regulatory mutant (KP5060, GlnA$^-$, GlnR$^-$, Hut$^-$, Nif$^-$) is deficient in ammonium transport (Kleiner, 1982). Studies in *E. coli* revealed that both NtrA (GlnF) and NtrC (GlnG) are required to activate synthesis of the ammonium carrier, while NtrB (GlnL) plays a role in its repression (Jayakumar et al., 1986; Servin-Gonzalez and Bastarrachea, 1984). As the ammonium carrier of *A. brasilense* is found to be repressed by fixed nitrogen (Hartmann and Kleiner, 1982), we investigated the regulatory role of the *A. brasilense ntrA*, *ntrB*, and *ntrC* genes in ammonium uptake.

The *ntrC* gene product in *Azospirillum brasilense* is involved in the regulation of nitrate utilisation, switch-off of nitrogenase by ammonia, and to a lesser extent, *nifA* expression. Similarity of the *Azospirillum brasilense* NtrBC polypeptides with other NtrBC polypeptides suggests that, after phosphorylation by NtrB, NtrC acts as a transcriptional activator for genes encoding enzymes subjected to nitrogen control (Liang et al., 1993). The *ntrA* gene product encodes the σ^{54} (RpoN), which is involved in the recognition of -24/-12 type promoters.

Ammonium transport was measured using [C^{14}]methylammonium as a radioactive ammonium analogue. In a study by Milcamps et al. (1996), results suggest that the *ntrABC* genes are necessary for the active uptake of ammonium. This seems to be in conflict with the observation that these mutants grow normally on medium with a low ammonium concentration (Liang et al., 1993; Milcamps et al., 1996).

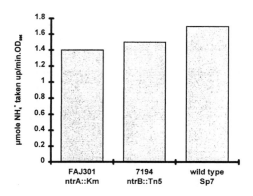

Figure 1. Ammonium uptake in wild type Sp7 strain, and in *ntrA* and *ntrB* mutant strains.

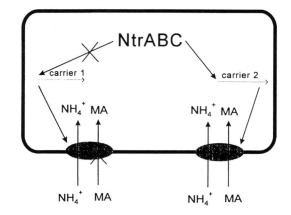

Figure 2. Model for ammonium and methylammonium (MA) transport in *Azospirillum brasilense*.

A selective ammonium electrode (F2322NH$_4$ ammonium selectrode, Radiometer, Copenhagen) was subsequently used to measure directly the ammonium uptake in minimal medium (Figure 1). As expected, the mutants are still able to take up ammonium. The loss of methylammonium (MA) uptake in *ntr* mutants can be explained by assuming the existence of (at least) two 2 types of ammonium carriers in *A. brasilense*. Only one of them, which is regulated by the *ntr* system, can transport [C^{14}]methylammonium (Figure 2). The simultaneous existence of an inducible methylammonium uptake system and a constitutive ammonium uptake system has also been suggested for *Rhodobacter sphaeroides* (Cordts and Gibson, 1987), *R. capsulata* (Cordts and Gibson, 1987), and *Anacystis nidulans* (Boussiba et al., 1984).

The findings for *A. brasilense*—that the inducible methylammonium carrier has considerably higher affinity for ammonium than for methylammonium and that methylammonium is toxic for *Azospirillum* spp. and cannot serve as a carbon or nitrogen source—indicate that ammonium must be the natural substrate for this carrier (Hartmann and Kleiner, 1982).

The reduced levels of glutamine synthetase activity observed in the *A. brasilense ntrC* and *ntrB* mutant strains (de Zamaroczy et al., 1993; Zhang et al., 1994; Milcamps et al., 1996) are not likely to be responsible for the complete loss of [C^{14}]methylammonium transport in the *A. brasilense ntrC* and the *ntrB* mutant strains. It was shown in *Escherichia coli* that mutants affected in glutamine synthetase with 2% of the wild type activity, have a [C^{14}] methylammonium transport activity of about 70% of the wild type (Jayakumar et al., 1986).

Glutamine synthetase mutants (Gauthier and Elmerich, 1977; Christiansen-Weniger and van Veen, 1991) also lack methylammonium transport, but this can be a consequence of their ammonium excretion. The excreted ammonium has a much higher affinity than methylammonium for the carrier. Hence, the radioactive-labelled methylammonium will not be transported into the cells. In *ntr* mutants, no ammonium excretion was observed.

Cloning and sequencing of *A. brasilense* Sp7 *amtA* gene

To further study the role of the ammonium carrier, attempts were made to isolate the gene coding for this carrier. Marini et al. (1994) have cloned and sequenced three ammonium transport proteins in yeast (*Sacharromyces cerevisiae*): *mep1*, *mep2*, and *mep3* (personal communication). The first plant ammonium transporter was isolated by functional complementation of a *mep1mep2 S. cerevisiae* mutant strain (Ninneman et al., 1994). This led to the cloning and sequencing of the *Arabidopsis thaliana amt1* (ammonium transport) gene. The Mep1p, Mep2p, and Amt1p proteins are very similar to five proteins of unknown function identified in various bacterial species including *Bacillus subtilis*, *Corynebacterium glutamicum*, *Mycobacterium smegmatis*, and *Rhodobacter capsulata*. One of these proteins, the NrgAp protein encoded by the *nrgA* gene of *Bacillus subtilis*, was identified for its high expression during nitrogen-limited growth (Wray et al., 1994). This *nrgA* gene is located upstream of *nrgB*, encoding a 13-kDa protein that shares similarity with the *E. scherichia coli* P$_{II}$ regulatory protein (GlnB).

Meletzus et al. (1995) found in *Azotobacter vinelandii* an *nrgA* gene downstream of *glnB*. The same situation appeared in *E. coli*, where a putative ammonium transporter encoded by the *amtB* gene was found downstream of the *glnK* gene encoding an PII homolog (van Heeswijk et al., 1996). Recently, the *glnKamtB* operon was also isolated *in Rhizobium etli* (Patriarca et al., 1996).

After alignment of the deduced amino acid sequences of these reported ammonium transporter genes, four regions with conserved amino acids were used to design degenerated polymerase chain reaction primers. The combination of two of these primers yielded an amplification product of the expected size (344 bases). This product was cloned and sequenced. The sequence was homologous with the expected region of the known ammonium transporter genes. After hybridisation on a genomic *A. brasilense* bank, an *Eco*RI fragment was isolated and subcloned. An 1.6-kb *Eco*RI-*Sal*I fragment containing the first 1,249 bases of the *A. brasilense amtA* (ammonium transporter) gene was sequenced. The rest of the gene was sequenced from the *Eco*RI clone using the 3' terminal part of the known sequence as primer. DNA sequence analysis revealed an open reading frame (ORF) with homology to the known ammonium transporter genes. The highest similarity was found with the *E. coli amtB* gene: 50% identical amino acids and 16% similar amino acids. Figure 3 shows the dendrogram of the alignment.

Figure 4 shows the organisation of the sequenced DNA region. The (G+C) content of the entire sequence is 67.2%, consistent with the high (G+C) content of *A. brasilense* DNA. The (G+C) content in the third position of the codons is 91.9%. The potential ATG start codon is preceded by a putative ribosome binding site (Shine-Dalgarno sequence). The ORF is followed by a palindromic sequence (ΔG(25°): −21.6 kCal). The predictive algorithm of Klein et al. (1985), shown by comparative analysis to offer low ambiguity and high accuracy in determining integral membrane segments (Fasman and Gilbert, 1990), predicted 12 transmembrane segments and classified the protein as integral protein.

Occurrence of signalling between plant roots and *A. brasilense*

The first chemotaxis (chemo-receptor) gene in *A. brasilense* was identified by comparing protein pro-

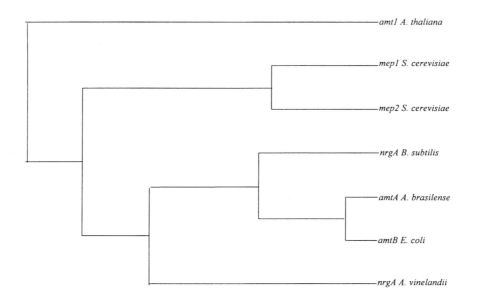

Figure 3. Dendrogram of the alignment of deduced amino acid sequencesnes of ammonium transporter genes and homologous genes.

files generated by two-dimensional polyacrylamide gel electrophoresis (2D-PAGE) (Van Bastelaere et al., 1996). The most obvious induced protein on these gels was an acidic 40-kDa protein. Proteolytic cleavage of the 40-kDa protein by the *Staphylococcus aureus* V8 protease, followed by amino acid sequencing of internal peptides, resulted in two amino acid sequences. Data base searches with these peptides revealed striking similarity between these regions of the induced *A. brasilense* 40-kDa protein and the ChvE protein of *Agrobacterium tumefaciens* (Huang et al., 1990; Shimoda et al., 1993).

ChvE is a periplasmic sugar-binding protein, which is homologous to the *E. coli* galactose- and ribose-binding proteins. It is necessary for *vir* gene induction in response to sugars. Besides its role in tumorigenicity on the host plant, the ChvE protein also functions in uptake of sugars and chemotaxis of *A. tumefaciens* towards sugars (Cangelosi et al., 1990; Shimoda et al., 1993).

Using an internal restriction fragment of the *chvE* gene of *A. tumefaciens*, a *chvE* homologous gene of *A. brasilense* was cloned and sequenced. This gene was designated *sbpA* (sugar binding protein A). Upstream of this *sbpA* gene, another ORF open reading frame was found. This ORF shows homology with the *A. tumefaciens* GbpR (galactose-binding protein regulator), the transcriptional regulator of *chvE* (Doty et al., 1993).

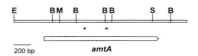

Figure 4. Physical and genetic map of the sequenced DNA region containing the *A. brasilense* Sp7 *amtA* gene. E=*Eco*RI; B=*Bgl*I; M=*Sma*I; S=*Sal*I; ← and → = PCR primers used to amplify a 344-bp internal sequence.

Induction of *sbpA* in *A. brasilense* was studied with the translational *sbpA::gusA* fusion in pLAFR3: pFAJ115. The induction pattern corresponds to the pattern described for *A. tumefaciens* (Doty et al., 1993) (Table 1). The role of the SbpA protein in chemotaxis towards sugars and in uptake has been investigated using a *sbpA* insertion mutant of Sp245. SbpA clearly mediates chemotaxis towards D-galactose, L-arabinose, and D-fucose, while chemotaxis towards D-fructose is SbpA independent.

With C^{14} labelled D-galactose, uptake of D-galactose into a *sbpA* insertion mutant of *A. brasilense* Sp245 was studied. Uptake of D-galactose into the mutant (Figure 5) was much slower compared to the wild type. However, the uptake was not completely abolished. This suggests the existence of at least two uptake systems for D-galactose in *A. brasilense*. One of these uptake systems is the binding-protein dependent system, using SbpA as a periplasmic sugar-binding protein. These systems are known as high-

Table 1. Induction of the *sbpA::gusA* fusion in *A. brasilense*. β-glucuronidase activity was measured 24 hours after addition of the sugars to minimal (MMAB) medium. Without sugars added, no detectable β-glucuronidase activity could be measured. ND: not determined

Sugar	Sp7 (wild type)	Sp245 (wild type)
L-arabinose	+++	+++
D-fucose	+++	+++
D-galactose	+++	+++
2-deoxy-D-glucose	–	–
6-deoxy-D-glucose	–	–
D-galacturonic acid	–	–
D-glucose	–	–
D-glucuronic acid	–	–
inositol	–	–
D-mannose	–	–
D-xylose	–	–
D-fructose	–	–
wheat root exudates	+	ND

sugar uptake

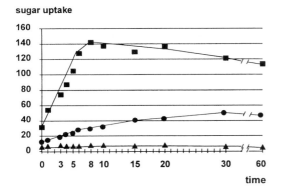

Figure 5. Uptake of ^{14}C-galactose by *A. brasilense* Sp245 ■ and *A. brasilense* Sp245 *sbpA*::Km mutant ● grown in MMAB+10 mM D-galactose; ▲: cells were grown in MMAB. Sugar uptake is expressed as 10^{-12} mol per mg protein. Uptake is measured at different time intervals (indicated in minutes) after addition of D-galactose.

affinity uptake systems. Apparently, galactose can also be transported into the cell via a low-affinity uptake system. It has been shown in *E. coli* that at least six different uptake mechanisms for D-galactose exist (Furlong, 1987).

Conclusions

Until now, all characterised *A. brasilense* ammonium-excreting mutants seem to be defective in their ammonium assimilatory enzymes. In order to investigate another possible ammonium excretion mechanism, the regulation of an ammonium carrier enabling *A. brasilense* to absorb ammonium from dilute solutions was studied. When ammonium uptake is measured with [C14]methylammonium, no activity is observed in *ntrA*, *ntrB*, and *ntrC* regulatory mutants. This indicates that the *ntr* system may regulate the activity of an ammonium carrier. Although no uptake is measured, *ntr* mutants are still able to grow on medium with low ammonium concentration. This suggests the existence of a second ammonium transport system that has no or very low affinity for [C14]methylammonium. A gene homologous to known ammonium transporter genes was cloned and sequenced in *A. brasilense*.

In *A. brasilense*, a ChvE homologous protein was found to be part of a binding protein-dependent, high-affinity inducible uptake system for D-galactose and required for chemotaxis towards particular sugars, such as D-galactose, L-arabinose, and D-fucose. This is the first report on a gene involved in chemotaxis in *Azospirillum*. The gene might play an important role in *Azospirillum* ecology, since it is specifically expressed in the presence of plant root exudates.

Acknowledgements

We thank C. Elmerich for kindly providing us the *A. brasilense ntrBC* mutants and A. Milcamps for the *A. brasilense ntrA* mutant. *Agrobacterium* strains and plasmids were kindly provided by Prof. E W Nester. This research was financed by grants from the Fonds voor Wetenschappelijk Onderzoek (FGWO no 3.0095.93 to J Vanderleyden), the Flemish Government (GOA Vanderleyden, 1993), and BIOTECH (contract BIO2-CT93-0053) of the European Union. A V Dommelen is a recipient of the Belgian "Nationaal Fonds voor Wetenschappelijk onderzoek".

References

Boussiba S, Dilling W and Gibson J 1984. Methylammonium transport in *Anacystis nidulans* R-2. J. Bacteriol. 160, 204–210.
Boussiba S and Gibson J 1990 Ammonia translocation in cyanobacteria. FEMS Microbiol. Rev. 88, 1–14.
Bozouklian H, Fogher C and Elmerich C 1986. Cloning and characterization of the *glnA* gene of *Azospirillum brasilense* Sp7. Ann. Inst. Pasteur/Microbiol. 137B, 3–18.
Cangelosi G A, Ankenbauer R G and Nester E W 1990 Sugars induce the *Agrobacterium* virulence genes through a periplasmic binding protein and a transmembrane signal protein. Proc. Natl. Acad. Sci. USA 87, 6708–6712.

160

Christiansen-Weniger C, and van Veen J A 1991 NH_4^+-Excreting *Azospirillum brasilense* mutants enhance the nitrogen supply of a wheat host. Appl. Environ. Microbiol. 57, 3006–3012.

Cordts M L and Gibson J 1987 Ammonium and methylammonium transport in *Rhodobacter sphaeroides* and *R. capsulata*. J. Bacteriol. 169, 1632–1638.

de Zamaroczy M, Paquelin A and Elmerich C 1993 Functional organization of the glnB-glnA cluster of *Azospirillum brasilense*. J. Bacteriol. 175, 2507–2515.

Doty S L, Chang M and Nester E W 1993 The chromosomal virulence gene, *chvE*, of *Agrobacterium* tumefaciens is regulated by a LysR family member. J. Bacteriol. 175, 7880–7886.

Fasman G D and Gilbert W A 1990 The prediction of transmembrane protein sequences and their conformation: an evaluation. Trends Biochem. Sci. 15, 89–92.

Furlong C E 1987 Osmotic-shock-sensitive transport systems. *In Escherichia coli* and *Salmonella typhimurium*, cellular and molecular biology. Ed. N.C. Neidhardt, F.C., pp 768–796. American Society for Microbiology. Washington, D.C.

Gauthier D and Elmerich C 1977 Relationship between glutamine synthetase and nitrogenase in *Spirillum lipoferum*. FEMS Microbiol. Lett. 2, 101–104.

Hartmann A and Kleiner D 1982 Ammonium (methylammonium) transport by *Azospirillum* spp. FEMS Microbiol. Lett. 15, 65–67.

Hartmann A et al. 1984 Advances in nitrogen fixation research. C. Veeger et al. eds. The Hague. pp 227.

Huang, M W, Cangelosi G A, Halperin W and Nester E W 1990 A chromosomal *Agrobacterium tumefaciens* gene required for effective plant signal transduction. J. Bacteriol. 172, 1814–1822.

Jayakumar A, Schulman I, MacNeil D and Barnes E M Jr 1986 Role of the *Escherichia coli* glnALG operon in regulation of ammonium transport. J. Bacteriol. 166, 281–284.

Klein P, Kanehisa M and DeLisi C 1985 The detection and classification of membrane-spanning proteins. Biochim. Biophys. Acta 15, 468–76.

Kleiner D 1982 Ammonium (methylammonium) transport by *Klebsiella pneumoniae*. Biochim. Biophys. Acta 688, 702–708.

Kleiner D 1985 Bacterial ammonium transport. FEMS Microbiol. Rev. 32, 87–100.

Liang Y Y, Arsène F and Elmerich C 1993 Characterization of the *ntrBC* genes of *Azospirillum brasilense* Sp7: their involvement in the regulation of nitrogenase synthesis and activity. Mol. Gen. Genet. 240, 188–196.

Machado H B, Funayama S, Rigo L U and Pedrosa F O 1990 Excretion of ammonium by *Azospirillum brasilense* mutants resistant to ethylenediamine. Can. J. Microbiol. 57, 549–553.

Marini A-M, Vissers S, Urrestarazu A and André B 1994 Cloning and expression of the *mep1* gene encoding an ammonium transporter in *Saccharomyces cerevisiae*. EMBO Journal 13, 3456–3463.

Meletzus D, Doetsch N, He A, Green L, Rudnick P, Yan D and Kennedy C 1995 Genetic characterization of ammonium sensing and signal transduction in *Azotobacter vinelandii*. Nitrogen Fixation: Fundamentals and Applications. Eds. I.A. Tikhonovich et al. Kluwer Academic Publishers, the Netherlands. pp 220.

Milcamps A, Van Dommelen A, Stigter J, Vanderleyden J and de Bruijn F 1996 *Azospirillum ntrA* gene is involved in nitrogen fixation, nitrate assimilation, ammonium uptake and flagella biosynthesis. Can. J. Microbiol. 42, 467–478.

Ninneman O, Jauniaux J-C and Frommer W B 1994 Identification of a high affinity NH_4^+ transporter from plants. The EMBO Journal 13, 3464–3471.

Okon Y 1994 Azospirillum-plant root associations. Boca Raton, Florida, CRC Press.

Patriarca E J, Taté R, Riccio A, Chiurazzi M, Merrick M and Iaccarino M 1996 The *Rhizobium etli amtB* gene coding for a NtrC-dependent ammonium transporter. Abstract of the 8th International Congress Molecular Plant-Microbe interactions. Eds. G. Stacey, B. Mullin, P.M. Gresshoff. poster H-18.

Servin-Gonzales L and Bastarrachea F 1984 Nitrogen regulation of the synthesis of the high affinity methylammonium transport system of *Escherichia coli*. J. Gen. Microbiol. 130, 3071–3077.

Shanmugam K T and Valentine R C 1975 Microbial production of ammonium ion from nitrogen. Proc. Nat. Acad. Sci. USA 72, 136–139.

Shimoda N, Toyoda-Yamamoto A, Aoki S and Machida Y 1993 Genetic evidence for an interaction between the VirA sensor protein and the ChvE sugar-binding protein of *Agrobacterium tumefaciens* J. Biol. Chem. 268, 26552–26558.

Singh H N, Singh R K and Sharma R 1983 An L-methionine-D,L-sulfoximine-resistant mutant of the cyanobacterium *Nostoc muscorum* showing inhibitor resistant γ-glutamyl-transferase, defective glutamine synthetase and producing extracellular ammonia during N_2 fixation. FEBS Lett. 154, 10–14.

Thomas S P, Zaritsky A and Boussiba S 1990 Ammonium excretion by an L-methionine-D,L-sulfoximine-resistant mutant of the rice field cyanobacterium *Anabaena siamensis*. Appl. Environ. Microbiol. 56, 3499–3504.

Van Bastelaere E, Vermeiren H, Van Dommelen A, Keijers V, Proost P and Vanderleyden J 1997 Characterization of a sugar-binding protein from *Azospirillum brasilense* mediating chemotaxis to and uptake of sugars. Mol. Microbiol. (*in press*).

Van Heeswijk W C, Hoving S, Molenaar D, Stegeman B, Kahn D and Westerhoff H V 1996 An alternative PII protein in the regulation of glutamine synthetase in *Escherichia coli*. Mol. Microbiol. 21, 133–146.

Wray L V JR, Atkinson M R and Fisher S H 1994 The nitrogen-regulated *Bacillus subtilis* nrgAB operon encodes a membrane protein and a protein highly similar to the *Escherichia coli* glnB-encoded PII protein J. Bacteriol. 176, 108–114.

Zhang Y, Burris R H, Ludden P W and Roberts G P 1994 Post-translational regulation of nitrogenase activity in *Azospirillum brasilense ntrBC* mutants: ammonium and anaerobic switch-off occurs through independent signal transduction pathways. J. Bacteriol. 176, 5780–5787.

Guest editors: J K Ladha, F J de Bruijn and K A Malik

Plant and Soil **194**: 161–169, 1997.

Chitin recognition in rice and legumes

Gary Stacey[1] and Naoto Shibuya[2]

[1] *Center for Legume Research, Department of Microbiology and Department of Ecology and Evolutionary Biology, The University of Tennessee, Knoxville, TN 37996-0845, USA* and* [2] *Department of Biotechnology, National Institute of Agrobiological Resources, Ministry of Agriculture, Forestry and Fisheries, Tsukuba, Ibaraki, Japan*

Key words: chitin, elicitor, receptor, rhizobium, rice, legume

Abstract

This review focuses on a comparison of plant reception of chitin oligosaccharides by legumes and rice. Chitin oligosaccharides (dp=6-8) released from fungal pathogens induce plant defense reactions in rice, while lipo-chitin oligosaccharides (dp=4-5) induce the development of a new plant organ, the nodule, in legumes during infection by rhizobia. The former situation is pathogenic and the latter situation beneficial to the plant. However, these two systems do share some common features. We hypothesize that rice and legumes, as well as other plants, may possess members of an evolutionarily conserved family of chitin binding proteins. These proteins may play an important role in chitin reception and subsequent signal transduction. However, data support the idea that legumes may possess a second chitin binding receptor that shows a greater specificity for the lipo-chitin nodulation signals. The presence of this second receptor may be one of the key factors that distinguishes plants capable of nodulation by rhizobia (e.g., soybean) from those that cannot be nodulated (e.g., rice).

Introduction

Agonist recognition by a receptor protein leading to the activation of a signal transduction pathway is a key event in many important cellular processes (e.g., organogenesis). The molecular mechanisms of sensory perception in plants are largely unknown. Investigations of plant-microorganism interactions provide a useful method to dissect the molecular steps involved in agonist-receptor recognition and the following signal transduction pathway. This review focuses on a comparison of plant reception of chitin oligosaccharides by legumes and rice. Chitin oligosaccharides (dp=6-8) which could be released from fungal pathogens induce plant defense reactions in rice, while lipo-chitin oligosaccharides (dp=4-5) induce the development of a new plant organ, the nodule, in legumes during infection by rhizobia. The former situation is pathogenic and the latter situation beneficial to the plant. However, these two systems do share some common features (Table 1).

Higher plants initiate various defense reactions when they are attacked by pathogens such as fungi,

bacteria and viruses (reviewed in Dixon and Harrison, 1990; Ryan and Farmer, 1991). These responses can be triggered by specific elicitors that include oligo-/polysaccharides produced by the pathogenic microbes. One such example is oligochitin fragments (dp=6-8) that can be generated from a cell wall polymer of various fungi and are potent elicitors of defense responses in rice (that protects the plant from fungal infection). In contrast, rhizobia, Gram-negative soil bacteria, produce lipo-chitin molecules (dp=4-5) that induce de novo organogenesis on legume roots leading to the formation of a nodule in which the bacteria reside (reviewed in Dénarié et al., 1996). In this case, the interaction between bacterium and plant host is beneficial, leading to a nitrogen fixing symbiosis, and the plant does not mount a defensive reaction.

Nodule formation can be viewed as a model system for the general study of plant organogenesis. A nodule is a true organ with a specific temporal and spatial developmental pattern and internal cell specialization (e.g., infected vs. non-infected cells). A number of recent studies have provided circumstantial evidence to support the notion that lipo-chitin nod signals, that induce nodule formation, are really mimics of endoge-

---

* FAX No: +14239744007. E-mail: GStacey@utk.edu

Table 1. Comparison of chitin binding activity in plants

Trait	Legume	Rice	Tomato
Chitin oligomer size	*lipo-chitin oligomer* dp= 4-5	*chitin oligomer* dp= 7-8	*chitin oligomer* dp = 4
chitin binding protein, molecular weight	*alfalfa*, unknown *D. biflorus,* 46 kDa (Etzler and Murphy, 1996)	75 kDa (Ito et al., 1996)	unknown
Dissociation constant	*alfalfa:* NFBS1, Kd= 72 nM NFBS2, Kd= 2 nM (Bono et al., 1996) *D. biflorus*: unknown	Kd= 29 nM (Shibuya et al., 1996)	Kd= 23 nM (Baureithel et al., 1994)
Cellular location	*alfalfa:* root, membrane fraction (Bono et al., 1996) *D. biflorus:* root, soluble (Etzler and Murphy, 1996)	plasma membrane (suspension culture cells) (Shibuya et al., 1993, 1996)	microsomal membranes (suspension culture cells) (Baureithel et al., 1994)
Cellular response elicited by optimal chitin oligomer	*lipo-chitin oligomer induces* (reviewed in Dénarié et al., 1996): - medium alkalinization - membrane depolarization (Ehrhardt et al., 1992) - calcium spiking (Ehrhardt et al., 1996) - gene expression (Minami et al. - nodule morphogenesis	*chitin oligomer induces*: - medium alkalinization cytoplasmic acidification, ion flux (Kuchitsu et al., unpubl.) -membrane depolarization (Kuchitsu et al., 1993) - gene expression 1996c; Nishizawa et al., unpubl.) -protein phosphorylation (Kuchitsu et al., 1993) - jasmonic acid biosynthesis (Nojiri et al., 1996) - phytoalexin production (Yamada et al., 1993)	*chitin oligomer induces*: - medium alkalinization - protein phosphorylation (Baureithel et al., 1994)

nous plant growth regulators (e.g., De Jong et al., 1993; Schmidt et al., 1993; Spaink et al., 1993; Truchet et al., 1989). However, at present, such an idea is speculation. Precedence comes from research on other plant growth regulators; e.g., gibberellin produced by bacteria and actinomycetes (Atzorn et al., 1988; Katznelson and Cole, 1965). If indeed lipo-chitin compounds play an important role in plant growth and development, then study of the plant response to such compounds may have importance far beyond the study of nodule development. Recently, the DG42 protein of *Xenopus laevis*, a homologue of the rhizobial NodC protein (a chitin synthase, Geremia et al., 1994) expressed at a specific stage in frog development, has been shown to synthesize chitin oligomers (Semino and Robbins, 1995). Indeed, homologues of NodC and DG42 have now been found in zebra fish and mice (Semino et al., 1996). In the case of zebra fish, chitin synthase activity is primarily expressed in the late gastrula stage of embryo development. These data suggest that both plants and animals may have the ability to synthesize and respond to chitin oligomers. Therefore, chitin reception by cells may be an evolutionarily conserved function that plays an important role in developmental processes.

Chitin oligosaccharides elicit plant defense responses in rice

A. Defense responses induced by chitin oligosaccharides

Chitin oligomers (oligochitin, *N*-acetylchitooligosaccharides), that can be generated from fungal cell walls by endochitinase, can induce defense responses or related cellular responses in many monocots and some dicots (Barber et al., 1989; Ishihara et al., 1996; Kaku et al., 1996; Felix et al., 1993; Ren and West, 1992; Roby et al., 1987). Yamada et al. (1993) showed that purified chitin fragments (dp=7-8) could induce phytoalexin biosynthesis in suspension-cultured rice cells at nM concentrations . The same chitin fragments have been shown to induce various cellular responses in rice including, transient depolarization of membrane potential (Kuchitsu et al., 1993a), ion flux (Kuchitsu et al., unpubl.), protein phosphorylation (Kuchitsu et al., 1993b), transient generation of reactive oxygen species (Kuchitsu et al., 1995) and jasmonic acid (Nojiri et al., 1996), and transient expression of several unique "early responsive" genes, as well as typical defense-

related genes such as phenylalanine ammonia lyase (PAL) (Minami et al., 1996) and chitinase (Nishizawa et al., unpubl.). The structural requirement for the chitin fragments for each of these responses was found to be the same suggesting that all of these cellular responses may be mediated through a single receptor. The rapid and transient nature of some of these cellular responses suggests their involvement in the signal transduction cascade. Chitin oligosaccharides have also been shown to induce various defense-related cellular responses in wheat (Barber et al., 1989), barley (Kaku et al., submitted), oat (Ishihara et al., 1996), melon (Roby et al., 1987) and tomato cells (Felix et al., 1993).

B. Putative receptors for elicitor signals

To understand signal perception and transduction in plant-microbe interactions, it is critically important to know the detailed properties of the receptor molecules which perceive the microbe-produced signal. However, the knowledge of such receptor molecules in plants is extremely limited. For example, only the presence of high-affinity binding sites, putative receptor molecules, for some well characterized elicitors (i.e., inducers of a plant defense reaction) have been reported in the membrane preparations for several plants. A high affinity binding protein for the hepta-β-glucoside elicitor has been detected in microsomal/plasma membrane preparations from soybean (Cosio et al., 1988; Cosio et al., 1990a; Cheong and Hahn, 1991; Cosio et al., 1990b). Mithöfer et al. (1996) recently reported the purification of this binding protein to homogeneity. The presence of high-affinity binding sites for a glycopeptide elicitor (Basse et al., 1993) and a peptide elicitor (Nürnberger et al., 1994) have also been reported from microsomal membrane preparations of suspension-cultured tomato cells and parsley cells, respectively. At present, a gene for a bonafide receptor protein has not been isolated. The well publicized isolation of genes encoding for plant resistance to bacterial and fungal pathogens may indeed represent the first such genes isolated (reviewed in Martin 1996). However, firm data showing that these genes encode receptors has yet to be published.

C. High-affinity binding protein for chitin oligosaccharide elicitor

An ^{125}I-labeled tyramine conjugate of *N*-acetylchitooctaose was used as a radioactive ligand to identify the presence of a high-affinity binding site in microsomal (Shibuya et al., 1993), as well as plasma membrane preparations (Shibuya et al., 1996), of suspension cultured rice cells. Scatchard plot analysis of binding assay results indicated the presence of an independent, non-interactive binding site (Kd = 5.4 nM). Aqueous two phase partitioning localized the binding activity to the plasma membrane. Studies of the binding specificity of this site revealed that the binding specificity correlated well with the specificity for the N-acetylchitooligosaccharide elicitor shown by rice cells. For example, larger N-acetylchitooligosaccharides, such as an octamer, showed several ten thousand-fold higher affinity for this binding site compared to a trimer.

Affinity labeling of plasma membrane preparations was used to identify the binding protein for the N- acetylchitooligosaccharide elicitor. ^{125}I-Labeled N-acetylchitooctaose conjugates were synthesized with 2-(4-azidophenyl)ethylamine and 2-(4-aminophenyl)ethylamine (APEA) for photoaffinity labeling and affinity cross-linking, respectively. Autoradiographic analysis of the SDS-PAGE gels revealed that a single, 75 kDa protein band was labeled by both ligands (Ito et al., 1996). The incorporation of radioactivity was saturable with increasing concentration of the ligand. The labeling of the 75 kDa protein was inhibited with unlabeled N-acetylchitooctaose with a half maximal concentration of 30 nM but was not inhibited with N-acetylchitotriose nor deacetylated octamer, chitooctaose, at 25 μM. These characteristics of the 75 kDa protein corresponded well with that of the high-affinity binding site detected in the plasma membrane preparation.

The 75 kDa binding protein was successfully solubilized from the plasma membrane preparation by use of several detergents. Approximately one-third of the binding activity was solubilized using Triton-X100. The solubilized fraction showed saturable binding (Kd = 90 nM) with the ^{125}I-labeled tyramine conjugate of N-acetylchitooctaose. Affinity cross-linking with the ^{125}I-labeled APEA conjugate of N-acetylchitooctaose showed the presence of the same 75 kDa binding protein in the solubilized fraction. The 75 kDa protein was affinity purified to apparent homogeneity using a N-acetylchitoheptaosyl-lysil-agarose column (Shibuya et al., 1996). The purified fraction contained a binding affinity toward the radioactive ligand which was specifically inhibited with unlabeled N-acetylchitooligosaccharides. These results show that the 75 kDa binding protein, a putative receptor

molecule for chitinoligosaccharide elicitor, has been purified. Cloning of the corresponding gene and the generation of transgenic plants containing such a gene will clarify whether the 75 kDa protein really functions as a receptor for the elicitor.

Lipo-chitin nodulation signals induce de novo organogenesis in legume roots

Rhizobia are Gram- bacteria that fix nitrogen in a symbiotic association with leguminous plants. The ability of rhizobia to infect legume roots and to initiate nodule formation is determined, at least in part, by a family of unique lipo-oligosaccharides called nod signals (reviewed in Dénarié et al., 1996). Nod signals are produced by the bacteria and are lipo-chitin oligosaccharides of three to five β 1,4 linked-N-acetylglucosamine residues which are N-acylated on the terminal, non-reducing glucosaminosyl residue. These molecules can be further modified by the addition of a glycosyl residue or other modifying group, such as sulfate, carbamoyl, acetyl, and/or glycerol residue (reviewed in Dénarié et al., 1996). The nod signals are synthesized by enzymes encoded by the bacterial nod genes. Purified or synthetic nod signals are able to induce the deformation of root hairs, formation of pre-infection threads, and the induction of root cortical cell division (Dénarié et al., 1996). On Medicago, Sesbania, Glycine soja, and Lotus preslii, the compatible nod signals can trigger the formation of nodule-like structures (Dénarié et al., 1996). Nod signals are also able to induce the expression of infection-related early nodulin genes in pea (Horvath et al., 1993), Medicago truncatula (Pichon et al., 1992), and Glycine soja (Minami et al., 1996a,b).

A. Structure/function studies of nod signal action.

The lipo-chitin molecules produced by rhizobia are biologically active at concentrations below 1 nM (Dénarié et al., 1996). Because of this, it has been postulated that specific plant receptors must exist that recognize the nod signals produced by the compatible rhizobium. Thus, it is generally claimed that the plant response to nod signals is very specific with regard to chemical structure. Indeed, only four lipo-chitin molecules were found capable of inducing root hair deformation (HAD) or root cortical cell division (NOI) on G. soja roots (Stokkermans et al., 1995). However, careful reading of the literature suggests that nod signal recognition by legume roots may be more complex. For example, whereas the response of *Vicia sativa* roots to nod signal addition was found to show a strict structural specificity with regard to nodule primordium formation, several structurally distinct lipo-chitin molecules were able to induce a HAD response (Spaink et al., 1991; van Brussel et al., 1992). Horvath et al. (1993) showed ENOD12 mRNA accumulation was induced upon treatment of pea roots with purified nod signals from *R. leguminosarum* bv. viciae (the compatible symbiont) and *R. meliloti* (an incompatible symbiont). Journet et al. (1994) reported the specific induction of ENOD12 mRNA expression in alfalfa only when roots were treated with NodRmV(16:2Δ2,9;SO$_3$) nod signal. Lipo-chitin molecules lacking the sulfate substituent were 1000-fold less active and chitin tetramer was inactive up to a concentration of 1 mM. These various studies present a rather confusing picture with regard to the specificity of the plant response to nod signal addition.

Ardourel et al. (1994) offered an explanation for the various results by suggesting that different steps in the nodulation process may vary with regard to their stringency for nod signal structure. For example, *R. meliloti* mutants (e.g., *nodF/nodL* double mutant) that produced nod signals lacking O-acetylation or the appropriate C16 poly unsaturated fatty acid were able to induce cell wall tip growth in trichoblasts (giving rise to root hairs) and were also able to elicit cell division of inner cortical cells. However, unlike the *R. meliloti* wild type (producing sulfated nod signals with a C16Δ2,9 fatty acid), these mutants were unable to induce infection thread formation and to penetrate the host root. The nod signals purified from these mutants, as well as those from the wild type, possessed similar ability to induce a HAD response on alfalfa roots, but differed in their ability to induce nodule primordia. The results of such studies led Ardourel et al. (1994) to propose that nod signal recognition by alfalfa involves two separate events with differing chemical specificity.

Recent structure/function studies have led to the formulation of a model for soybean nodulation in which at least two independent chitin/ lipo-chitin recognition events are required for nodule development (Minami et al., 1996a,b). According to this model the first recognition event is rapid and relatively non-specific. For example, fucosylation of the *B. japonicum* lipo-chitin nod signals, previously shown to be essential for the induction of nodule primordia (Stacey et al., 1994), is not required for the rapid induction of the nodulin ENOD40 (Minami et al., 1996a). Indeed,

the addition of chitin pentamer was found to induce the expression of this gene. This recognition event may be mediated by a general chitin receptor that is evolutionarily conserved among all plants, including legumes and rice. In contrast, a specific lipo-chitin nodulation signal was found to be required to induce later nodulation events. Therefore, it was postulated that a specific lipo-chitin molecule receptor was involved in this recognition step. Minami et al. (1996b) showed that the induction of the nodulin ENOD2 required the addition of two, structurally distinct nod signals. ENOD2 induction required the addition of the specific lipo-chitin nod signal produced by *Bradyrhizobium japonicum*, normal symbiont of soybean, while the other signal could be either a lipo-chitin or chitin oligomer (i.e., dp = 5). These data suggest that nodule organogenesis requires the action of both recognition systems acting cooperatively.

B. Biochemical studies of nod signal binding.

Recently, Bono et al. (1996) reported the presence of two, distinct lipo-chitin nod signal binding sites in membrane preparations from *Medicago varia*. One of the binding sites, termed nod factor binding site 1 (NFBS1), exhibited a $K_d = 72$ nM and, therefore, resembled the nod signal binding site previously reported in membrane preparations of *M. truncatula* (Bono et al., 1995). This binding site was shown to be specific for lipo-chitin molecules. However, competition studies showed that binding was independent of both the O-acetyl and sulphyl modifications of the chitin chain and was also not dependent on the unsaturation of the fatty acid. Previous studies had shown that these features were important for the biological activity of *Rhizobium meliloti*, normal symbiont of *Medicago*, lipo-chitin nod signal (reviewed in Dénarié et al., 1996). Similar binding activity as that of NFBS1 was found in membrane preparations from tomato. Therefore, these studies are consistent with the idea that a non-specific chitin binding activity is evolutionarily conserved among legumes and, at least, some other plants. The second nod signal binding site found in membrane preparations of *M. varia* was found to have a $K_d = 1.9$ nM (Bono et al., 1996). The specificity of this binding activity has yet to be determined. If this binding site has high specificity for the specific substituents found on the *R. meliloti* lipo-chitin nod signal, then it would be a prime candidate for the specific nod signal receptor hypothesized above. This work of Bono et al. (1995, 1996) is consistent with the model, based on lipo-chitin structure/function studies, that two receptors may be involved in nod signal recognition in plants. If it is also true that one of these receptors is evolutionarily conserved among all plants, then it may be the presence of the second, specific nod signal receptor that distinguishes legumes from other plants. However, at this point, these statements are speculation and much additional research needs to be done.

C. Role of lectins in nodulation

Research on the role of lectins in legume nodulation has had a long and controversial history. However, the work of Diaz et al. (1989) brought renewed interest in the role of lectins in nodulation. In these experiments, it was shown that transfer of the pea lectin gene to clover conferred the ability to the transgenic plants to be nodulated by *R. leguminosarum* bv. viciae, the normal symbiont of pea. These experiments suggested that lectin is an important determinant in nodulation specificity. However, it is difficult to believe that this effect of the pea lectin is due to its ability to act as a receptor for the nod signal. Any receptor protein must have at least two functions; one, the ability to recognize the agonist and two, the ability to transduce the signal that results from agonist recognition. Although one can envision pea lectin binding to the carbohydrate portion of the nod signal, this well characterized protein has no obvious means by which to transduce the signal resulting from such interaction.

Previously, Quinn and Etzler (1987) reported the purification and characterization of a root lectin (DB46) from the roots of the legume, *Dolichos biflorus*. Recently, Etzler and Murphy (1996) reported the surprising finding that this lectin has the ability to bind to the lipo-chitin nod signal of *B. japonicum*. *D. biflorus* is nodulated by *B. japonicum* but the nodules formed do not fix nitrogen (Day and Stacey, unpubl.). This work suggests that DB46 could be involved in lipo-chitin nod signal reception in legume roots. DB46 was found to exist as a monomer in solution, as opposed to the tetrameric subunit structure of the *D. biflorus* seed lectin and the dimeric structure of the 58 kDa stem and leaf lectin (Etzler, 1994). Surprisingly, DB46 has been shown to possess phosphatase activity that is stimulated upon binding of the hapten (Etzler and Murphy, 1996). Thus, DB46 appears to have the ability to bind to the lipo-chitin nodulation signal and an enzymatic activity that could be postulated to transduce the signal to other cellular components. However,

much work remains to show that this lectin is involved in nod signal recognition.

Conclusions

Rice is the most important staple food for more that 2 billion people in Asia and hundreds of millions in Africa and Latin America. The predicted increase in world population will lead to a concomitant need for greater rice production. Nitrogen supply is critical for attaining yield potential. In rice, it takes 1 kg of nitrogen to produce 15-20 kg of grain. It has been estimated that to increase rice production to match growing population trends will require a doubling of the 10 million tons of nitrogen fertilizer currently used per year for rice production. An exciting possibility would be to attain some of this needed nitrogen through the development of nitrogen fixing systems in rice. Such an idea remains a remote possibility but recent research has documented the ability of rhizobium to interact with rice roots, as well as other plants (e.g., *Arabidopsis*, Gough et al., 1996). For example, Reddy et al. (1995) have recently demonstrated the ability of specific rice cultivars to induce the transcription of the *nod* genes of *Bradyrhizobium japonicum*. The products of these genes are required for soybean infection and biosynthesize the lipo-chitin nodulation signals. The invasion of rice and *Arabidopsis* roots by rhizobia has been demonstrated (Gough et al., 1996; Ladha et al., 1996). During legume nodulation, specific plant genes (i.e, nodulins)are induced that are required for the normal function of the nodule. Recently, homologues of the nodulin ENOD93 have been found in rice (Reddy et al., 1996a). These data provide evidence that rhizobia and rice have the ability to interact and that rice possesses some of the gene functions necessary for nodule formation. Transgenic rice have been constructed that express the nodulin ENOD12 gene, isolated from alfalfa, fused to the β-glucuronidase (GUS) reporter gene (Reddy et al., 1996b; Terada et al., 1996). Although rice does not form nodules in response to rhizobial infection, ENOD12-GUS expression was stimulated by rhizobium inoculation of these rice plants, as well as by the addition of purified lipo-chitin nodulation signals (Reddy et al., 1996b; Terada et al., 1996). These surprising results presented by two independent groups suggest that rice may possess a receptor protein(s) that recognizes lipo-chitin nodulation signals.

If such a receptor exists it may be related to the 75 kDa chitin receptor that has been purified from rice suspension culture cells (Shibuya et al., 1996). However, this receptor seems to prefer chitin oligomers of dp=7-8, while the nod signals are oligomers of dp=4-5. Thus, this receptor may be very inefficient in recognizing the rhizobial nod signals. Alternatively, rice may possess an alternative receptor that can recognize the nod signals. However, biochemical experiments have failed to detect such a receptor. Structure/function studies of nod signal recognition in legumes, as well as biochemical studies, support the notion that legumes may possess two chitin receptors with differing chemical specificity. One receptor is somewhat non-specific and can apparently respond to chitin oligomers (dp=4-5), as well as the lipo-chitin nodulation signals. This receptor may be a member of an evolutionarily conserved family of chitin binding proteins that may include the rice 75 kDa chitin binding protein. The postulated nod signal receptor in legumes would appear to have a higher specificity for the lipo-chitin nodulation signals and may play a key role in determining nodulation specificity. It may be that plant species that cannot be nodulated by rhizobium lack this specific receptor. The cloning of such a receptor and the construction of transgenic plants (e.g., rice) expressing this receptor may result in a plant with a greater capability of responding to rhizobial inoculation. The DB46 lectin isolated from *Dolichos biflorus* (Etzler and Murphy, 1996) is one candidate for this receptor.

However, at this point, these ideas are mere speculation based on only a handful of studies. Much work remains to be done to understand chitin reception in rice and legumes, as well as other plants. Yet, the above discussion does point out that increased understanding of the basic recognition events behind plant-microorganism interactions can lead to rational schemes to extend such interactions to other plants. Future research will determine whether new information will lead to practical successes.

References

Ardourel M, Demont N, Debelle F, Maillet F, de Billy F, Prome J-C, Dénarié J and Truchet G 1994 *Rhizobium meliloti* lipooligosaccharide nodulation factors: different structural requirements for bacterial entry into target root hair cells and induction of plant symbiotic developmental processes. Plant Cell 6, 1357–1374.

Atzorn R A, Crozier C T, Weeler C T and Sanberg G 1988 Production of gibberelins and indole-3-acetic acid by *Rhizobium phaseoli* in relation to nodulation of *Phaseolus vulgaris* roots. Plant 175, 532–536.

Barber M S, Bertram R E and Ride, J P 1989 Chitin oligosaccharides elicit lignification in wounded wheat leaves. Physiol. Mol. Plant Pathol. 34, 3–12.

Basse C W, Fath A and Boller T 1993 High affinity binding of a glycopeptide elicitor to tomato cells and microsomal membranes and displacement by specific glycan suppresors. J. Biol. Chem. 268, 14724–14731.

Baureithel K, Felix G and Boller T 1994 Specific, high affinity binding of chitin fragments to tomato cells and membrane. J. Biol. Chem. 269, 17931–17938.

Bono J J, Riond J, Nicolaou K C, Bockovich N J, Estevez V A, Cullimore J V and Ranjeva R 1995 Characterization of a binding site for chemically synthesized lipo-oligosaccharidic nod*Rm* factors in particulate fractions prepared from roots. Plant J. 7, 253–260.

Bono J J, Gressent F, Niebel A, Cullimore J V and Ranjeva R 1996 Biochemical characterization of nod factor binding sites in *Medicago* roots and cell suspension cultures. *In* Biology of Plant-Microbe Interactions. Eds. G Stacey, B Mullin and P M Gresshoff. pp. 99–104. Int. Soc. Molecular Plant-Microbe Int., St. Paul, MN.

Cheong J J and Hahn M G 1991 A specific, high-affinity binding site for the hepta-β-glucoside phytoalexin elicitor in soybean. Plant Cell 3, 137–147.

Cosio E G, Popperl H, Schmidt W E and Ebel J 1988 High-affinity binding of fungal β-glucan fragments to soybean (*Glycine max* L.) microsomal fractions and protoplasts. Eur. J. Biochem. 173, 309–315.

Cosio E G, Frey T, Verdyn R, van Boom J and Ebel J 1990a High-affinity binding of a synthetic heptaglucoside and fungal glucan phytoalexin elicitors to soybean membranes. FEBS Lett. 271, 223–226.

Cosio E G, Frey T and Ebel J 1990b Solubilization of soybean membrane binding sites for fungal β-glucans that elicit phytoalexin accumulation. FEBS Lett. 264, 235–238.

Dénarié J, Debellé F and Promé J-C 1996 Rhizobium lipo-oligosaccharide nodulation factors. Ann. Rev. Biochem. 65, 503–535.

De Jong A J, Heidstra R, Spaink H P, Hartog M V, Meijer E A , Hendriks T, Schiavo F L, Terzi M, Bisseling T, Van Kammen A and de Vries S C 1993 *Rhizobium* lipooligosaccharides rescue a carrot somatic embryo mutant. Plant Cell 5, 615–620.

Diaz C, Melchers L S , Hooykaas P J J, Lugtenberg B J J and Kijne J W 1989 Root lectin as a determinant of host specificity in the *Rhizobium*-legume symbiosis. Nature 338, 579–581.

Dixon R A and Harrison M J 1990 Activation, structure, and organization of genes involved in microbial defense in plants. Adv. Genetics 28, 165–234.

Ehrhardt D W, Atkinson E M and Long S R 1992 Depolarization of alfalfa root hair membrane potential by *Rhizobium meliloti* Nod factors. Science 256, 998–1000.

Ehrhardt D W, Wais R and Long S R 1996 Calcium spiking in plant root hairs responding to rhizobium nodulation signals. Cell 85, 673–681.

Etzler M E 1994 Isolation and characterization of subunits of DB58, a lectin from the stems and leaves of *Dolichos biflorus*. Biochem. 33, 9778–9783.

Etzler M E and Murphy J B 1996 Do legume vegetative tissue lectins play roles in plant-microbial interactions? *In* Biology of Plant-Microbe Interactions. Eds. G Stacey, B Mullin, and P M Gresshoff. pp 105–110. Int. Soc. Molecular Plant-Microbe Int., St. Paul, MN.

Felix G, Regenass M and Boller T 1993 Specific perception of sub-nanomolar concentrations of chitin fragments by tomato cells: induction of extracellular alkalinization, changes in protein phosphorylation, and establishment of a refractory state. Plant J. 4, 307–316.

Geremia R A, Mergaert P, Geelen D, Van Montagu M and Holsters M 1994 The NodC protein of *Azorhizobium caulinodans* is an N-acetylglucosaminyltransferase. Proc. Natl. Acad. Sci. (USA) 91, 2669–2673.

Gough C, Webster G, Vasse J, Galera C, Batchelor C, O'Callahan K, Davey M, Dénarié J and Cocking E 1996 Specific flavonoids/stimulate intercellular colonization of non-legumes by *Azorhizobium caulinodans*. In Biology of Plant-Microbe Interactions Eds. G Stacey, B Mullin and P Gresshoff, Int. Soc. Mol. Plant-Microbe Int., Minneapolis, MN.

Horvath B, Heidstra R, Lados M, Moerman A, Spaink H P, Promé J-C, Van Kammen A and Bisseling T 1993 Lipo-oligosaccharides of *Rhizobium* induce infection-related early nodulin gene expression in pea root hairs. Plant J. 4, 727–733.

Ishihara A, Miyagawa H, Kuwahara Y, Ueno T and Mayama S 1996 Involvement of Ca^{2+} ion in phytoalexin induction in oats. Plant Sci. 115, 9–16.

Ito Y, Kaku H and Shibuya N 1996 Identification of a high-affinity binding protein for *N*- acetylchitooligosaccharide elicitor in the plasma membrane of suspension-cultured rice cells by affinity labeling. (Submitted).

Journet E P, Pichon M, Dedieu A, de Billy F, Truchet G and Barker D G 1994 *Rhizobium meliloti* nod factors elicit cell-specific transcription of the ENOD12 gene in transgenic alfalfa. Plant J. 6, 241–249.

Kaku H, Shibuya N, Xu P, Aryan A P and Fincher G B 1996 *N*-Acetylchitooligosaccharides elicit expression of a single (1-3)-β-glucanase gene in suspension-cultured cells from barley (*Hordeum vulgare*). (Submitted).

Katznelson H and Cole S G 1965 Production of gibberelin-like substances by bacteria and actinomycetes. Can. J. Microbiol. 11, 733–741.

Kuchitsu K, Kikuyama M and Shibuya N 1993 *N*-acetylchito-oligosaccharides, biotic elicitor for phytoalexin production, induce transient membrane depolarization in suspension-cultured rice cells. Protoplasma 174, 79–81.

Kuchitsu K, Komatsu S, Hirano H and Shibuya N 1993 Induction of protein phosphorylation by *N*-acetylchitooligosaccharide elicitor in suspension-cultured rice cells. Plant Cell Physiol. 35, s90.

Kuchitsu K, Kosaka T, Shiga T and Shibuya N 1995 EPR evidence for generation of hydroxyl radical triggered by *N*-acetylchitooligosaccharide elicitor and a protein phosphatase inhibitor in suspension-cultured rice cells. Protoplasma 188, 138–142.

Ladha J K, So' R, Angeles O R, Hernandez R, Ramos M C and Reddy P M 1996 Rhizobial invasion and induction of phenotypic changes in rice roots are independent of nod factors. 8th Int. Symp. Mol. Plant-Microbe Int., Knoxville, TN, (July 14-19, 1996), abstract.

Martin G 1996 Plant resistance genes. In Plant-Microbe Interactions, Vol. 1. Eds. G Stacey and N Keen. pp. 1–32. Chapman & Hall Publ. Co., New York, N.Y.

Minami E, Kouchi H, Cohn J R, Ogawa T and Stacey G 1996a Expression of the early nodulin, ENOD40, in soybean roots in response to various lipo-chitn signal molecules. Plant J. 10, 23–32.

Minami E, Kouchi H, Carlson R W, Cohn J R, Kolli V K, Day R B, Ogawa T and Stacey G 1996b Cooperative action of lipo-chitin nodulation signals on the induction of the early nodulin, ENOD2, in soybean roots. Mol. Plant-Microbe Int. 9, 129–136.

Minami E, Kuchitsu K, He D Y, Kouchi H, Midoh N, Ohtsuki Y and Shibuya N 1996c Two novel genes rapidly and transiently

activated in suspension-cultured rice cells by treatment with N-acetylchitoheptaose, a biotic elicitor for phytoalexin production. Plant Cell Physiol. 37, 563–567.

Mithöfer A F, Lottspeich F and Ebel J 1996 One-step purification of the β-glucan elicitor binding protein from soybean (*Glycine max* L.) Roots and characteriztion of an anti-peptide antiserum. FEBS Lett. 381, 203–207.

Nojiri H, Sugimori M, Yamane H, Nishimura Y, Yamada A, Shibuya N, Kodama O, Murofushi N and Omori T 1996 Involvement of jasmonic acid in elicitor-induced phytoalexin production in suspension-cultured rice cells. Plant Physiol. 110, 138–142.

Nürnberger T, Nennsteil D, Jabs T, Sacks W R, Hahlbrock K and Scheel D 1994 High affinity binding of a fungal oligopeptide elicitor to parsley plasma membranes triggers multiple defense responses. Cell 78, 449–460.

Pichon M, Journet E-P, Dedieu A, de Billy F, Truchet G and Barker D G 1992 *Rhizobium meliloti* elicits transient expression of the early nodulin gene ENOD12 in the differentiating root epidermis of transgenic alfalfa. Plant Cell 4, 1199–1211.

Quinn J M and Etzler M 1987 Isolation and characterization of a lectin from the roots of *Dolichos biflorus*. Arch. Biochem. Biophys. 258, 535–544.

Reddy P M, Ramos M N C, Hernandez R J and Ladha J K 1995 Rice-rhizobial interactions. 15 N.Am.Conf. Symb. N_2 Fixation, N.C. State Univ., Raleigh, N.C., USA (Aug. 13-17, 1995), abstract.

Reddy P M, Kouchi H, Hata S, and Ladha J K 1996a Homlogs of GmENOD93 from rice. 8th Int. Symp. Mol. Plant-Microbe Int., Knoxville, TN, (July 14-19,1996), abstract.

Reddy P M, Torrizio L, Ramos M C, Datta S K, and Ladha J K 1996b Expression of *MtENOD12* promoter driven GUS in transformed rice. 8th Int. Symp. Mol. Plant-Microbe Int., Knoxville, TN, (July 14-19,1996), abstract.

Ren Y Y and West C A 1992 Elicitation of diterpene biosynthesis in rice (*Oryza sativa* L.) by chitin. Plant Physiol. 99, 1169–1178.

Roby D, Gadelle A and Toppan P 1987 Chitin oligosaccharides as elicitors of chitinase activity in melon plants. Biochem. Biophys. Res. Comm. 143, 885–892.

Ryan C A and Farmer E E 1991 Oligosaccharide signals in plants: a current assessment. Ann. Rev. Plant Physiol. 42, 651–674.

Savoure A, Magtyar Z, Pierre M, Brown S, Schultze M, Dudits D, Kondorosi A and Kondorosi E 1994 Activation of the cell cycle machinery and the isoflavonoid biosynthesis pathway by active *Rhizobium meliloti* Nod signal molecules in *Medicago* microcallus suspensions. EMBO J. 13, 1093–1102.

Schmidt J, Rohrig H, John M, Wieneke U, Stacey G, Koncz C and Schell J 1993 Alteration of plant growth and development by *Rhizobium nodA* and *nodB* genes involved in the synthesis of oligosaccharide signal molecules. Plant J. 4, 651–658.

Semino C E and Robbins P W 1995 Synthesis of "Nod"-like chitin oligosaccharides by the *Xenopus* developmental protein DG42. Proc. Natl. Acad. Sci. (USA) 92, 3498–3501.

Semino C E, Specht C A, Raimondi A and Robbins P W 1996 Homologs of the *Xenopus* developmental gene DG42 are present in zebrafish and mouse and are involved in the synthesis of Nod-like chitin oligosaccharides during early embryogenesis. Proc. Natl. Acad. Sci. USA 93, 4548–4553.

Shibuya N, Kaku H, Kuchitsu K and Maliarik M J 1993 Identification of a novel high-affinity binding site for N-acetylchitooligosaccharide elicitor in the microsomal membrane fraction from suspension-cultured rice cells. FEBS Lett. 329, 75–78.

Shibuya N, Ebisu N, Kamada Y, Kaku H, Cohn J and Ito Y 1996 Localization and binding characteristics of a high-affinity binding site for *N*-acetylchitooligosaccharide elicitor in plasma membrane from suspension-cultured rice cells suggest a role as a receptor for the elicitor signal at the cell surface. Plant Cell Physiol. 37, 894–898.

Shibuya N, Ito Y and Kaku H 1996 Perception of oligochitin (N-acetylchitooligosaccharide) elicitor signal in rice. *In* Biology of Plant-Microbe Interactions. Eds. G Stacey, B Mullin and P M Gresshoff. pp. 83–88. Int. Soc. Mol. Plant-Microbe Int., St. Paul, MN.

Spaink H P, Sheeley D M, van Brussel A A N, Glushka J, York W S, Tak T, Geiger O, Kennedy E P, Reinhold V N and Lugtenberg B J J 1991 A novel, highly unsaturated, fatty acid moiety of lipo-oligosaccharide signals determines host specificity of *Rhizobium leguminosarum*. Nature 354, 125–130.

Spaink H P, Wijfjes A H M, vanVliet T B, Kijne J W and Lugtenberg B J J 1993 Rhizobial lipo-oligosaccharide signals and their role in plant morphogenesis; are analogous lipophilic chitin derivatives produced by the plant? Aust. J. Plant Physiol. 20, 381–392.

Stacey G, Luka S, Sanjuan J, Banfalvi Z, Nieuwkoop A J and Carlson R W 1994 *nodZ*, a unique host-specific nodulation gene, is involved in the fucosylation of the lipo-oligosaccharide nodulation signal of *Bradyrhizobium japonicum*. J. Bacteriol. 176, 620–633.

Stokkermans T J W, Ikeshita S, Cohn J, Carlson R W, Stacey G, Ogawa T and Peters N K 1995 Structural requirements of synthetic and natural product lipo-oligosaccharides to induce nodule primoridia on *Glycine soja*. Plant Physiol. 108, 1587–1595.

Terada R, Bauer P, Schultze M, Kondorosi E, Kondorosi A, Potrykus I and Sautter C 1996 The early nodulin gene promoter MsENOD12B is expressed in transgenic rice. 8th Int. Symp. Mol. Plant-Microbe Int., Knoxville, TN, (July 14-19,1996), Abstract.

Truchet G, Barker D G, Camut S, de Billy F, Vasse J and Huguet T 1989 Alfalfa nodulation in the absence of *Rhizobium*. Mol. Gen. Genet. 219, 65–68.

Van Brussel A A N, Bakhuizen R, Van Spronson P C, Spaink H P, Tak T, Lugtenberg B J J and Kijne J 1992 Induction of pre-infection thread structures in the leguminous host plant by mitogenic lipo-oligosaccharides of *Rhizobium*. Science 257, 70–72.

Yamada A, Shibuya N, Kodama O and Akatsuka T 1993 Induction of phytoalexin formation in suspension-cultured rice cells by N-acetylchitooligosaccharides. Biosci. Biotech. Biochem. 57, 405–409.

Guest editors: J K Ladha, F J de Bruijn and K A Malik

Plant and Soil **194**: 171–184, 1997.
© 1997 *Kluwer Academic Publishers. Printed in the Netherlands.*

The role of phytohormones in plant-microbe symbioses

A.M. Hirsch[1,2], Y. Fang[1], S. Asad[1,4] and Y. Kapulnik[3]
[1]*Department of Molecular, Cell and Developmental Biology,* [2]*Molecular Biology Institute, University of California, Los Angeles, CA 90095-1606, USA* and* [3]*Institute of Field and Garden Crops ARO, The Volcani Center, Bet Dagan 50-250, Israel.* [4]*Present address: National Institute for Biotechnology and Genetic Engineering, Faisalabad, Pakistan*

Key words: auxin, cytokinin, ethylene, plant hormones, plant-microbe symbiosis

Abstract

Plant hormones, especially auxin, cytokinin, and ethylene, have long been implicated in nodule development. In addition, plant hormones have been shown to have increased concentrations in mycorrhizal associations. We show that the early nodulin (*ENOD*) genes can be used as indicators for the status of endogenous hormones in symbiotic root tissues. Transcripts for *ENOD2* and *ENOD40* genes are shown to accumulate in uninoculated, cytokinin-treated alfalfa roots, even in roots of the non-nodulating alfalfa mutant MN1008, which is unresponsive to *Rhizobium meliloti* inoculation and to Nod factor treatment. Transcripts for these *ENOD* genes also accumulate in mycorrhizal roots of alfalfa. A model describing the involvement of cytokinin and auxin in stimulating cell divisions in the inner cortex which leads to nodule formation is presented.

Introduction

The formation of a nitrogen-fixing root nodule is a complex developmental event requiring the exquisite coordination of gene expression of two very distinct organisms, one a prokaryote and the other a eukaryote. The developmental pathway giving rise to nodulation has been under intense scrutiny for a number of years, and major progress has been made especially with respect to the bacterial partner. However, the plant partner has been much more recalcitrant in part because the genetic tools which have been applied so successfully to the prokaryote have not been routinely used for the plant. Nevertheless, this is rapidly changing because of the recent interest in studying host plants that are mutated in stages of the nodulation pathway.

An important question that can be asked in light of such studies is: what are the antecedents of the nitrogen-fixing symbiosis? Is the legume nitrogen-fixing nodule a modified root, for example, as the actinorhizal nodule clearly is, or some other type of modified plant organ, for example, a storage organ? Or is it something very different, such as an organ *sui generis* as suggested by Libbenga and Bogers (1974)? As an extension of this hypothesis, could the *Rhizobium*-legume interaction have evolved from an even older symbiosis, like that of endomycorrhizae as suggested by LaRue and Weeden (1994)?

Our goal in this article is to look at the molecules that serve as mediators of a "molecular conversation" in an attempt to see whether there are conserved elements among the various symbiotic associations. We start by looking at Nod factor, a recently identified signal molecule produced by the rhizobia, and then turn our attention to the phytohormones. We will keep in mind the overall picture of legume and non-legume symbioses, both nitrogen-fixing and mycorrhizal, to determine whether there are any common features linking them together. Finding answers to questions of this type may give us leads for designing future experiments.

The conversation between symbiotic microbes and their host plants involves small molecules

Nod factor

Perhaps one of the most exciting findings in the studies of legume nodulation in the last few years was the identification of Nod factor, the end-product of

* FAX No: +13102065413.
E-mail: hirsch@biovxl.biology.ucla.edu

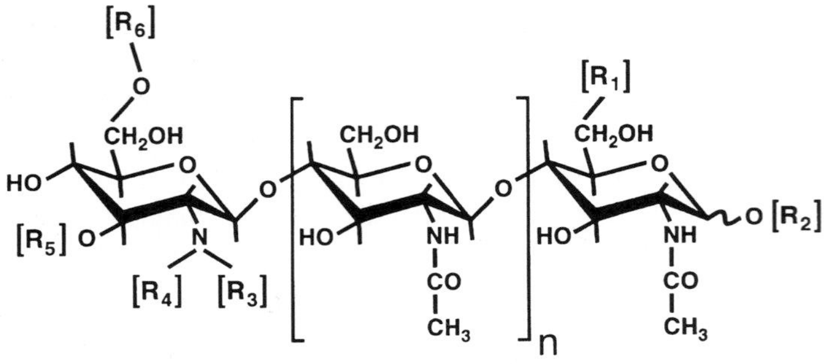

Figure 1. Generalized chemical structure of Nod factor. n refers to the number of glucosamine residues. R_1 can be H, sulfate, fucose, methylfucose, sulfo-methylfucose, acetyl-methylfucose, or D-arabinose. R_2 is either H or glycerol, whereas R_3 is either H or a methyl group. R_4 is an acyl group and is usually 18 or 16C; the degree of unsaturation varies. R_5 is either H or a carbamoyl group, and R_6 can be H, or an acetyl or carbamoyl group.

the expression of *Rhizobium meliloti nod* genes (Lerouge et al., 1990) (Figure 1). The major Nod factor of *R. meliloti*, NodRm-IV (Ac, S) is a sulfated β 1,4 tetra D-glucosamine with three acetylated amino groups. A C16 fatty acid with two double bonds occupies the non-reducing end of the molecule, while the reducing end contains a sulfate. There is also an O-acetyl group at carbon 6 of the terminal sugar of the non-reducing end (Truchet et al., 1991). Nod factors have also been chemically characterized from other rhizobia (see review by van Rhijn and Vanderleyden, 1995). All of the Nod factors are N-fatty acid oligoglucosamines with variations in the length and degree of unsaturation of the acyl group as well as showing differences in the substitutions on the oligoglucosamine backbone. In addition, tri, penta- and even hexaglucosamine chains as well as the more common tetraglucosamine oligomer are detected in Rhizobiaceae (Schultze et al., 1992; López-Lara et al., 1995). In addition, many rhizobia produce more than one Nod factor-type molecule; perhaps a combination of Nod factors is important for host recognition.

Nod factor has been defined as a primary morphogen for the legume nitrogen-fixing symbiosis because it elicits not only deformation of treated root hairs, but also cortical cell divisions that lead to nodule primordium formation. However, higher concentrations of Nod factor are required for the latter response (Truchet et al., 1991). One of the major questions now facing the legume-*Rhizobium* research community is the mechanism of action whereby Nod factor elicits plant responses. Several models have been proposed suggesting that Nod factor binds to a plasma membrane-bound receptor (Hirsch, 1992; Ardourel et al., 1994). Ehrhardt et al. (1992) and Kurkdjian (1995)

have shown that adding purified Nod factor to root hairs leads to membrane depolarization within minutes after application, and Ehrhardt et al. (1996) have demonstrated that exogenous Nod factor leads to rapid spiking of the root hair calcium levels in treated root hairs. These studies suggest that the root hair membrane is the site of Nod factor action.

However, some recent experiments demonstrate that the tetraglucosamine itself, without its fatty acid, after being ballistically targeted into plant cells, elicits cell divisions and nodule primordium formation (Spaink et al., 1995), suggesting that an intracellular receptor may exist. In addition, a number of Nod factor-degrading enzymes are produced by the host legume, indicating that the vast majority of lipochitooligosaccharide molecules are degraded into smaller constituents (Staehelin et al., 1995). Thus, the questions of mechanism of action and how the signal is perceived and transduced are still unanswered.

In addition to the legume-*Rhizobium* symbiosis, there are two other very important plant-microbe interactions that involve some type of molecular signaling: the nitrogen-fixing symbiosis between *Frankia* and actinorhizal plants, and the phosphate-acquiring mycorrhizal associations that exist between a number of diverse fungi and a broad spectrum of plant hosts. Very little is known about these molecular conversations, with respect to either the ligand or the receptor(s). Although attempts have been made to find *nod* gene homologs in *Frankia* (Reddy et al., 1988; Chen et al., 1992), so far there is no evidence that a Nod factor-like molecule is produced, implying that some other type of signaling interaction may be taking place. These two symbioses represent the areas where the next breakthroughs need to be made, especially in light of the finding that the signal transduction pathways between legume symbiosis and mycorrhizae may be conserved (van Rhijn et al., 1997; see later section).

Plant hormones

The various phytohormones, which can be considered as more "traditional" signal molecules, in the sense that they have been known for a longer time than Nod factors, are also important for plant growth and development. Plant hormones have been assumed to be part of the nodulation process ever since Thimann (1936) reported that pea nodules contain elevated levels of auxin. Although there have been more studies on the involvement of plant hormones in legume nodule development than of any other symbiosis, it is likely that the plant hormones also play a role in nodulation of non-legumes. Earlier, we briefly reviewed some of the literature about the different classes of phytohormones and their involvement in nodulation (Hirsch and Fang, 1994). In this report, we will concentrate on auxin and cytokinin because these two hormones frequently interact with one another, often times in antagonistic ways. For example, cytokinins applied exogenously to seedling roots or excised roots repress lateral root formation whereas exogenous auxin stimulates lateral root development in seedling roots, excised roots, and roots of mature plants (see references in Torrey, 1986).

Although the evidence for hormones having an important role in nodulation is striking (Table 1), it is difficult to determine which partner produces the hormone. As these are *phyto*hormones, it is likely that the plant up-regulates their synthesis in the production of a new organ. However, the production of plant hormones by rhizobia is well documented (see Torrey, 1986). Rhizobia synthesize both auxins and cytokinins but the genes responsible for cytokinin production are as yet unknown. Mutants defective in IAA synthesis have been described, but none of them are Nod$^-$, suggesting that auxin production by rhizobia is not essential for nodule morphogenesis (see references in Hirsch and Fang, 1994). Thus, rigorous testing of the involvement of rhizobial-produced hormones on nodulation is not possible at this time.

As far as the *Frankia*-actinorhizal plant symbiosis is concerned, *Frankia* appear to produce both auxins and a cytokinin (Stevens and Berry, 1988; Berry et al., 1989). However, none of the genes have been identified so their relationship to other hormone-synthesizing genes in bacteria is completely unknown.

The major questions now revolve around exactly *how* phytohormones are involved mechanistically in the nodulation pathway as well as *which* of the symbiotic partners is the source of the hormone(s). The latter question may be less important than the first because there are examples of nodules that form without an inductive stimulus from *Rhizobium*. These are the so-called "spontaneous" nodules originally described by Truchet et al. (1989) for alfalfa. This phenotype has been described as Nar$^+$ (*N*odulation in the *A*bsence of *R*hizobium) and is the consequence of a mutation in a dominant gene (Caetano-Anollés et al., 1992). Surprisingly, this phenotype has not been described for legumes other than alfalfa, suggesting that it is not a common mutation in legumes. In any case, expression of the Nar gene may result in the production of some factor (hormone?) that allows the constitutive formation of *bona fide* nodules. Similar to nitrogen-fixing nodules, these nodules have a discrete nodule meristem, peripheral vascular bundles, and express the early nodulin gene *ENOD2* (Truchet et al., 1989; Hirsch et al., 1992).

What evidence links either cytokinin or auxin or a balance between the two hormones to the nodulation pathway?

Several reports have suggested that the balance of hormones, particularly that between auxin and cytokinin, is part of the nodulation stimulus, but a priori it is not obvious in which direction the balance is shifted. There are no simple, accurate methods to measure hormone concentration on a cell-to-cell basis. Allen et al. (1953) treated several legumes with compounds that function as auxin transport inhibitors (ATI) and found that nodule-like structures were formed. These structures resulted from an inhibition of lateral root elongation and the subsequent fusion of closely spaced root primordia. Hirsch et al. (1989) determined that alfalfa pseudonodules induced by the ATI *N*-(1-naphthyl)phthalamic acid (NPA) contained transcripts for the early nodulin gene *ENOD2*, suggesting that this "nodule-specific" gene was developmentally rather than symbiotically regulated. This study further indicated that the early nodulin genes could be used as diagnostic makers for changes in hormone balance. The nodule-like structures more closely resembled roots than nodules in that they had a single central vascular bundle and no distal meristem (Hirsch et al., 1989). Later, Scheres et al. (1992) determined that the *ENOD12* gene was expressed in pseudonodules

Table 1. Cytokinins and nodulation-related events in legumes and non-legumes

Nodulation-related event	Reference
Cytokinins induce pseudonodules on tobacco roots	Arora et al., 1959
Cytokinin induces pseudonodules on roots of *Alnus glutinosa*	Rodriguez-Barrueco and Bermudez de Castro, 1973
Cytokinins detected in nodules of *Phaseolus vulgaris*	Puppo et al., 1974
Cytokinins detected in nodules of *Vicia faba*	Henson and Wheeler, 1976
Identification of cytokinins in pea nodules	Syono and Torrey, 1976
Cytokinins detected in nodules of *Myrica gale*	Rodriguez-Barrueco et al., 1979
Cytokinins detected in nodules of *Phaseolus mungo*	Jaiswal et al., 1981
High levels of cytokinins measured in pea nodules	Badenoch-Jones et al., 1987
Cytokinins induce *ENOD2* expression in *Sesbania rostrata*	Dehio and deBruijn, 1992
Nod⁻ *R. meliloti* carrying *Agrobacterium trans*-zeatin synthase gene induce nodules and *ENOD2* gene expression in alfalfa	Cooper and Long, 1994
Cytokinins induce pseudonodules on siratro roots	Relic et al., 1994
Cytokinins induce *ENOD12* expression in alfalfa	Bauer et al., 1996

formed on Afghanistan pea in response to ATIs. Wu et al. (1996) found that even *non-nodulating* mutants of sweetclover (*Melilotus alba* Desr.) after NPA treatment developed pseudonodules which contained transcripts that hybridize to *A2ENOD2*.

Suggestions that the auxin:cytokinin balance is pushed into the direction of cytokinin comes from several additional studies (Table 1). Exogenous cytokinin has been found to induce pseudonodules on tobacco, a plant that normally is not nodulated (Arora et al., 1959). The pseudonodules originated from cortical cells, an origin consistent with that of legume nodules rather than of lateral root primordia. However, the pseudonodules did not exhibit any tissue specialization and consisted solely of large, highly vacuolated parenchyma cells (Arora et al., 1959).

There is considerable data showing that nodule development ensues in a zone of the root where the root hairs are susceptible to infection (Bhuvaneswari et al., 1980). The cortical cells in this region of the uninoculated root are mitotically quiescent but often are tetraploid due to endoreduplication. In contrast, the cells of the pericycle are diploid. The levels of cytokinin in the root hair-emerging zone are generally low. Short and Torrey (1972) determined that the highest levels of cytokinin were in the area 1 mm behind the apex. There was a decrease in cytokinin concentration 1-5 mm behind the apex, and very little cytokinin in older sections of the root.

Torrey (1961) found that tetraploid cortical cells could be stimulated to divide when kinetin was added to the culture medium in which pea root segments excised 10-11 mm behind the root tip had been grown. Few diploid cells of the central cylinder divided under these conditions. Thus, from outside to inside of the root, a gradient of cells which differ in their sensitivity to cytokinin exist. Libbenga et al. (1973) studied this transverse gradient by stripping the cortex from the stele and growing the cortical explants on a defined medium containing auxin alone. When cytokinins or stelar extracts were added to the medium, cell divisions were stimulated, usually opposite protoxylem points. This stelar extract has been identified as uridine (Smit et al., 1995).

Actinorhizal nodule development can also be induced by plant hormones. Rodriguez-Barreco and Bermudez de Castro (1973) induced pseudonodules on the roots of *Alnus glutinosa* with cytokinin. Various phytohormones have also been detected in actinorhizal nodules, namely, auxin (Wheeler et al., 1979), gibberellic acid (Henson and Wheeler, 1977) and cytokinins (Henson and Wheeler, 1977).

Involvement of Ppytohormones in arbuscular-mycorrhizal associations

Mycorrhizal formation enhances the growth and survival of numerous plant species (Hyman, 1980). Many of the observed plant growth responses may be regulated in part by alterations in endogenous phytohormone levels. Host-AM fungus associations result in elevated levels of ABA, cytokinins, or gibberellin-like substances (Allen et al., 1980, 1982) in different plant organs. Recently, a correlation between altered zeatin riboside (ZR) levels and growth was described by Drüge and Schönbeck (1992) to support their conclusion that the improved growth of AM-plants was caused by the enhanced cytokinin production by mycorrhizal roots. However, does the elevated level of cytokinins play a role in the establishment of the symbiosis or are these phytohormones produced in response to fungal colonization? For a more comprehensive discussion on the role of phytohormones in the function and biology of mycorrhizae, the reader is referred to a review by Beyrle (1995).

Regardless of their origin, cytokinins typically stimulate protein and chlorophyll synthesis as well as cell division and expansion in plants (van Stader and Davey, 1979). In the symbiosis, and as a result of the AM-fungal colonization, enhancement of cytokinin levels in mycorrhizal *Bouteloua gracilis* plants was reported (Allen et al., 1980). A higher cytokinin level was found in leaves compared to the levels in roots (Allen et al., 1982). It has also been pointed out that AM fungal colonization increases the flux of cytokinins from roots to shoots in *Citrus* (Dixon et al., 1988), suggesting that AM fungal colonization may alter both the production and allocation of cytokinin in the host plant.

The enhanced root and leaf cytokinin levels in mycorrhizal associations might come about in response to a number of causes: *(i)* cytokinin production by the fungus and subsequent translocation into the plant; *(ii)* an inhibition of cytokinin degradation by compounds produced by the fungus or the plant; or *(iii)* a stimulation of cytokinin production by the plant as a result of improved nutrition or in response to a signal from the fungus. Edriss et al. (1984) demonstrated that the enhancement of cytokinin production in *Citrus* mycorrhizae was associated with AM fungal infection rather than with increased P uptake, arguing against the former. Nevertheless, it is still not clear whether nutrients other than P are important or whether cytokinin is actually required for the establishment or for later stages of the symbiosis.

Several soil microorganisms, e.g., *Azotobacter* (Taller and Wong, 1989), *Rhizobium* (Sturtevant and Taller, 1989), and *Azospirillum* (Okon and Itzigsohn, 1995) among others, are known to produce plant growth regulators. In some cases, the alteration of plant

growth responses to a particular class of growth regulators was suggested as a consequence of inoculation with these bacteria (Brown, 1972). Auxin, cytokinin, GA, and B-vitamin production by ectomycorrhizal fungi has been reported (Slankis, 1973; Strzelczyk et al., 1977; Crafts and Miller, 1974). Also, Barea and Azcon-Aguilar (1982) demonstrated that germinated spores of the endomycorrhizal fungus *Glomus mosseae* synthesize cytokinin and gibberellin-like substances *in vitro*. However, it is unclear at the present whether other AM fungi share these properties.

In conclusion, the AM symbiosis results in an increase in the internal cytokinin levels of the infected tissue and improves the fluxes of this phytohormone to other plant organs, independently of the nutrient status of the host plant, albeit by mechanism(s) not yet understood. Whether such mechanism(s) are common to all AM fungal-plant interactions or whether they exhibit similarities with the induction of other plant hormones (e.g., auxins and gibberellins) remains to be demonstrated.

In this paper, we investigate the involvement of cytokinins in plant-microbe associations by treating mutant or wild-type alfalfa roots or roots from transgenic plants carrying *MsENOD40* promoter-Gus constructs, with cytokinin or with rhizobia genetically engineered to produce cytokinin. Alfalfa roots colonized with the arbuscular-mycorrhizal (AM) fungus *Glomus intraradices* were also studied.

Materials and methods

Plants, treatments, and growth conditions

Three-day old alfalfa seedlings (*Medicago sativa* cv. Iroquois) were transferred to Magenta jars containing Jensen's medium with nitrogen (Vincent, 1970) and various concentrations of different plant hormones. The phytohormones tested were: α-napthalene acetic acid (NAA); 2,4-dichlorophenoxyacetic acid (2,4-D); gibberellic acid (GA$_3$); abscisic acid (ABA); 6-benzylaminopurine (BAP), and kinetin. The concentrations used in initial tests were 10^{-9} to 10^{-5} M for all the hormones, except for 2,4-D which was utilized as high as 10^{-4} M. Three separate experiments for each phytohormone were performed. Roots were also flood-inoculated with wild-type *Rhizobium meliloti* (Rm1021) which had been grown on RDM (*Rhizobium* Defined Medium; Vincent, 1970). Plants were also treated with purified Nod factor [PNF; NodR-mIV(Ac,S)] at 10^{-8} M.

For some experiments, healthy stems of *M. sativa* cv. Saranac and the non-nodulating mutant derived from cv. Saranac, MN1008, were rooted as cuttings in Magenta jars containing sterilized vermiculite watered with 1/4 strength Hoagland's medium plus N (Machlis and Torrey, 1956). After roots were regenerated (ca. 2 weeks), the plants were transferred to sterilized Magenta jars containing Jensen's medium minus N (Vincent, 1970) with the appropriate cytokinin dilution. Some plants were transferred to square Petri dishes (Labtek) containing solidified Jensen's medium minus N and spot-inoculated with either *Rhizobium meliloti* strain SL44/pTZS, which is a Nod$^-$ mutant (ΔnodDABC) of *R. meliloti* carrying the *trans*-zeatin synthase gene of *Agrobacterium tumefaciens* (Cooper and Long, 1994). Plants were spot-inoculated following a previously published protocol (Dudley et al., 1989).

In some experiments, 1-mm^2 1.5% agar blocks containing concentrations of BAP from 10^{-4} to 10^{-7} M were placed on the root at the root-hair emerging zone. Control plants were sham-inoculated with blocks containing Jensen's medium in 1.5% agar. India ink was used to mark the point of agar block or bacterial placement. After varying times, sections of the root adjacent to the marked spot, ca. 5 mm in length, were fixed in FAA for further analysis.

All plants were grown in a Conviron growth cabinet where they were maintained under 16 h 21 °C day/8 h 19 °C night with a relative humidity of 80%. The bottoms of the Magenta jars were covered with aluminum foil to exclude light and the containers were left slightly ajar to avoid ethylene accumulation. Tissue was harvested into liquid nitrogen at the termination of the experiments, and stored at −70 °C until RNA was isolated.

For the mycorrhizal studies, alfalfa (*M. sativa* L. cv. Gilboa) seedlings were grown in sand under microbiologically controlled *Rhizobium*-free conditions as described previously (Volpin et al., 1994). Before planting, surface-sterilized spores of *Glomus intraradices* (Schenck and Smith) were layered 4 cm or 8 cm, respectively, below the soil surface to supply 10 to 40 spores per seedling. After each harvest, the roots were immediately frozen in liquid nitrogen and kept at −70 °C until assayed. Mycorrhizal colonization was estimated colorimetrically by measuring glucosamine released from fungal chitin (Hepper, 1976). All assays were conducted in two or three replicates; each replicate containing roots from 12 seedlings. Analysis of variance was tested for significant ($p < 0.05$) differences, and when appropriate, standard error (SE) values were calculated.

RNA isolation, RNA transfer blot analysis, and in situ hybridization

These procedures were performed as previously described (McKhann and Hirsch, 1993; McKhann and Hirsch, 1994). RNA was isolated from the treated roots and subjected to electrophoresis. Twenty μg of total RNA were loaded per lane for the RNA transfer blot analyses. The RNA was subsequently transferred to nylon membranes for hybridization experiments with probes derived from *MsENOD40*-2 (Asad et al., 1994) and *A2ENOD2* (Dickstein et al., 1988) cDNAs. An *Msc27* cDNA clone (Kapros et al., 1992) was used to normalize RNA loading.

Alfalfa transformation and Gus staining

Five *MsENOD40* genomic clones were isolated by screening an alfalfa genomic library with the *MsENOD40*-2 cDNA clone (Asad et al., 1994). The putative promoter regions were subcloned into the HindIII-BamHI site of the binary vector pBI101.3 (Clontech). The resultant plasmids were electroporated into *A. tumefaciens* strains LBA4404, which was then used to transform alfalfa (*Medicago sativa* L.) cv. Regen. Transformation and plant regeneration procedures were performed as previously described (Hirsch et al., 1995).

Cuttings from transgenic plants were rooted in 1/4 strength complete Hoagland's medium for 3 to 4 weeks. Individual rooted plants were transferred into 50-mL Falcon tube (Fisher) containing 50 mL of nitrogen-free Jensen's medium. Six to 7 d after transfer, the plants were either inoculated with *R. meliloti* strain 1021, treated with purified Nod factor at 10^{-8}M, or with phytohormones at 10^{-6} M. Gus staining was carried out as described by Jefferson (1987).

Results

The MsENOD40 Gene is expressed in uninoculated roots treated with cytokinin

To monitor whether or not cytokinin was an inductive stimulus for *MsENOD40* gene expression, we treated

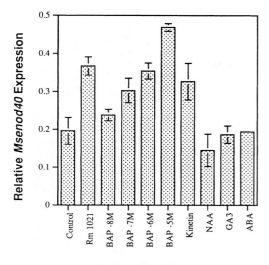

96 hr. after Treatment

Figure 2. Densitometric analysis of *MsENOD40* gene expression relative to *Msc27* in alfalfa seedling roots following Rm1021 inoculation or different phytohormone treatments for 4 d. 10^{-6} M IAA, 10^{-6} M GA$_3$, or 10^{-6} M ABA did not result in an increased accumulation of *MsENOD40* transcripts after this time period.

alfalfa roots with several concentrations of the different plant hormones (Figure 2). We found that the cytokinin benzylaminopurine (BAP) at concentrations ranging from 10^{-8} to 10^{-5} m induced *MsENOD40* and *MsENOD2* gene expression. BAP at 10^{-6} M induced *MsENOD40* gene expression to levels comparable to that of Rm1021 (Figure 2). This response appeared to be specific to cytokinins, as the only other phytohormone which induced *MsENOD40* gene expression was kinetin. All of the cytokinins are adenine derivatives. Other nucleotide bases, such as uridine (recently identified as the stele factor; Smit et al., 1995) did not induce *MsENOD40* gene expression (data not shown).

We then proceeded to localize the sites of *MsENOD40* gene expression in cytokinin-treated alfalfa roots using *in situ* hybridization. The roots were spot-inoculated either with 10^{-6} M BAP dissolved in a 1 mm^2 agar block or with a drop of SL44/pTZS bacteria as described in the Materials and methods. In both treatments, by *in situ* hybridization, we detected *MsENOD40* transcripts in the cells of the cortex and the epidermis, and also in the pericycle (Figure 3b and c). These are the same sites where *MsENOD40* is expressed in alfalfa roots upon *R. meliloti* inoculation (Asad et al., 1994; Crespi et al., 1994). Figure 3a is a sense control.

MsENOD40 and MsENOD2 are expressed in nodulation mutants of alfalfa that do not respond to R. meliloti.

The alfalfa mutant MN1008 undergoes neither root hair deformation nor cellular divisions in response to wild-type *R. meliloti* (Dudley and Long, 1989), suggesting that the mutation potential is in a receptor for Nod factor. When RNA from BAP-treated roots from rooted cuttings of MN1008 plants were examined by northern blot analysis, both *MsENOD40* and *MsENOD2* transcripts were detected (BAP; Figure 4). Both genes were also expressed in the parental cultivar Saranac (Figure 4).

M. sativa cv. MN1008 roots were spot-inoculated with Nod$^-$/pTZS$^+$ bacteria as well as with agar blocks containing different concentrations of BAP to determine the site of localization of *MsENOD40* transcripts. Twenty to 27 dpi with Nod$^-$/pTZS$^+$ rhizobia, small uninfected sites of cell division activity were observed on MN1008 roots (Figure 3d). No such sites were observed on mutant alfalfa roots spot-inoculated with BAP-containing agar blocks, even 27 d after treatment. The small regions of cell division activity in MN1008 roots included both cortical and pericycle divisions; vascular tissue differentiated into the base of the small nodule-like structure, but no further. *MsENOD40* transcripts were detected by *in situ* hybridization in epidermal cells, enlarged outer cortical cells and mid-cortical cells (arrows, Figure 3e), or in cells which appear to be derived from the pericycle (data not shown). No transcripts were detected in cells of a fully developed outgrowth, however (Figure 3d). We concluded that these sites of localized cell division were arrested nodules.

Although agar blocks containing BAP, even with concentrations as high as 10^{-4} M, did not routinely elicit cell divisions on treated alfalfa roots, microscopic examination of the roots indicated that histological changes occurred. Both epidermal and outer cortical cells expanded in response to BAP concentrations ranging from 10^{-7} to 10^{-4} M (Figure 3f).

Neither cell divisions, *MsENOD40* or *MsENOD2* gene expression, nor or any other histological changes were detected in control roots spot-inoculated with Jensen's medium (data not shown).

Isolation of the MsENOD40 gene promoter

Using a full-length alfalfa cDNA clone (*MsENOD40-2*; Asad et al., 1994), we screened an alfalfa genomic

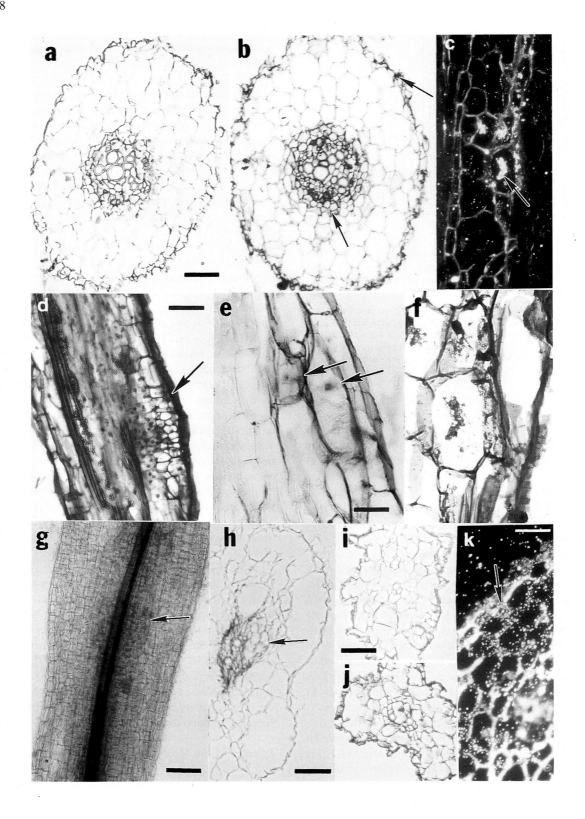

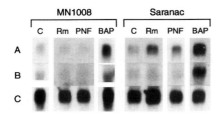

Figure 3. *MsENOD40* (A) and *MsENOD2* (B) genes can be induced by purified Nod factor or cytokinin treatment in wild-type and non-nodulating alfalfa mutant roots. Rooted cuttings were transferred to Jensen's medium without nitrogen and treated as indicated for 4 d. Twenty μg of total root RNA per lane were used for analysis. (C) *Msc27* was used to standardize the amount of RNA loaded. C, untreated control plants; Rm, wild-type *Rhizobium meliloti* strain 1021; PNF, 10^{-8} M purified Nod factor; BAP, 10^{-6} M benzylaminopurine. This figure is a composite from two independent experiments. Replicates gave identical results.

library to obtain phage DNAs that contained putative promoter elements. Southern analysis had previously indicated that there were probably two *ENOD40* genes in alfalfa (Asad et al., 1994). Five genomic clones were isolated, and two were found to be identical. Subsequent analysis of the four remaining clones showed that there were two distinct promoters; they were only 40% similar to one another at the 5'-distal ends. The two promoters were designated as *MsENOD40*-1 and *MsENOD40*-2 (Fang and Hirsch, manuscript submitted).

The transcriptional start site for each of the promoters was determined, and they were then transcriptionally fused to the reporter gene *uidA* (*gusA*). The constructs were introduced into *A. tumefaciens* by electroporation, and transformation of alfalfa (*Medicago*

sativa L. cv. Regen) proceeded as previously described (Hirsch et al., 1995). More than 50 independent transgenic lines were generated for each of the promoters.

We found that the full-length promoter-Gus constructs were expressed in the same tissues as *MsENOD40* transcripts were detected by *in situ* hybridization methods (Asad et al., 1994; Crespi et al., 1994). The *MsENOD40*-2 promoter was found to be stronger than the *MsENOD40*-1 promoter. Gus staining was detected within a few hours for the former promoter whereas roots containing constructs with the *MsENOD40*-1 promoter required staining for 10-12 h. For both promoters, Gus staining was found in the nodule meristem and in the pericycle of the nodule vascular bundles. However, Gus product indicating the expression of the *MsENOD40*-2 construct was found in non-symbiotic tissues even in uninoculated roots, e.g., in non-emergent lateral root primordia, and in the root pericycle. On the other hand, the *MsENOD40*-1 promoter appeared to be induced upon inoculation and Gus product was found in the root hairs, both outer and inner cortical cells, usually in the pericycle adjacent to the site where the nodule was to develop, and finally in the nodule primordium itself (Fang and Hirsch, manuscript submitted).

Ten different lines of plants containing the different promoter-constructs were then used for studies on cytokinin induction of the *MsENOD40* promoter. We found that both *MsENOD40* promoters were induced by cytokinin treatment two-fold over the untreated controls (Fang and Hirsch, manuscript submitted). Again, the *MsENOD40*-2 promoter was stronger than the *MsENOD40*-1 promoter. *MsENOD40*-2 promoter-Gus plants had Gus product localized to the stele,

Figure 3. Light micrographs of treated roots of alfalfa. (a) Transverse section of an alfalfa root treated with Nod$^-$/pTZS$^+$ bacteria for 9 d. *MsENOD40* sense control. Scale bar = 60 μm. (b) Transverse section of an alfalfa root treated with Nod$^-$/pTZS$^+$ bacteria for 9 d. *MsENOD40* expression as shown by the blue color (digoxigenen-labeled probe) is detected in the epidermis (arrow) and in the stele, specifically in the pericycle (arrow) and to some extent in the cortex (dividing nuclei). Same magnification as (a). (c) Longitudinal section showing the epidermis and cortex of an alfalfa root treated with 10^{-6} M BAP in agar for 20 d. *MsENOD40* transcripts as detected by an ^{35}S-labeled antisense probe (white dots are silver grains) are found over the nuclei of the cortical cells (arrow). Same magnification as (a). (d) A small region of cell division activity (arrow) on an MN1008 root collected 27 d after spot-inoculation with Nod$^-$/pTZS$^+$ bacteria. Scale bar = 60 μm. (e) Longitudinal section of an MN1008 root collected 20 d after spot-inoculation with Nod$^-$/pTZS$^+$ bacteria. *MsENOD40* transcripts are localized by the blue-purple color in epidermal and cortical cells (arrows). Note that the nucleus has moved from a peripheral to a central position within the expanded cortical cell. Digoxigenin-labeled probe. Scale bar = 30 μm. (f) Outer epidermis and cortex of an alfalfa root treated with 10^{-6} M BAP in agar for 20 d. Enlargement of (c). Same magnification as (e). (g) Part of an entire transgenic (*MsENOD40*-2 promoter-Gus construct) alfalfa root stained for Gus after treatment with 10^{-6} M BAP for 4 d. Blue color is detected in the cortex and in the stele. A region of cell division activity in the inner cortex is indicated by the arrow. (h) Transverse section of the same root shown in (g). Blue color indicating Gus product is found in the stele and in inner cortical cell derivatives (arrow). The blue color in the cortex does not show in the section. Scale bar = 60 μm. (i) Transverse section of a mycorrhizal root. *MsENOD40* sense control. Scale bar = 120 μm. (j) Transverse section of a mycorrhizal root showing blue color in most of the tissues of the root. *MsENOD40* antisense. Digoxigenin-labeled probe. Same magnification as (i). (k) Transverse to oblique section of a mycorrhizal root. *MsENOD40* transcripts (detected as white dots) are over most of the cortical cells. ^{35}S-labeled probe. Scale bar = 40 μm.

the pericycle, and the cortex following treatment with 10^{-6} M BAP for 3 days (Figure 3g). Occasionally, cell divisions were induced in the inner cortex following cytokinin treatment (arrow, Figure 3g). When these roots were sectioned, blue color was detected in the stele and in the inner cortical cell derivatives (Figure 3h). The blue color of the outer cortical cells was lost during the tissue processing for paraffin embeddment. The *MsENOD40*-1 promoter-Gus transgenic roots gave a similar response but the blue color of the outer cortical cells was not as intense as that just described for the *MsENOD40*-2 promoter. Thus, the cytokinin inducibility of *MsENOD40* is established, thereby confirming the idea that this gene can be considered a marker for increased cytokinin levels in the plant tissues.

The MsENOD40 gene is expressed in mycorrhizal roots

We examined RNA from alfalfa roots that had been inoculated either with *G. intraradices*, or left uninoculated. Hybridization to a *MsENOD40* probe occurred only in the roots that were colonized with *G. intraradices* and not in control roots (data not shown) (van Rhijn et al., 1997). *In situ* hybridization studies localized *MsENOD40* transcripts to the outer and inner cortical cells of the root as well as to the pericycle (Figure 3j and k). Figure 3i is a sense control.

Discussion

Several early nodulin (*ENOD*) genes, have been found to be induced by cytokinin–*ENOD2*, *ENOD12A*, and *ENOD40* (Dehio and deBruijn, 1992; Bauer et al., 1996; Hirsch and Fang, 1994; van Rhijn et al., 1997). The increase in transcript accumulation for these three early nodulin genes is thus likely to reflect the endogenous status of cytokinin in the inoculated root.

In this report, we have demonstrated that cytokinin induces *MsENOD40* gene expression not only in uninoculated alfalfa roots, but also in a non-nodulating mutant of alfalfa. This demonstrates that *MsENOD40* expression and Nod factor perception are uncoupled in the non-nodulating alfalfa MN1008, which is blocked in perceiving and transducing the first signal, i.e., Nod factor. However, MN1008 is not blocked in those stages that lead to nodule organogenesis. *MsENOD40* is expressed and aborted nodules are formed. One interpretation of these results is that a change in endogenous hormone balance occurs downstream of the perception of Nod factor. Nod factor perception could lead to an alteration of endogenous hormone(s) levels and/or sensitivities which then triggers early nodulin gene expression. This implies that in wild-type alfalfa Nod factor and exogenous cytokinin intercept at the same point in the signal transduction cascade although each may utilize different mechanisms to reach this common point.

However, other research points to the involvement of auxin in triggering nodulation as evidenced by the expression of auxin-induced genes in response to Nod factor or *Rhizobium*. Auxin rather than cytokinin levels could change after *Rhizobium* inoculation or Nod factor application, or alternatively Nod factor and auxin, instead of cytokinin, could work through the same signal transduction pathway. Van de Sande et al. (1996) have implicated auxin resistance as part of the response of tobacco protoplasts transformed with one of the conserved regions of the soybean *ENOD40* cDNA. Protoplasts lacking this *GmENOD40* construct are unable to divide in the presence of high auxin concentrations, whereas tobacco protoplasts carrying the construct divide even up to 13.8 μM NAA. These authors propose that *ENOD40* encodes a small peptide that modulates the action of auxin.

Röhrig et al. (1995) found that very low concentrations of synthetic Nod factors alleviated the requirement for auxin and cytokinin for continued division of tobacco protoplasts. Also, in these studies synthetic Nod factor treatment was found to induce the expression of genes which are auxin- responsive. The -90-base pair region of the CaMV 35S RNA promoter was fused to Gus and used to measure transient expression after synthetic Nod factor application; this promoter region is responsive to auxin, jasmonic acid, and salicylic acid. Similarly, the *axi1* gene, which is both auxin- and cytokinin responsive, is expressed in response to synthetic Nod factor and derivatives.

The auxin responsive promoter *GH3* (Hagen et al., 1991) linked to Gus as well as constructs consisting of chalcone synthase (CHS) genes fused to a Gus reporter gene have been used to gauge the auxin levels of white clover (*Trifolium repens*) in response to *Rhizobium* inoculation (Larkin et al., 1996) and to target molecules such as Nod factor, auxin, NPA, and flavonoids ballistically into clover root cells (Mathesius et al., 1997). *GH3* expression linked to Gus is detected in inner cortical cells as early as 10 h post-inoculation, but not in the nodule primordium approximately 70 h after inoculation. Seven dpi, the nodule primordium cells which will differentiate into vascular tissue stain blue and later, Gus activity is detected in the nodule meristem suggesting that auxin levels are elevated in these cells.

Thus, there is evidence for the involvement of both auxin and cytokinin in the elicitation of nodule development. To understand how these two growth regulators change in the root in response to *Rhizobium* inoculation, we first must review the status of the phytohormones in the uninoculated root. Auxin transport in the root is thought to proceed basipetally from the shoot and then acropetally in the root via the vascular parenchyma of the stele. A considerable amount of auxin accumulates in the root tip and it is likely that some of this auxin is actually synthesized by the root tip. Immunolocalization of auxin in the root tip of *Zea mays* shows that the auxin levels in certain root tip cells–quiescent center, root cap cells, vascular tissue and outer cortical cells are elevated compared to others–root cap meristem region, inner cortex, and epidermis (Kerk and Feldman, 1995). Uninoculated transgenic clover roots exhibit *GH3*-Gus expression in the phloem region of the stele and in the root apical meristem as well as the pericycle just before and during the formation of a lateral root, indicating that auxin is accumulating in these cells (B G Rolfe, pers. comm.).

Figure 5 summarizes the changes in auxin and cytokinin concentrations after inoculation based on our studies and those of Mathesius et al. (1997). Twenty-four to 50 h after inoculation, auxin accumulates transiently in the inner cortical cells that are opposite a protoxylem point (Mathesius et al., (1997)) (Figure 5). Although the source of auxin is unknown, it is likely to be derived from the vascular tissues. However, another possibility is that there is a localized accumulation of auxin due to a block in auxin transport. This block in auxin transport could come about due to a localized increase in flavonoid production (Hirsch, 1992). Flavonoids are known to function as auxin transport inhibitors (Jacobs and Rubery, 1988). Evidence supporting this hypothesis comes from Mathesius et al. (1997) who used a *CHS* promoter-Gus construct in transgenic clover roots to detect localized changes in the activity of this promoter in response to inoculation.

What about cytokinin? Based on our studies with the *MsENOD40*-1 promoter-Gus fusions, cytokinin levels are elevated in cells opposite a protoxylem point soon after inoculation also (Fang and Hirsch, manuscript submitted). The response of the *MsENOD40*-2 promoter to exogenous cytokinin replicates this result (see Figure 3g and h). Unlike auxin, however, cytokinin accumulates in a horizontal gra-

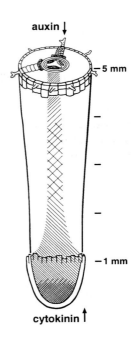

Figure 4. Diagram of a indeterminate nodule-forming legume root showing the localization of *GH3*-Gus activity [right to left slanted lines indicating auxin levels; based on the results of Mathesius et al., (1997)] and *MsENOD40*-Gus activity (left to right stanted lines indicating cytokinin levels; based on this report and Fang and Hirsch, manuscript submitted) approximately 48 hours after inoculation with *R. meliloti*. The inner cortical cells, where both auxin and cytokinin concentrations are elevated, are stimulated to divide. Auxins and cytokinins (based on Gus reporting) overlap also in the root tip and in the stele approximately 2 to 3 mm behind the root apical meristem.

dient extending from the root hair cells, extending through the outer cortical cells and to the inner cortical cells; Gus product is found in all these cells (summarized in Figure 5). The higher level of cytokinins observed in these cells may be reflective of or causal for the change in cell cycle from G_1 to G_2 (Yang et al., 1994). Nevertheless, the outer cortical cells do not divide even though they have been activated. Instead, they form a phragmoplast or preinfection thread, a mass of centralized cytoplasm through which the infection thread penetrates (van Brussel et al., 1992), whereas the inner cortical cells divide and establish the nodule primordium. Thus, the cells which show an increase in both cytokinin and auxin levels based on the detection of Gus product from the two different promoters are stimulated to divide. In contrast, the outer cortical cells, which do not accumulate auxin, remain undivided.

Figure 5, which is based on the hypothesis that both auxin and cytokinin are required for the cell divisions lead to the formation of the nodule primordium, allows us to make some predictions. If both auxin and cytokinin levels must increase for the elicitation of cell divisions, then the pattern for determinate nodules would be such that *GH3*-Gus and *ENOD40*-Gus together would first be expressed in the outer cortical cells and then in the inner cortical cells. Moreover, mycorrhizal roots which express *ENOD40*, but which do not exhibit any cell divisions, should exhibit little or no *GH3*-Gus expression in the root zones where AM-fungi are present. We measured the concentration of cytokinins by radioimmunoassay and found that ZR concentration within the mycorrhizal roots was significantly higher than the concentrations in control roots (van Rhijn et al., 1997). Thus, elevated cytokinin concentration and *MsENOD40* gene expression are coupled in this symbiosis as well.

If these predictions hold, then we will have an understanding of some of the triggers for cell division leading to nodule formation. However, we will still not understand how Nod factor sets this entire process into motion. This is a challenge for the future.

Acknowledgements

Thanks are extended to James B Cooper (UC-Santa Barbara) for making the Nod⁻/pTZS⁺ bacterial strain available to us and informing us of his unpublished data on MN1008 producing "bumps" after inoculation with this strain. We thank David Ehrhardt and Sharon R Long (Stanford University) for purified Nod factor, Don Barnes and Carroll Vance (U Minn.) for seeds of alfalfa MN1008 and its parental line, and Agway, Inc. (Syracuse, NY) for alfalfa cv. Iroquois seeds. We are grateful to Tom LaRue (Boyce Thompson Institute) for his helpful comments on a previous version of this manuscript. Gratitude is extended to Lew Feldman (UC Berkeley) and Barry G Rolfe (Australian National University) for insights and unpublished data. We also thank Margaret Kowalczyk for Figure 5, and Shelly Thai, Yousun Kim, Clive Spray, and members of our laboratory for their help. This research was funded in part by NSF 90-23888 to AMH and a Pakistan Participant Training Program grant (USAID) to SA.

References

Allen E K, Allen O N and Newman A S 1953 Pseudonodulation of leguminous plants induced by 2-bromo-3,5-dichlorobenzoic acid. Amer. J. Bot. 40, 429–435.

Allen M F, Moore T S Jr and Christensen M 1980 Phytohormone changes in *Bouteloua gracilis* infected by vesicular-arbuscular mycorrhizae. I. Cytokinin increases in the host plant. Can. J. Bot. 58, 371–374.

Allen M F, Moore T S Jr and Christensen M 1982 Phytohormone changes in *Bouteloua gracilis* infected by vesicular-arbuscular mycorrhizae. II. Altered levels of gibberellin-like substances and abscisic-acid in the host plant. Can. J. Bot. 60, 468–471.

Arora N, Skoog F and Allen O N 1959 Kinetin-induced pseudonodules on tobacco roots. Amer. J. Bot. 46, 610–613.

Ardourel M, Demont N, Debellé F, Maillet F, de Billy F, Promé J-C, Dénarié J and Truchet G 1994 *Rhizobium meliloti* lipooligosaccharide nodulation factors: different structural requirements for bacterial entry into target root hair cells and induction of plant symbiotic developmental responses. Plant Cell 6, 1357–1374.

Asad S, Fang Y, Wycoff K and Hirsch A M 1994 Isolation and characterization of cDNA and genomic clones of *MsENOD40*; transcripts are detected in meristematic cells. Protoplasma. 183, 10–23.

Badenoch-Jones J, Parker C W and Letham D S 1987 Phytohormones, *Rhizobium*-mutants, and nodulation in legumes VII. Identification and quantification of cytokinins in effective and ineffective pea root nodules using radioimmunoassay. J. Plant Growth Regul. 6, 97–111.

Bauer P, Ratet P, Crespi M D, Schultze M and Kondorosi A 1996 Nod factors and cytokinins induce similar cortical cell division, amyloplast deposition and *MsEnod12A* expression patterns in alfalfa roots. Plant J. 10, 91–106.

Barea J M and Azcon-Aguilar C 1982 Production of plant growth-regulating substances by vesicular-arbuscular mycorrhizal fungus *Glomus mosseae*. Appl. Environ. Microbiol. 43, 810–813.

Berry A M, Kahn R K S and Booth M C 1989 Identification of indole compounds secreted by *Frankia* HFPArI3 in defined culture medium. Plant Soil 118, 205–209.

Bhuvaneswari T V, Turgeon G B and Bauer W D 1980 Early events in the infection of soybean (*Glycine max* L. Merr.) by *Rhizobium japonicum*. I. Localization of infectible root cells. Plant Physiol. 66, 1027–1031.

Brown M E 1972 Plant growth substances produced by microorganisms of soil and rhizosphere. J. Appl. Bacteriol. 35, 443–451.

Caetano-Anollés G, Joshi P A and Gresshoff P M 1992 Nodulation in the absence of *Rhizobium*. *In* Current Topics in Plant Molecular Biology. vol. 1, Plant Biotechnology and Development. Ed. P M Gresshoff. pp 61–70. CRC Press, Boca Raton, FL.

Chen L, Cui Y, Qin M, Wang Y, Bai X and Ma Q 1992 Identification of a *nodD*-like gene in *Frankia* by direct complementation of a *Rhizobium nodD*-mutant. Mol. Gen. Genet. 233, 311–314.

Cooper J B and Long S R 1994 Morphogenetic rescue of *Rhizobium meliloti* nodulation mutants by *trans*-zeatin secretion. Plant Cell. 6, 215–225.

Crafts C B and Miller C O 1974 Detection and identification of cytokinins produced by mycorrhizal fungi. Plant Physiol. 54, 586–588.

Crespi M D, Jurkevitch E, Poiret M, d-Aubenton-Carafa Y, Petrovics G, Kondorosi E and Kondorosi A 1994 *enod40*, a gene expressed during nodule organogenesis, codes for a non-translatable RNA involved in plant growth. EMBO J. 13, 5099–5112.

Dehio C and deBruijn F J 1992 The early nodulin gene *SrEnod2* from *Sesbania rostrata* is inducible by cytokinin. Plant J. 2, 117–128.

Dickstein R, Bisseling T, Reinhold V N and Ausubel F M 1988 Expression of nodule-specific genes in alfalfa root nodules blocked at an early stage of development. Genes Dev. 2, 677–687.

Dixon R K, Garrett H E and Cox G S 1988 Cytokinins in the root pressure exudate of *Citrus jambbiri* Lush colonized by vesicular-arbuscular mycorrhizae. Tree Physiol. 4, 9–18.

Drüge U and Schönbeck F 1992 Effect of vesicular-arbuscular mycorrhizal infection on transpiration, photosynthesis and growth of flax (*Linum usitatissimum* L.) in relation to cytokinin levels. J. Plant Physiol. 141, 40–48.

Dudley M E and Long S R 1989 A non-nodulating alfalfa mutant displays neither root hair curling nor early cell division in response to *Rhizobium meliloti*. Plant Cell 1, 65–72.

Dudley M E, Jacobs T W and Long S R 1987 Microscopic studies of cell divisions induced in alfalfa roots by *Rhizobium meliloti*. Planta (Berl.) 171, 289–301.

Edriss M H, Davis R M and Burger D W 1984 Influence of mycorrhizal fungi on cytokinin production in sour orange. J. Amer. Soc. Hort. Sci. 109, 587–590.

Ehrhardt D W, Atkinson E M and Long S R 1992 Depolarization of alfalfa root hair membrane potential by *Rhizobium meliloti*. Planta (Berl.) 171, 289–301.

Ehrhardt D W, Wais R and Long S R 1996 Calcium spiking in plant root hairs responding to *Rhizobium* nodulation signals. Cell 85, 673–682.

Hagen G, Martin G, Li Y and Guilfoyle T J 1991 Auxin-induced expression of the soybean GH3 promoter in transgenic tobacco plants. Plant Molec. Biol. 17, 567–579.

Henson I E and Wheeler C T 1976 Hormones in plants bearing nitrogen-fixing root nodules: The distribution of cytokinins in *Vicia faba* L. New Phytol. 76, 433–439.

Henson I E and Wheeler C T 1977 Hormones in plants bearing nitrogen fixing root nodules: Cytokinin levels in roots and root nodules of some nonleguminous plants. Z. Pflanzenphysiol. 84, 179–182.

Hepper C M 1976 A colorimetric method for estimating vesicular arbuscular infection in roots. Soil Biol. Biochem. 9, 15–18.

Hirsch A M 1992 Developmental biology of legume nodulation. New Phytol. 122, 211–237.

Hirsch A M, Bhuvaneswari T V, Torrey J G and Bisseling T 1989 Early nodulin genes are induced in alfalfa root outgrowths elicited by auxin transport inhibitors. Proc. Natl. Acad. Sci. USA. 86, 1244–1249.

Hirsch A M, Brill L M, Lim P O, Scambray J and van Rhijn P 1995 Steps toward defining the role of lectins in nodule development in legumes. Symbiosis 19, 155–173.

Hirsch A M and Fang Y 1994 Plant hormones and nodulation: What's the connection? Plant Mol. Biol. 26, 5–9.

Hirsch A M, McKhann H I and Löbler M 1992 Bacterial-induced changes in plant form and function. Inter. J. Plant Sci. 153, S171–S181.

Hyman D S 1980 Mycorrhiza and crop production. Nature (London). 287, 487–488.

Jacobs M and Rubery P H 1988 Naturally-occurring auxin transport regulators. Science 241, 346–349.

Jaiswal V, Rizvi S J H, Mukerji D and Mature S N 1981 Cytokinins in root nodules of *Phaseolus mungo*. Ann. Bot. 48, 301–305.

Jefferson R A 1987 Assaying chimeric genes in plants. The Gus gene-fusion system. Plant Mol. Rep. 5, 387–405.

Kapros T, Bögre L, Németh K, Bakö L, Györgyey J, Wu S C and Dudits D 1992 Differential expression of histone H3 gene variants during cell cycle and somatic embryogenesis in alfalfa. Plant Physiol. 98, 621–625.

Kerk N M and Feldman L J 1995 A biochemical model for the initiation and maintenance of the quiescent center: implications for organization of root meristems. Development 121, 2825–2833.

Kneen B E and LaRue T A 1988 Induced symbiosis mutants of pea (*Pisum sativum*) and sweetclover (*Melilotus alba annua*). Plant Sci. 58, 177–182.

Kurkdjian A C 1995 Role of the differentiation of root epidermal cells in Nod factor (from *Rhizobium meliloti*)-induced root-hair depolarization of *Medicago sativa*. Plant Physiol. 107, 783–790.

Larkin P J, Gibson J M, Mathesius U, Weinman J J, Gartner E, Hall E, Tanner G J, Rolfe B G and Djordjevic M A 1996 Transgenic white clover. Studies with the auxin-responsive promoter *GH3* in root gravitropisms and lateral root development. Transg. Res. 5, 1–11.

LaRue T A and Weeden N F 1994 The symbiosis genes of the host. *In* Proc. 1st European Nitrogen Fixation Conference. Eds. G B Kiss and G Endre. pp 147–151. Officina Press, Szeged, Hungary.

Lerouge P, Roche P, Faucher C, Maillet F, Truchet G, Promé J-C, and Dénarié J 1990 Symbiotic host-specificity of *Rhizobium meliloti* is determined by a sulphated and acylated glucosamine oligosaccharide signal. Nature 344, 781–784.

Libbenga K R and Bogers R J 1974 Root-nodule morphogenesis. *In* The Biology of Nitrogen Fixation. Ed. A Quispel. pp. 430–472. North-Holland Publish. Co. Amsterdam.

Libbenga K R, van Iren F, Bogers R J and Schraag- Lamers M F 1973 The role of hormones and gradients in the initiation of cortex proliferation and nodule formation in *Pisum sativum* L. Planta (Berl.) 114, 29–30.

López-Lara I M, van der Drift K M G M, van Brussel A A N, Haverkamp J, Lugtenberg B J J, Thomas- Oates J E and Spaink H P 1995 Induction of nodule primordia on *Phaseolus* and *Acacia* by lipo-chitin oligosaccharide nodulation signals from broad-host-range *Rhizobium* strain GRH2. Plant Molec. Biol. 29, 465–477.

MacDonald E M S, Akiyoshi D E and Morris R O 1981 Combined high performance liquid chromatography radioimmunoassay for cytokinins. J. Chromatogr. 214, 101–109.

Machlis L, and Torrey J G 1956 Plants in Action: A Laboratory Manual of Plant Physiology. Freeman, San Francisco

Mathesius U, Schlaman H R M, Meijer D, Lugtenberg B J J, Spaink H P, Weinman J J, Roddam L F, Sautter C, Rolfe B G and Djordjevic M A 1997 New tools for investigating nodule initiation and ontogeny: spot inoculation and microtargeting of transgenic white clover roots shows auxin involvement and suggests a role for flavonoids. *In* Advances in Molecular Genetics of Plant-Microbe Interactions. vol. 4. Eds. G Stacey, B Mullin and P M Gresshoff. pp 353–358. Kluwer Academic Publishers, Dordrecht.

McKhann H I and Hirsch A M 1993 *In situ* localization of specific mRNAs in plant tissues. *In* Methods in Plant Molecular Biology and Biotechnology. Eds. B R Glick and J E Thompson. pp. 179–205. CRC Press, Boca Raton.

McKhann H I and Hirsch A M 1994 Isolation of chalcone synthase and chalcone isomerase cDNAs from alfalfa (*Medicago sativa* L.): highest transcript levels occur in young roots and root tips. Plant Mol. Biol. 24, 767–777.

Okon Y and Itzigsohn R 1995 The development of *Azospirillum* as commercial inoculum for improving crop yield. Biotech. Adv. 13, 415–424.

Puppo A, Rigaud J and Barthe P 1974 Sur la presence de cytokinines dans les nodules de *Phaseolus vulgaris* L. C. R. Acad. Sci. Paris, Ser. D. 279, 2029–2032.

184

Reddy A, Bochenek B and Hirsch A M 1992 A new *Rhizobium meliloti* symbiotic mutant isolated after introducing *Frankia* DNA sequence into a *nodA*::Tn*5* strain. Molec. Plant-Microbe Inter. 5, 62–71.

Relic B, Talmont F, Kopcinska J, Golinowski W, Promé J-C and Broughton W J 1994 Biological activity of *Rhizobium* sp. NGR234 Nod-factors on *Macroptilium atropurpureum*. Molec. Plant-Microbe Inter. 6, 764–774.

Rodriguez-Barreuco C and Bermudez de Castro F 1973 Cytokinin-induced pseudonodules on *Alnus glutinosa*. Physiol. Plant. 29, 277–280.

Rodriguez-Barreuco C, Miguel C and Palni L M S 1979 Cytokinins in root nodules of the nitrogen-fixing non-legume *Myrica gale* L. Z. Pflanzenphysiol. 95S, 275–278

Röhrig H, Schmidt J, Walden R, Czaja I, Mikasevics E, Wieneke U, Schell J and John M 1995 Growth of tobacco protoplasts stimulated by synthetic lipo-chitooligosaccharides. Science 269, 841–884.

Scheres B, McKhann H I, Zalensky A, Löbler M, Bisseling T and Hirsch A M 1992 The PsENOD12 gene is expressed at two different sites in Afghanistan pea pseudonodules induced by auxin transport inhibitors. Plant Physiol. 100, 1649–1655.

Schultze M, Quiclet-Sire G, Kondorosi E, Virelizier H, Glushka J N, Endre G, Géro S D and Kondorosi A 1992 *Rhizobium meliloti* produces a family of sulfated lipooligosaccharides exhibiting different degrees of plant host specificity. Proc. Natl. Acad. Sci. USA. 89, 191–196.

Short K C and Torrey J G 1972 Cytokinins in seedling roots of pea. Plant Physiol. 49, 155–160.

Spaink H P, Bloemberg G V, Wijfjes A H M, Ritsema T, Geiger O, López-Lara I M, Harteveld M, Kafetzopoulous D, van Brussel A A N, Kijne J W, Lugtenberg B J J, van der Drift K M G M, Thomas-Oates J E, Potrykus I and Sautter C 1994 The molecular basis of host specificity in the *Rhizobium leguminosarum*-plant interaction. *In* Advances in Molecular Genetics of Plant-Microbe Interactions. vol. 3. Eds. M J Daniels, J A Downie, and A E Osbourn. pp 91–98. Kluwer Academic Publ., Dordrecht.

Staehelin C, Schultze M, Kondorosi E and Kondorosi A 1995 Lipo-chitooligosaccharide nodulation signals from *Rhizobium meliloti* induce their rapid degradation by the host plant alfalfa. Plant Physiol. 108, 1607–1614.

Stevens G A and Berry A M 1988 Cytokinin secretion by *Frankia* sp. HFPArI3 in defined medium. Plant Physiol. 87, 15–16.

Strzelczyk E, Sitek J M and Kowalski S 1977 Synthesis of auxin from tryptophan and tryptophan-precursors by fungi isolated from mycorrhizae of pine (*Pinus sylvestris* L.). Acta Microbiol. Pol. 26, 255–264.

Sturtevant D B and Taller B J 1989 Cytokinin production by *Bradyrhizobium japonicum*. Plant Physiol. 89, 1247–1252.

Syono K and Torrey J G 1976 Identification of cytokinins of root nodules of the garden pea, *Pisum sativum* L. Plant Physiol. 57, 602–606.

Taller B J and Wong T-Y 1989 Cytokinins in *Azotobacter vinelandii* culture medium. Appl. Environ. Microbiol. 5, 266–267.

Thimann K V 1936 On the physiology of the formation of nodules on legume roots. Proc. Natl. Acad. Sci. USA. 22, 511–513.

Torrey J G 1961 Kinetin as trigger for mitosis in mature endomitotic plant cells. Exp. Cell Res. 23, 291–299.

Torrey J G 1986 Endogenous and exogenous influences on the regulation of lateral root formation. *In* New Root Formation in Plants and Cuttings Ed. M B Jackson. pp 31–66. Martinus Nijhoff Publishers, Dordrecht.

Truchet G, Barker D G, Camut S, de Billy F, Vasse J and Huguet T 1989 Alfalfa nodulation in the absence of *Rhizobium*. Molec. Gen. Genet. 219, 65–68.

Truchet G, Roche P, Lerouge P, Vasse J, Camut S, deBilly F, Promé J-C and Dénarié J 1991 Sulphated lipooligosaccharide signals from *Rhizobium meliloti* elicit root nodule organogenesis in alfalfa. Nature 351, 670–673.

Van Brussel A A N, Bakhuizen R, van Spronsen P C, Spaink H P, Tak T, Lugtenberg B J J and Kijne J W 1992 Induction of pre-infection thread structures in the leguminous host plant by mitogenic lipo-oligosaccharides of *Rhizobium*. Science 257, 70–72.

Van de Sande K, Pawlowski K, Czaja I, Wieneke U, Schell J, Schmidt J, Walden R, Matvienko M, Wellink J, van Kammen A, Franssen H and Bisseling T 1996 Modification of phytohormone response by a peptide encoded by *ENOD40* of legumes and a nonlegume. Science 273, 370–373.

Van Rhijn P and Vanderleyden J 1995 The *Rhizobium*-plant symbiosis. Micro. Rev. 59, 124–142.

Van Rhijn P, Fang Y, Galili S, Shaul O, Atzmon N, Winiger S, Eshead Y, Kapulnik Y, Lum M, Li Y, To V, Fujishige N and Hirsch A M 1997 Signal transduction pathways in forming arbuscular-mycorrhizae and *Rhizobium*-induced nodules may be conserved based on the expression of early nodulin genes in alfalfa mycorrhizae. Proc. Natl. Acad. Sci. (USA) (*In press*).

Van Stader J and Davey J E 1979 The synthesis, transport and metabolism of endogenous cytokinins. Plant Cell Environ. 2, 93–106.

Vincent J M 1970 A Manual for the Practical Study of Root-Nodule Bacteria. IBP Handbook No. 15. Blackwell, Oxford.

Volpin H, Elkind Y, Okon Y and Kapulnik Y 1994 A vesicular arbuscular mycorrhizal fungus (*Glomus intraradix*) induces a defense response in alfalfa roots. Plant Physiol. 104, 683–689.

Wu C, Dickstein R, Cary A J and Norris J H 1996 The auxin transport inhibitor *N*-(1-naphthyl)phthalamic acid elicits pseudonodules on nonnodulating mutants of white sweetclover. Plant Physiol. 110, 501–510.

Yang W-C, de Blank C, Meskiene I, Hirt H, Bakker J, van Kammen A, Franssen H and Bisseling T 1994 *Rhizobium* Nod factors reactive the cell cycle during infection and nodule primordium formation, but the cycle is only completed in primordium formation. Plant Cell 6, 1415–1426.

Guest editors: J K Ladha, F J de Bruijn and K A Malik

Plant and Soil **194**: 185–192, 1997.

The impact of molecular systematics on hypotheses for the evolution of root nodule symbioses and implications for expanding symbioses to new host plant genera

Susan M. Swensen[1] and Beth C. Mullin[2]

[1]*Department of Biology, Ithaca College, Ithaca, NY 14850, USA and* [2]*The Department of Botany and the Center for Legume Research, The University of Tennessee, Knoxville, TN 37996, USA*

Key words: actinorhizal plants, evolution, nitrogen fixation, phylogenetic hypotheses

Abstract

Current taxonomic schemes place plants that can participate in root nodule symbioses among disparate groups of angiosperms. According to the classification scheme of Cronquist (1981) which is based primarily on the analysis of morphological characters, host plants of rhizobial symbionts are placed in subclasses Rosidae and Hamamelidae, and those of *Frankia* are distributed among subclasses Rosidae, Hamamelidae, Magnoliidae and Dilleniidae. This broad phylogenetic distribution of nodulated plants has engendered the notion that nitrogen fixing endosymbionts, particularly those of actinorhizal plants, can interact with a very broad range of unrelated host plant genotypes. New angiosperm phylogenies based on DNA sequence comparisons reveal a markedly different relationship among nodulated plants and indicate that they form a more coherent group than has previously been thought (Chase et al., 1993; Swensen et al., 1994; Soltis et al., 1995). Molecular data support a single origin of the *predisposition* for root nodule symbiosis (Soltis et al., 1995) and at the same time support the occurrence of multiple origins of symbiosis within this group (Doyle, 1994; Swensen, 1996; Swensen and Mullin, In Press).

Introduction

As long as humans have observed the natural world around them, they have attempted to identify and classify the organisms that they observe. In general, natural systems of classification, that is those that relied on the fundamental biological differences and similarities among and between groups, have been preferred over artificial systems. In the mid 18th century in an attempt to establish a natural system of classification, Linneaus set forth his Sexual System of classification in which all known genera of angiosperms were organized according numbers of carpels and stamens (Linneaus, 1753). With publication of the Origin of Species (Darwin, 1859) and gradual acceptance of Darwin's theory of evolution, there existed a rationale for natural systems of classification based on the evolutionary relatedness of organisms. As the concept of evolution was embraced, naturalists, and more recently systematists, have strived to create classification systems that best reflect the evolutionary relatedness of taxa rather than what might be superficial phenotypic similari-

ties. As many systematists will agree, it is not always easy to determine which characteristics of a group of organisms best reflect their evolutionary relatedness. For this reason, there are frequently multiple systems of classification set forth for any group of organisms, each system reflecting the best judgment of the proposer. For example, there is no consensus modern day classification scheme for the angiosperms. The classification schemes of Cronquist (1981), Dahlgren (1980), Takhtajan (1980), and Thorne (1992), although possessing many common elements, differ in the placement of a number of taxa, because of their differing interpretation of the presence or absence of particular characters or groups of characters. Cronquist (1988) has referred to taxonomy as an "artful science" acknowledging the subjectivity involved in the interpretation of taxonomic data. Species belong to genera, genera are placed within families, families within orders, orders within subclasses and classes, classes within divisions, and divisions within kingdoms. No group can be left out of a taxonomic scheme because of insufficient data. It is the job of the taxonomist to place it somewhere, no matter how tenuous the placement. It is with this background in mind that we discuss the placement of actinorhizal host plants within the classification schemes of angiosperms.

Current classification of actinorhizal plants

Approximately 194 plant species distributed among 24 genera and eight families of dicotyledonous angiosperms are known to participate in nitrogen-fixing symbiosis with *Frankia* (Benson and Silvester, 1993). Traditional taxonomic schemes based predominantly on morphological characters assign the actinorhizal host plant families to disparate orders and subclasses among the angiosperms. For example, Cronquist (1988) places actinorhizal genera in seven higher plant orders distributed among four of the six subclasses of angiosperms (Figure 1). Such taxonomic schemes have led to the belief that many actinorhizal host plants are only very distantly related to one another. The only two features previously known to unite actinorhizal host plants have been the ability to be nodulated by *Frankia* and possession of a woody or suffrutescent habit. The actinorhizal genera *Cercocarpus, Cowania, Chamaebatia, Dryas* and *Purshia* are securely placed within the family Rosaceae; *Ceanothus, Colletia, Discaria, Retanilla, Kentrothamnus, Talguenea* and *Trevoa* within the Rhamnaceae, which are generally accepted as being closely allied with *Elaeagnus, Hippophaë* and *Shepherdia* of the Elaeagnaceae, all within the Subclass Rosidae. *Alnus* (Betulaceae), *Casuarina, Allocasuarina, Gymnostoma* and *Ceuthostoma* (Casuarinaceae), and *Myrica* and *Comptonia* (Myricaceae) are considered by Takhtajan and Cronquist to be in the Hamamelidae although considered to be "higher hamamelids." However, the "higher hamamelids" of Cronquist are considered by Thorne to be closer to the rosids that the hamamelids (Thorne, 1992). The placement of the two remaining actinorhizal genera *Datisca* (Datiscaceae) and *Coriaria* (Coriariaceae) have been the most problematic. The Datiscaceae have been allied with as many as six different families distributed among the subclasses Rosidae and Dilleniidae, and *Coriaria* has been placed within the orders Ranunculales (Cronquist, 1981), the Sapindales (Dahlgren, 1980; Thorne, 1992) and the Rutales (Takhtajan, 1980) distributed among the subclasses Rosidae and Magnoliidae. With such disagreement among systematists, it was evident that new data

were needed in order to resolve the phylogenetic affinities of the actinorhizal host plants.

Molecular phylogenies

It is well established that amino acid and nucleic acid sequence data have proven to be powerful characters for establishing phylogenetic hypotheses. Among plants the nucleotide sequence of *rbcL*, the chloroplast gene coding for the large subunit of ribulose-1,5-bisphosphate carboxylase/oxygenase (Rubisco) has proved to be very useful in constructing phylogenetic trees because it is highly enough conserved to be useful to assess distant relationships and yet it contains enough variation to be useful to distinguish among most genera. An analysis of *rbcL* nucleotide sequence data from members of the Datiscaceae indicated that this family was not monophyletic and that the genus *Datisca* was affiliated more closely with members of the families Begoniaceae and Cucurbitaceae than with the two non-actinorhizal members of the Datiscaceae, *Octomeles* and *Tetrameles* (Swensen et al., 1994). In addition, this analysis provided evidence for a much closer relationship between *Datisca* and *Coriaria* than had been previously proposed, and pointed to a possible affinity with the "higher hamamelids."

By 1993 the database of *rbcL* nucleotide sequences had grown to the point where it was possible to construct a global molecular phylogeny of the angiosperms (Chase et al., 1993). The phylogenetic trees resulting from these analyses confirmed many of the relationships previously determined for angiosperm groups. However, two major rearrangements affected the placement of actinorhizal genera. The analysis of Chase et al. indicated that Subclass Dilleniidae did not form a natural group, and taxa currently assigned to it had phylogenetic affinities among the five other subclasses. A second major finding was that the so-called "higher hamamelids" in Subclass Hamamelidae grouped in the molecular analysis with plants in Subclass Rosidae. In fact, all of the actinorhizal plants included in this analysis were grouped within Rosid I of the three Rosid clades found in this analysis (Figure 2). Although only a few actinorhizal genera were represented in this study, it was evident that in general actinorhizal genera were more closely related to one another than had previously been thought.

The addition of more actinorhizal taxa and their putative relatives to the *rbcL* database has led to more refined phylogenetic trees that group all known

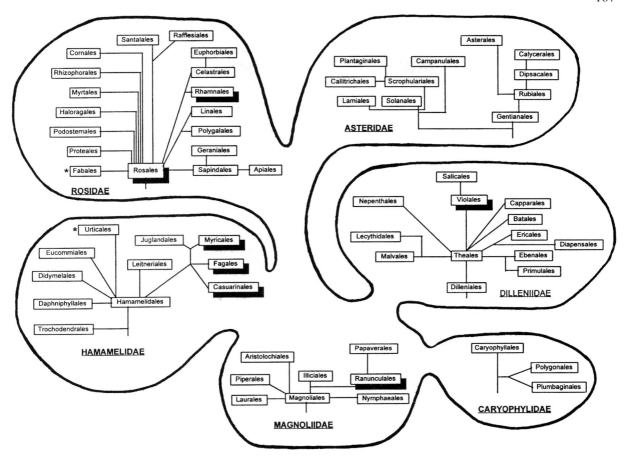

Figure 1. Putative relationships among the subclasses and orders of dicots according to Cronquist, 1988. This figure is a modified composite figure compiled and redrawn from separate figures in Cronquist's The Evolution and Classification of Flowering Plants, Second Edition, Copyright 1988, reprinted by permission from The New York Botanical Garden. Orders that contain actinorhizal genera are highlighted. Orders that contain genera nodulated by rhizobia are marked with an asterisk.

actinorhizal taxa, nonsymbiotic relatives, as well as legumes and *Parasponia* within a single clade. This symbiotic nitrogen fixing root nodule clade falls within Rosid I shown in Figure 2 (Soltis et al., 1995). This remarkable finding indicates that only one small group of angiosperms possesses the genetic predisposition to host nitrogen-fixing symbionts. In the case of legumes the symbiont is either *Rhizobium*, *Bradyrhizobium* or *Azorhizobium*, in the case of *Parasponia*, *Bradyrhizobium*, and in the case of the actinorhizal plants the symbiont is the actinomycete *Frankia*.

Evidence for multiple origins of symbiosis

In both legumes and actinorhizal plants a variety of evidence points to the occurrence of multiple origins of nodulation within each group (Sprent, 1994; Doyle, 1994; Swensen, 1996). Figure 3 is a phylogenetic tree based on *rbcL* nucleotide sequence data from many of the actinorhizal genera and their close relatives. Several different scenarios are possible. It is possible that the ability to be nodulated by *Frankia* was gained by an ancestral plant at the point in the tree of the asterisk. This event would have to have been followed by a series of losses of the ability to nodulate (15 losses would be required), leading to the extant actinorhizal plants. In addition, in one case, loss of the ability to be nodulated by *Frankia* would have had to be followed

188

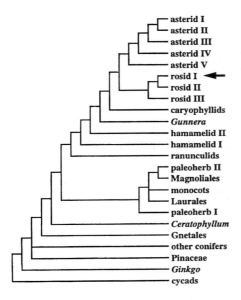

Figure 2. Summary of the major clades identified in the strict consensus of 3900 equally parsimonious trees based on *rbcL* sequences for 499 taxa. Reprinted from Chase et al., 1993 with permission from The Missouri Botanical Garden. The figure has been modified with an arrow pointing to the Rosid clade which contains all known plants that participate in nitrogen-fixing root nodule symbiosis.

by a gain by *Parasponia* in the ability to be nodulated by *Bradyrhizobium*. It is also possible that eight separate gains in the ability to nodulate occurred, with no losses. Based simply on the principle of parsimony, eight gains are more likely to have occurred than one gain and 15 losses. However, nodulation appears to be a complex process involving many genes, and it is certainly reasonable to assume that it would be easier to lose the ability to be nodulated by interfering with the functioning of any one of the essential genes than it would be to gain that ability.

There are four major clades that contain the actinorhizal genera, in some cases interspersed with non-actinorhizal taxa. Various biological characteristics of the nodules of plants within each clade support the hypothesis that symbiosis arose in each as a result of a separate gain in the ability to be nodulated. Plants within Clade I are nodulated by intercellular penetration (IP), do not have nodule roots, have microsymbionts with elliptical non-septate vesicles and have been shown to host frankiae of similar genetic backgrounds which have not been able to be cultured from the nodule. Analysis of the 16S RNA genes of *Dryas* and *Purshia* show that they are also closely related to unisolated microsymbionts from *Ceanothus* in Clade

II, and *Coriaria* and *Datisca* in Clade III (Benson et al., 1996; Normand et al., 1996). Plants within Clade I appear only recently in the fossil record (Pleiocene) and all are North American genera (Müller, 1981).

Plants within Clade II are also infected by IP and do not have nodule roots. There is a striking difference in this clade, however, between microsymbionts of host plants within the Rhamnaceae and those within the Elaeagnaceae. No infective and effective microsymbionts have been isolated from plants within the Rhamnaceae. In contrast to this, many *Elaeagnus* isolates have been obtained and are known to differ from the microsymbiont of *Ceanothus* of the Rhamnaceae, but to be similar to the microsymbiont found within the nodules of *Discaria*, also within the Rhamnaceae, by sequence analysis of 16S RNA (Benson et al., 1996). Some of the *Elaeagnus* isolates are infective only on members of the Elaeagnaceae (Host Specificity Group 4), others are infective on *Elaeagnus* and *Myrica* (HSG 3) and some are infective on *Elaeagnus*, *Myrica* and *Alnus* (flexible strains) (Miller and Baker, 1986; Lumini and Bosco, 1996). Although the orders Rhamnales and Elaeagnales appear in the fossil record during roughly the same period, the actinorhizal members of the Rhamnaceae are predominantly South American genera while those of the Elaeagnaceae have a much broader distribution. Based on the difference in the microsymbionts, it is quite conceivable that these two groups established symbioses independently of one another.

Clade III contains the two actinorhizal genera *Datisca* and *Coriaria* interspersed among non-actinorhizal genera. These two host plants share a number of unique nodule characteristics. Both are infected by IP, infective and effective isolates have not been made from either plant, both are infected by closely related microsymbionts (Mirza et al., 1994a,b,c), that are at the same time closely related to the microsymbiont from *Dryas* of Clade I (Normand et al., 1996), they share a unique nodule morphology (Calvert et al., 1979; Newcomb and Pankherst, 1982; Akkermans et al., 1983) and both demonstrate an adaptive response to changing oxygen levels (Tjepkema et al., 1988; Silvester and Harris, 1989).

All of the actinorhizal members of Clade IV are infected by root hair infection, and many *Frankia* isolates have been obtained from this group. Three members of the clade, *Casuarina*, *Allocasuarina* and *Myrica*, have a layer of cells within the nodule, specialized to provide an oxygen diffusion barrier and these same genera also have higher levels of nodule hemoglobin

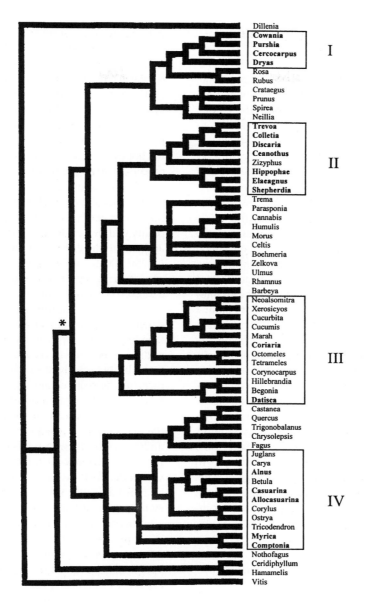

Figure 3. Strict consensus tree summarizing the major clades found in a phylogenetic analysis of 70 *rbcL* sequences from actinorhizal plants (bold type) and non-actinorhizal relatives. Modified from Swensen, 1996 and reproduced with permission from The Botanical Society of America.

(Silvester et al., 1990) as well as nodule roots. *Myrica* is considered to be a promiscuous host because it can be nodulated by most *Frankia* strains; *Alnus*, although not considered to be promiscuous, is nodulated by *Frankia* strains from several *Alnus*-infective genogroups. It has been shown on the basis of 16S RNA sequence analysis that the fairly homogeneous *Casuarina* microsymbionts originate within an *Alnus*-infective clade (Cournoyer et al., 1993; Normand et

al., 1996). Casuarinales appear later in the fossil record (lower Paleocene) than Betulaceae (lower part of the Upper Cretaceous), and their relative position in *rbcL* tree, as well as the estimated time of divergence of their microsymbionts, supports the view that the evolution of the host plants influenced that of the microsymbionts.

Genetic elements leading to the predisposition for nodulation

With the knowledge that a single angiosperm clade, appears to have a predisposition towards nodulation, new efforts should be mounted to identify the unique features of this clade that impart to it the ability to recognize and host diazotrophic bacteria. Studies of legume symbioses have revealed a number of important biochemical, developmental and genetic components necessary for successful symbioses but the presence of these components has not been assessed in non-nodulating members of the root nodulating clade (RNC) or in plants outside this clade. Because of the ease with which rhizobia can be manipulated, much more rapid progress has been made in the identification of bacterial determinants contributing to successful symbioses. Chemical signaling is known to be an integral part of symbiosis with legumes. Are there unique secondary metabolites in the RNC that foster interaction between the potential host and microsymbiont? Is there the absence of specific phytoalexins or antimicrobial compounds, or a mechanism for bypassing their production or masking their effect? Do members of this clade share any distinctive plant pathogens or any unique resistance to pathogens? Are there specific receptor molecules or other proteins such as nodulins in the RNC that are not present in sister clades? Do members of the RNC share any unique sensitivities to hormone levels? Are there unique physiological traits held in common by this group relative to their nitrogen uptake, transport and metabolism or to their photosynthetic efficiency or carbon partitioning?

Implications for establishing new symbiotic associations

There is a tremendous pressure worldwide to increase the biomass, decrease the cost and at the same time to lower the environmental impact of the production of food, fuel and fiber crops. Because nitrogen is the mineral element that most frequently limits plant productivity, much effort has been put into studies related to nitrogen supply, availability and retention in the soil, and nitrogen, uptake, utilization and recycling within plants. Plants that fix nitrogen symbiotically have a built-in supply of reduced nitrogen, which allows them to meet their nitrogen needs independent of nitrogen levels in the soil and therefore without the monetary, energy or environmental costs associated with the production and use of nitrogen fertilizer. For these reasons there are ongoing attempts to engineer new nitrogen-fixing symbiotic associations. Rice is a crop targeted for the establishment of such new nitrogen-fixing symbiotic associations because of its fundamental importance to the world food supply (Bennett and Ladha, 1992). Early reports that rhizobia could induce nodules on rice seedlings (Al-Mallah et al., 1989) raised the hopes of many working in this area but subsequent reports of nitrogen fixation within nodular structures remain controversial.

As cited above, phylogenetic trees based on molecular characters place all known root nodulating host plants within a relatively small group of angiosperms and lead to the conclusion that "...of the approximately 250,000 to 300,000 species and 380 families of angiosperms only one lineage of closely related taxa achieved the underlying genetic architecture necessary for root nodule symbiosis" (Soltis et al., 1995). The genetic background of known host plants is less diverse than has previously been thought and microsymbionts as diverse as *Frankia* and rhizobia have been limited in their ability to establish symbioses to this group. It may therefore be more difficult than previously envisioned to establish new symbiotic associations *outside* the single root nodule clade. On the other hand, evidence for the existence of a genetic predisposition towards root nodule symbiosis within this clade as well as evidence for the occurrence of multiple origins of nodulation within the clade would support the idea that it may be more feasible than previously thought to establish new symbioses *within* members of this clade.

Within the root nodule clade are a number of genera with members that are important in food or timber production. For example, within the family Rosaceae which includes Clade I actinorhizal genera are the genera *Malus* (apples), *Prunus* (almonds, apricots, cherries, peaches and plums), *Rubus* (raspberries, etc.) *Pyrus* (pears), *Fragaria* (strawberries). Within actinorhizal Clade III are the important genera *Cucurbita* (squash and pumpkin) and *Cucumis* (cantaloupe and cucumber). Within actinorhizal Clade IV are valuable timber species in the genera *Juglans*, *Quercus* and *Betula*. It is not implied that fruit and vegetable crops can replace the need for increased efficiency of production of staples such as rice and wheat, but simply that new symbioses may be more easily established with these plants than with those that cross the boundaries not only of family and order but also of the division between (Liliopsida) monocots and (Magnoliopsida) dicots.

A well-supported phylogenetic hypothesis such as that presented by Soltis et al. (1995) that places all known nodulating plants within a single clade will facilitate studies to determine the genetic, physiological, biochemical and developmental factors found within this clade that have allowed successful symbioses to be established. Once these attributes have been identified, it will be possible to proceed with a logical and orderly plan for extending symbioses to species both within and outside of this group of plants.

Acknowledgments

S.M.S. acknowledges support from National Science Foundation grant DEB-9306913 and Loren H. Rieseberg. B.C.M. acknowledges support from the United States Department of Agriculture National Research Initiative Competitive Grant 95-37305-2086.

References

Akkermans A D L, Hafeez F, Roelofsen W, Chaudhary A H and Baas R 1983 Ultrastructure and nitrogenase activity of *Frankia* grown in pure culture and in actinorhizae of *Alnus*, *Colletia* and *Datisca*. *In* Advances in Nitrogen Fixation Research. Eds. C Veeger and W E Newton. pp 311–319. Nijhoff/Junk, The Hague.

Al-Mallah M K, Davey M R, Cocking E C 1989 Formation of nodular structures on rice seedlings by rhizobia. J. Exp. Bot. 40, 473–478.

Bennett J and Ladha J K 1992 Introduction: feasibility of nodulation and nitrogen fixation in rice–potential and prospects. *In* Nodulation and Nitrogen Fixation in Rice. Eds. G S Khush and J Bennett. pp 1–11. International Rice Research Institute, Manila, Philippines.

Benson D R and Silvester W B 1993 Biology of *Frankia* strains, actinomycete symbionts of actinorhizal plants. Microbiol. Rev. 57, 293–319.

Benson D R, Stephens D W, Clawson M L and Silvester W B 1996 Amplification of 16S rRNA genes from *Frankia* strains in root nodules of *Ceanothus griseus*, *Coriaria arborea*, *Coriaria plumosa*, *Discaria toumatou*, and *Purshia tridentata*. Appl. Environ. Micro. 62, 2904–2909.

Calvert H E, Chaudhary A H, and Lalonde M 1979 Structure of an unusual nodule root symbiosis in a non-leguminous herbaceous dicotyledon. *In* Symbiotic Nitrogen Fixation in the Management of Temperate Forests. Eds. J C Gorden, C T Wheeler and D A Perry. p 474 Forest Research Laboratory, Oregon State University, Corvallis, OR.

Chase M, Soltis D E, Olmstead R G, Morgan D, Les D H, Mishler B D, Duvall M R, Price R, Hills H G, Qiu Y, Kron K A, Rettig J H, Conti E, Palmer J D, Manhart J R, Sytsma K J, Michaels H J, Kress W J, Donoghue J D, Clark W D, Hedrén M, Gaut B S, Jansen R K, Kim K-J, Wimpee C F, Smith J F, Furnier G R, Straus S H, Xiang Q, Plunkett G M, Soltis P S, Swensen S M, Eguiarte L E, Learn G H Jr, Barrett S C, Graham S and Albert V A 1993 Phylogenetics of seed plants: an analysis of nucleotide

sequences from the plastid gene *rbc*L. Annals of the Missouri Botanical Garden 80, 528–580.

Cournoyer B, Gouy M and Normand P 1993 Molecular phylogeny of the symbiotic actinomycetes of the genus *Frankia* matches host-plant infection processes. Mol. Biol. Evol. 10, 1303–1316.

Cronquist A 1981 An integrated system of classification of flowering plants. Columbia University Press, New York, NY.

Cronquist A 1988 The evolution and classification of flowering plants. New York Botanical Garden, Bronx, NY.

Dahlgren R M T 1980 A revised system of classification of the angiosperms. Botanical Journal of the Linnean Society 80, 91–124.

Darwin C 1859 On the origin of species by natural selection.

Doyle J J 1994 Phylogeny of the legume family: An approach to understanding the origins of nodulation. Ann. Rev. Ecol. Syst. 25, 325–349.

Linneaus C 1753 Species Plantarum. Lumini E and Bosco M 1996 PCR-restriction fragment length polymorphism identification and host range of single-spore isolates of the flexible *Frankia* sp. Strain UFI 132715. Appl. Environ. Micro. 62, 3026–3029.

Miller I M and Baker D D 1986 Nodulation of actinorhizal plants by *Frankia* strains capable of both root hair infection and intercellular penetration. Protoplasma 131, 82–91.

Mirza M S, Hahn D, Dobritsa S V, Akkermans A D L 1994a Phylogenetic studies on uncultured *Frankia* populations in nodules of *Datisca cannabina*. Can. J. Microbiol. 40, 313–318.

Mirza M S, Hameed S and Akkermans A D L 1994b Genetic diversity of *Datisca cannabina*-compatible *Frankia* strains as determined by sequence analysis of the PCR-amplified 16S RNA gene. Appl. Environ. Microbiol. 60, 2371–2376.

Mirza M S, Akkermans W M and Akkermans A D L 1994c PCR-amplified 16S rRNA sequence analysis to confirm nodulation of *Datisca cannabina* L. by the endophyte of *Coriaria nepalensis* Wall. Plant and Soil 160, 147–152.

Müller J 1981 Fossil pollen records of extant angiosperms. The Botanical Review 47, 1 1–142.

Newcomb W and Pankhurst C E 1982 Fine structure of actinorhizal nodules of *Coriaria arborea* (Coriariaceae). New Zealand Journal of Botany 20, 93–103.

Normand P, Orso S, Cournoyer B, Jeannin P, Chapelon C, Dawson J, Evtushenko L and Misra A K 1996 Molecular phylogeny of the genus *Frankia* and related genera and emendation of the family *Frankiaceae*. International Journal of Systematic Bacteriology 46, 1–9.

Silvester W B and Harris S L 1989 Nodule structure and nitrogenase activity of *Coriaria arborea* in response to varying pO_2. Plant and Soil 78, 245–258.

Silvester W B, Harris S L and Tjepkema J D 1990 Oxygen regulation and hemoglobin. *In* The Biology of *Frankia* and actinorhizal plants. Eds. C R Schwintzer and J D Tjepkema. pp 157–176. Academic Press, Inc., New York, NY.

Soltis D E, Soltis P S, Morgan D R, Swensen S M, Mullin B C, Dowd J M and Martin P G 1995 Chloroplast gene sequence data suggest a single origin of the predisposition for symbiotic nitrogen fixation in angiosperms. Proc. Natl. Acad. Sci. 92, 2646–2651.

Sprent J I 1994 Evolution and diversity in the legume-rhizobium symbiosis: Chaos theory? Plant and Soil 161, 1–10.

Swensen S M, Mullin B C, and Chase M W 1994 Phylogenetic affinities of the Datiscaceae based on an analysis of nucleotide sequences from the plastid *rbcL* gene. Systematic Botany 19(1), 157–168.

Swensen S M 1996 The evolution of actinorhizal symbioses: Evidence for multiple origins of the symbiotic association. Am. J. Bot. 83(11), 1503–1512.

192

Swensen S M and Mullin B C In Press Phylogenetic relationships among actinorhizal plants: The impact of molecular systematics and implications for the evolution of actinorhizal symbioses. Physiologia Plantarum.

Takhtajan A 1980 Outline of the classification of flowering plants (Magnoliophyta). The Botanical Review (Lancaster) 46, 225–359.

Thorne R T 1992 Classification and geography of the flowering plants. The Botanical Review (Lancaster) 58, 225–348.

Tjepkema J D, Schwintzer C R and Monz C A 1988 Time course of acetylene reduction in nodules of five actinorhizal genera. Plant Physiology 86, 581–583.

Guest editors: J K Ladha, F J de Bruijn and K A Malik

Plant and Soil **194**: 193–203, 1997.
© 1997 *Kluwer Academic Publishers. Printed in the Netherlands.*

Nif gene transfer and expression in chloroplasts: Prospects and problems

Ray Dixon[1], Qi Cheng[1], Gui-Fang Shen[2], Anil Day[3] and Mandy Dowson-Day[4]
[1]*Nitrogen Fixation Laboratory, John Innes Centre, Colney Lane, Norwich NR4 7UH, UK**, [2]*Biotechnology Research Center, Chinese Academy of Agricultural Sciences, Beijing 10081, China*, [3]*School of Biological Sciences, University of Manchester, Oxford Road, Manchester M13 9PT, UK* and [4]*School of Biological Sciences, University of Warwick, Coventry, CV4 7AL, UK*

Key words: chloroplast, genetic engineering, *nif* genes, nitrogenase, plant transformation, plastid

Abstract

The engineering of plants capable of fixing their own nitrogen is an extremely complex task, requiring the co-ordinated and regulated expression of 16 *nif* genes in an appropriate cellular location. We suggest that plastids may provide a favourable environment for *nif* gene expression provided that the nitrogenase enzyme can be protected from oxygen damage. Using the non-heterocystous cyanobacteria as a model, we argue that photosynthesis could be temporally separated from nitrogen fixation in chloroplasts by restricting nitrogenase synthesis to the dark period. We report preliminary data on the introduction and expression of one of nitrogenase components, the Fe protein, in transgenic tobacco and *Chlamydomonas reinhardtii*. Finally we discuss potential avenues for further research in this area and the prospects for achieving the ultimate goal of expressing active nitrogenase in cereal crops such as rice.

Introduction

Ever since the *nif* gene cluster was transferred from *Klebsiella pneumoniae* to *Escherichia coli*, thus creating the first engineered diazotroph more than twenty years ago (Dixon and Postgate,1972), there has been an ongoing debate concerning the prospects for introduction and expression of nitrogen fixation genes in cereal crops. Discussions on the potential for engineering diazotrophic plants have been widespread and often argumentative, with opposing views as to whether research in this area should be supported. One question which is frequently debated concerns the evolution of nitrogen fixation and the potential incompatibility of this process within the eukaryotic environment. Why was this important biochemical process not retained by primitive eukaryotes? Moreover, if plant chloroplasts have evolved from parasitic endophytic cyanobacteria, why are plastids not diazotrophic? On the other hand, if diazotrophy is so advantageous to an organism, one might also question why only some prokaryotes fix nitrogen. Furthermore, if the legume nodule provides such a highly sophisticated nitrogen- fixing organ, why

have plants not evolved to dispense with the symbiont and maintain nitrogen fixation genes as part of their own genome? We do not have a simple answer to any of these questions and to date there appears to be no definitive reason why plants should not fix their own nitrogen (Postgate, 1992). Nevertheless, it is a daunting task to consider the engineering of nitrogen fixing plants, particularly as it requires the co-ordinated and regulated expression of 16 *nif* genes in an appropriate cellular location together with any additional genes which may be required to maintain nitrogenase in an active form. It is extremely unlikely that this very ambitious goal will be achieved in the short term. The prospects for rice are likely to be even more long-term because the initial experiments will most probably be carried out in "model plants" such as tobacco, which are more amenable to transformation. Nevertheless, we believe that it is possible to test the feasibility of expressing nitrogenase in plants in a logical and systematic manner and that significant advances can be made on the basis of large-scale international collaboration. Indeed the overall goal of such research may justify the setting up of an agricultural equivalent of the "Manhattan" project, although the scientific com-

---
* FAX no: +441603454970. E-mail: ray.dixon@bbsrc.ac.uk

munity would need to be convinced of the potential long-term benefits.

We have previously suggested that plant plastids may provide the location of choice for the expression of *nif* genes in higher plants (Merrick and Dixon,1984) since chloroplasts are similar to prokaryotes in terms of gene organisation and expression. Moreover the chloroplast offers a good environment for the incorporation of the output of nitrogen fixation as it is a major site for the uptake of ammonia by glutamine synthetase as well as amino acid biosynthesis. Although it can be argued that photosynthetic oxygen evolution will provide an unfavourable environment for nitrogenase within the plastid, it should be possible to obtain temporal separation of nitrogen fixation and photosynthesis by regulating the *nif* genes so that they are expressed only in the dark. Alternatively spatial separation could be achieved by expressing nitrogenase in non-photosynthetic tissue e.g. in root amyloplasts. In this review we will consider the genetic, physiological and biochemical requirements for synthesis of active nitrogenase and the potential problems associated with the expression of this enzyme in the plastid environment. We present some of our preliminary data on the feasibility of expressing one of the nitrogenase component proteins in plants and suggest future strategies for engineering diazotrophic plastids.

How many *nif* genes are required ?

Nitrogenase is a complex enzyme consisting of two oxygen-sensitive protein components, usually referred to as Fe protein and MoFe protein (Howard and Rees, 1994; Eady, 1995). Fe protein is a homodimer (encoded by *nifH*) which contains a single Fe_4S_4 cluster whereas MoFe protein is an $\alpha_2\beta_2$ heterotetramer (encoded by *nifDK*) containing 2 Mo atoms and 30 Fe atoms organised into two pairs of novel metalloclusters which are called P clusters and FeMo-co. The primary translation products of the nitrogenase structural genes *nifHDK* are not active and they require processing by a consortium of additional *nif* genes, the majority of which are involved in metallocluster biosynthesis (Dean et al., 1993). How many of these genes are essential for nitrogenase biosynthesis in foreign hosts? From comparative sequence analysis of nitrogen fixation genes in diazotrophs it is clearly apparent that each organism maintains a common core of genes (*nif H,D,K,Y,T,E,N,X,U,S,V,Z,W,M,B,Q*) whose products are probably essential for efficient biosynthesis of

nitrogenase (Dean and Jacobson,1992; Merrick,1993) (see Figure 1). However, mutations in some of these genes (notably *nifY, nifT, nifX, nifU, nifS, nifV, nifW, nifM* and *nifQ*) do not completely eliminate nitrogenase activity and there is evidence that homologues elsewhere on the genome may at least partially substitute for their function. There is a strong probability that these genes may have evolved from "housekeeping" equivalents and were recruited in order to provide optimum support for nitrogenase biosynthesis.

Recent determination of the crystallographic structures of the nitrogenase component proteins has provided an excellent basis for more detailed analysis of the catalytic mechanism of nitrogenase and the structure of the iron-molybdenum co-factor (FeMo-co), but the precise role of gene products in the assembly of metalloclusters and maturation of the nitrogenase polypeptides has not yet been clearly defined. However, progress in this direction has been made with the genes required for biosynthesis of FeMo-co (Dean et al., 1993). In vitro assays for FeMo-co biosynthesis have unambiguously identified the genes required for this process and it is now clear that *nifH* is necessary as well as *nifE, nifN, nifV* and *nifB*; hence their gene products will be absolutely essential for biosynthesis of Mo Fe protein in the chloroplast. Mutations in other genes required for FeMo-co synthesis, *nifY* and *nifX*, have a somewhat "leaky" phenotype in vivo and therefore their roles may be partially substituted by "housekeeping" equivalents. Indeed in *A.vinelandii* the function of *nifY* appears to be carried out by the gamma protein, which is not encoded by a *nif* gene (Allen et al., 1993; G P Roberts, personal communication.)

Another substantial advance in our knowledge of the biosynthesis of metalloclusters in nitrogenase has arisen from the purification and characterisation of the *nifS* product from *A.vinelandii*. NIFS is a pyridoxal phosphate containing enzyme which catalyses the specific desulphurisation of L-cysteine, producing an enzyme bound persulphide which serves as a sulphur donor for Fe-S cluster formation. NIFS activates the apo-Fe protein of nitrogenase in vitro, catalysing the formation of a de novo Fe_4S_4 cluster in the presence of cysteine and Fe (Zheng and Dean, 1994; Zheng et al., 1993). Whereas *nifS* is required to mobilise sulphur for iron-sulphur cluster formation there is also evidence to suggest that *nifU* is involved in the mobilisation of Fe. Purified NIFU contains one redox-active Fe_2S_2 cluster per subunit as well as additional cysteinyl ligands which may be involved in sequestering Fe (Fu et al., 1994). It is possible that the redox-active Fe_2S_2 clus-

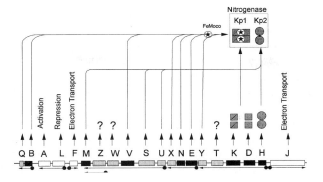

Figure 1. Map of the *nif* gene cluster in *Klebsiella pneumoniae*. The functions of each of the gene products are indicated schematically above each gene, which are shown as horizontal rectangles. Open boxes indicate those genes which are not "core" components and will probably not be required for synthesis of active nitrogenase in the chloroplast. Black boxes indicate essential "core" genes, whereas stippled boxes indicate genes whose functions might potentially be provided by "housekeeping" equivalents in the plant.

ters maintain bound Fe in the correct oxidation state for assembly into nitrogenase Fe-S clusters. While the in vitro experiments clearly show that NIFS mobilises sulphur for formation of the Fe_4S_4 cluster in the Fe protein, its role in the formation of the P clusters is unknown. However, functions equivalent to those of NIFS and NIFU could represent a universal pathway for Fe-S cluster formation since homologs have been found in the genomes of several other microorganisms and even humans. Indeed, it has been recently demonstrated that *Escherichia coli* contains a functional homologue of NIFS which supports assembly of the Fe-S cluster of dihydroxy- acid dehydratase (Flint, 1996). This finding is congruent with the observation that *nifS* and *nifU* are not required for biosynthesis of active Fe protein in *Escherichia coli* (Howard et al., 1986). Moreover, purified *A.vinelandii* NIFS functions to activate the apoprotein forms of the regulatory proteins FNR (Khoroshilova et al., 1995) and SOXR (Hilgado and Demple, 1996) via assembly of de novo Fe-S clusters.

The *nifM* gene product is required for activation and stability of the Fe protein. When *nifH* is expressed in either *E.coli* or yeast in the absence of *nifM* a very low level of dimeric Fe protein is synthesised (Berman et al., 1985; Paul and Merrick, 1989). The precise role of NIFM is unknown although it has been suggested that it could have a chaperone-like role in maintaining the apoFe protein in the correct conformation to accept the Fe_4S_4 cluster.

The role of non-*nif* genes in nitrogenase biosynthesis has been emphasised by the finding that molecular chaperones play a role in biogenesis of MoFe protein. The level of Mo-Fe protein is drastically reduced in an *E.coli groEL* mutant and overexpression of the *groE* operon in *K.pneumoniae* leads to more rapid accumulation of Mo-Fe protein. In accordance with these observations, GroEL transiently interacts with the NifD and NifK polypeptides during nitrogenase biosynthesis (Govezensky et al., 1991). Fortunately chloroplasts have homologues of GroEL and GroES termed ch-cpn60 and ch-cpn10 which function in a similar fashion to their bacterial counterparts (Viitanen et al., 1995).

Nif genes in addition to the "core" components are required for electron transport to nitrogenase and for regulation of nitrogenase synthesis (Merrick, 1993). These genes have probably evolved in order to integrate the process of nitrogen fixation into the overall physiology of the cell. For example, the *nifJ* gene which encodes a pyruvate-flavodoxin oxidoreductase is present in *K. pneumoniae* but not in *A.vinelandii*. NifJ, in conjunction with NifF, couples the oxidation of pyruvate to the reduction of nitrogenase Fe protein forming an electron transport pathway which is compatible with the expression of nitrogenase under anaerobic growth conditions in *Klebsiella*. In contrast, for example, in the obligately aerobic diazotroph *A. vinelandii*, electron transport to nitrogenase is most probably coupled to respiration via both flavodoxins and ferredoxins. A variety of electron donors can function with nitrogenase and since reducing power is abundant in the chloroplast it may not be necessary to provide additional electron carriers.

In addition to those genes specifically required for the biosynthesis and activity of nitrogenase, a number of regulatory gene are necessary for *nif* gene expression in bacteria. Transcription from most *nif* promoters requires a novel form of RNA polymerase holoenzyme containing a unique sigma factor, σ^N (σ^{54}) encoded by the *rpoN* gene, which is competent to bind its target promoters but is inactive in the absence of a specific transcriptional activator (Kustu et al., 1989). The latter function is provided by an enhancer binding protein, encoded by *nifA* which is ubiquitous among diazotrophic proteobacteria as an activator of *nif* gene transcription (Merrick, 1992). In considering the engineering of a *nif* gene cluster which can be expressed in the chloroplast environment it is not necessary to consider these regulatory gene products, although we might want to make use of the bacterial regulatory system if it could function in concert with the chloroplast polymerase (see below).

To summarise, synthesis of active nitrogenase in the plastid requires the 16 "core" *nif* genes listed above, although we may be able to dispense with some of these, particularly *nifU* and *nifS* which apparently have ubiquitous homologs. In addition mutations in *nifT,nifY,nifX,nifW* and *nifZ* all give "leaky" phenotypes and their function may also be complemented by other genes. As our knowledge of the biochemical functions of individual gene products increases we may be able to identify further "housekeeping" equivalents which can substitute for their function. On the other hand we may also have to consider the introduction of additional genes to support nitrogenase activity and protect the enzyme from oxygen damage (see below).

Physiological requirements for nitrogenase activity

Nitrogen fixation is an energy intensive process and as discussed above, requires a suitable reductant to support electron transport to nitrogenase. Fortunately, photosynthetically generated ATP and NADPH are plentiful in the chloroplast, although one must bear in mind that the energy demands of nitrogenase could compete with the demand for photosynthate provided by the Calvin cycle. This problem may be exacerbated by the requirement for synthesis of high levels of nitrogenase component proteins to support efficient nitrogen fixation, since nitrogenase is a relatively slow enzyme (Thorneley and Lowe, 1983). However it should be noted that the energy efficiency of nitrate utilisation is similar to that of nitrogen fixation (Postgate, 1981). Moreover, since rice is relatively low in protein, it has been suggested that lower rates of nitrogen fixation will be required to support yields compared with legumes (Reddy and Laddha, 1995).

Since nitrogenase is a metalloenzyme, its synthesis requires the assimilation of transition metal atoms namely iron and molybdenum. Iron is of course readily available in the chloroplast. Molybdenum is assimilated to support nitrogenase in nodules and is a relatively mobile element in plants, being translocated from roots to leaves as an essential component of the co-factor in nitrate reductase. However we are not aware of a molybdoenzyme in plastids. Alternatively, one could consider introducing alternative nitrogenases into plants such as the vanadium enzyme or the iron-only nitrogenase (Eady, 1995).

Both of the nitrogenase component proteins are extremely oxygen sensitive and are irreversibly denatured by oxygen so it will be necessary to either provide an anaerobic environment for the enzyme or protect it from oxygen damage. Free-living diazotrophs achieve the latter by a variety of different strategies, including "respiratory", "conformational" and "auto" protection (Hill, 1992). Cyanobacteria provide the best models for the reconciliation of photosynthesis with diazotrophy, using either a physical barrier to separate photosynthetic oxygen evolution from nitrogen fixation in heterocystous species or temporal separation of photosynthesis and nitrogen fixation in non-heterocystous forms (Gallon, 1992). It is possible that photorespiration could contribute to protection of nitrogenase from oxygen, but this would result in a significant drain of ATP and reductant. Conformational protection of nitrogenase, by introduction of the gene for the *Azotobacter vinelandii* Shethna (FeS II) protein (Moshiri et al., 1994) into plastids, would help to temporarily stabilise the enzyme in the presence of oxygen and prevent degradation. There are conditions in which nitrogenase itself can remove oxygen, resulting in autoprotection. In the presence of excess reductant, the Fe protein of *A.vinelandii* nitrogenase can actually reduce dioxygen to hydrogen peroxide and ultimately to water (Thorneley and Ashby, 1989). The generation of toxic activated oxygen species by nitrogenase could be deleterious to the chloroplast although reactive species will be removed by oxidant scavenging enzymes such as ascorbate peroxidase and glutathione reductase which are unique to the plastid. These enzymes could also potentially protect nitrogenase from oxygen damage. The engineering of a specialised compartment for nitrogenase such as the heterocyst would be an extremely complex and difficult task, particularly as our knowledge of heterocyst differentiation is incomplete, so we believe that non-heterocystous cyanobacteria provide the best models for the development of diazotrophic organelles. These organisms are able to cope with both photosynthesis and nitrogen fixation, even though nitrogenase is not spatially separated from photosynthetic oxygen evolution. Under laboratory conditions with alternating 12 hour light and dark periods, nitrogenase synthesis in *Gleothece* is initiated prior to the onset of the dark phase and nitrogenase activity in the dark is supported by ATP and reductant generated in the light. After several hours in the dark phase when the demand for ATP can no longer be sustained, nitrogenase becomes inactivated, its biosynthesis ceases and the enzyme is degraded (Gallon and Chaplin, 1988). However, this degradation may be specific to the algal enzyme since *Klebsiella pneumoniae* nitrogenase Fe protein is not degraded by *Gleothece* aerobic extracts (Gallon et al., 1995). As discussed above, degradation of nitrogenase in chloroplasts in the light could potentially be prevented by the presence of the Shethna protein.

What has been achieved so far?

In view of the complexity of this problem and the lack of funding opportunities it is not surprising that few laboratories have attempted to initiate research in this area. We argue however that it should be possible to examine the feasibility for expression of nitrogenase in plants using relatively simple model systems in which the expression and activity of only one of the nitrogenase component proteins is examined. The obvious choice for these experiments is the Fe protein, because it requires only the products of *nifH* and *nifM* for synthesis of active protein in *E.coli* (Howard et al., 1986), the biosynthesis of the Fe_4S_4 cluster being presumably carried out by non *nif*-encoded proteins. The activity of the Fe protein can be measured in cell-free extracts by in vitro complementation with purified MoFe protein.

Research in this area was initiated prior to the development of a routine transformation procedure for higher plant chloroplasts so the original strategy was to target *nif*-encoded polypeptides to the plastid from a nuclear located transgene. In addition, chloroplast transformation has been used to introduce *nifH* into the plastid genome of the unicellular green alga, *Chlamydomonas reinhardtii*.

(a) Import of the Fe protein into the tobacco chloroplast

In order to target the NIFH and NIFM polypeptides to the plant chloroplast their coding sequences were mutated to introduce a restriction site around the initiating ATG's to facilitate ligation to a chloroplast signal sequence. Consequently, the second codon of each protein was changed, resulting in the substitution of a proline for a threonine in NIFH and an asparagine to histidine change in NIFM. Neither of these mutations apparently influenced the activity of these polypeptides as determined by complementation assays with *nifH*⁻ and *nifM*⁻ mutant strains of *K. pneumoniae*. Surprisingly, it was found that when the coding sequence for the transit peptide of the small sub-unit of ribulose bisphophate carboxylase (SSU) were ligated to *nifH* or *nifM*, the precursor proteins (called prNIFH and prNIFM respectively) were also efficient at complementing the mutations (Dowson-Day et al., 1991).

The transit peptide *nif* fusions were tested in vitro to verify that they could be efficiently translated in a plant translation system as well as being imported into the chloroplast and cleaved to the native polypeptides. Transcripts derived from the *pr-nifH* and *pr-nifM* fusions expressed full-length fusion proteins in a wheat germ translation system and the in vitro translation products were both imported and correctly processed to their normal size in whole pea chloroplasts. As expected, both the imported NifH and NifM were present in the stromal fraction. Time course experiments revealed that both proteins were imported in vitro at equivalent rates to SSU and were equally stable. Moreover, import occurred anaerobically in the dark as well as in aerobic illuminated conditions (Dowson-Day et al., 1991).

The fusion pr-*nifH* and pr-*nifM* genes were individually placed under the control of the cauliflower mosaic virus (CaMV) 35S promoter and introduced in inverted orientation into a binary transformation vector for introduction into the nuclear genome of *Nicotiniana tobaccum*. Western blotting experiments on a number of transgenic plants indicated that the NifH and NifM polypeptides were expressed at a very low level. This low level of expression was improved somewhat by the incorporation of the tobacco mosaic virus (TMV) 5'Ω leader upstream of the *rbcS* transit peptide (Dowson-Day et al., 1993). The level of specific m-RNA expressed from these constructs in vivo, as measured either by PCR amplification of their corresponding c-DNAs or by primer extension analysis, was relatively high and therefore we presume that

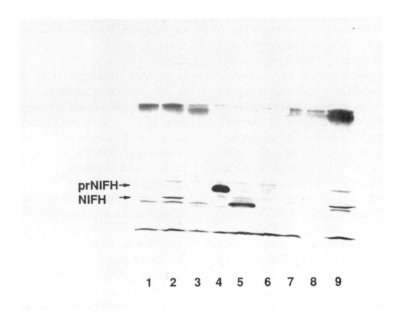

Figure 2. Western blots of transgenic *pr-nifH* (lanes 1-3) and pr-*nifH*, pr-*nifM* (lanes 7-9) tobacco leaf extracts probed with anti-NIFH antibody. Lane 4 contains an in vitro wheat germ translation reaction derived from the pr-*nifH* fusion, thus showing the SSU-NIFH pre-protein prior to cleavage. Lane 5 contains 5 ng of purified nitrogenase Fe protein. Lane 6 contains an extract from an untransformed tobacco plant.

the low level of the corresponding polypeptides is a consequence of either poor translation of the message or protein degradation. Nevertheless Western blots on isolated chloroplast fractions indicate that the NIFH polypeptide is imported in vivo into the chloroplast and is processed to the correct size (Figure 2). The chloroplast location of NIFH has been confirmed using immunogold localisation.

We have developed a very sensitive assay for the detection of nitrogenase Fe protein in leaf extracts. This assay measures hydrogen evolution by nitrogenase when MoFe protein is added to plant extracts and is sufficiently sensitive to detect Fe protein at a level of 0.01% of total soluble leaf protein. Unfortunately the level of Fe protein in the chloroplasts was below the limits of detection of the assay system.

Although these results are in some ways discouraging, they do provide important clues for future approaches. It would appear that the NIFH and NIFM are stable in chloroplasts, at least in vitro, and we favour the notion that these proteins are degraded in the cytoplasm prior to import. Although we cannot rule out the possibility that *nif* m-RNA is poorly translated in tobacco, the codon usage of *nifH* is similar to that of other highly expressed plant transgenes and the sequence is not AT rich. Now that chloroplast trans-

formation is routine in tobacco we believe it should be possible to increase the level of *nif* products by expressing *nif* genes directly in the chloroplast.

(b) Introduction of nifH into the chloroplast genome of Chlamydomonas reinhardtii

Lower plants provide an attractive model for testing the feasibility of *nif* gene expression in the plastid because these organisms already contain chloroplast genes which may be structurally and functionally related to nitrogenase component proteins (Figure 3). The chloroplast genomes of algae and lower plants for example, *Chlamydomonas, Marchantia*, and *Pinus* species contain a gene (designated *chlL* or *frxC*) encoding a protein that is 30% identical to nitrogenase Fe protein (NIFH) subunit (Ohyama et al., 1986; Lidholm and Gustafsson, 1991; Suzuki and Bauer, 1992). This gene is also present in cyanobacteria (Fujita et al., 1992) and in photosynthetic bacteria as *bchL* (Yang and Bauer, 1990). Interestingly, cysteine residues required for liganding the Fe_4-S_4 cluster in Fe protein are conserved in CHLL, raising the possibility that it may also be an oxygen sensitive protein. The sequence similarity between these proteins is maintained over their entire length and in addition to the environment surrounding

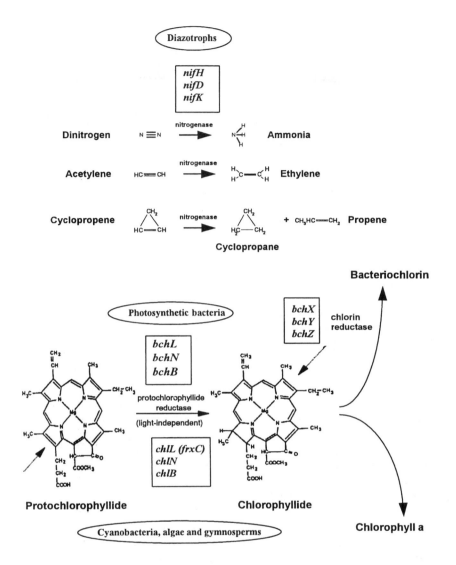

Figure 3. Potential functional similarities between nitrogenase and enzymes involved in the pathway of light-independent chlorophyll biosynthesis in photosynthetic bacteria, cyanobacteria, algae and gymnosperms. Nitrogenase reduces unsaturated substrates such as dinitrogen and acetylene but also reduces the double bond in cyclopropene, forming cyclopropane and propene. Protochlorophyllide reductase reduces the double bond of pyrrole ring D in the mg-tetrapyrrole. Another enzyme, chlorin reductase, which is unique to the photosynthetic bacteria, reduces ring B of the tetrapyrrole, thus providing a route to the synthesis of bacteriochlorophyll. Genes for the structural subunits of nitrogenase are *nifH*, *nifD* and *nifK* . Three genes required for protochlorophyllide reductase activity, *chlL* (*frxC* or *bchL*), *chlN* (*bchN*), and *chlB* (*bchB*), and three additional genes for chlorin reductase activity, *bchX*, *bchY* and *bchZ*, have been identified. The products of *chlL, bchL* and *bchX* show notable sequence similarity to NIFH, with the greatest conservation in residues required for binding the γ-phosphate of ATP and liganding the Fe₄S₄ cluster, and are refered to as "chlorophyll iron proteins" (Burke et al., 1993). The other potential components of protochlorophyllide reductase and chlorin reductase also show sequence conservation. CHLN also shows some sequence similarity to NIFD and NIFK.

the cysteine residues, the ATP binding pocket is also conserved. Analysis of the liverwort FRXC protein suggest that, like Fe protein , it binds ATP and is a dimer in solution (Fujita et al., 1989).

chlL is involved in the light-independent pathway of chlorophyll biosynthesis and has on disruption a "yellow in the dark" phenotype (Gid⁻) (Suzuki and Bauer, 1992). This phenotype is shared with the chloroplast genes *chlB* (Liu et al., 1993; Li et al., 1993) and *chlN* (Choquet et al., 1992) which are required for the activity of light-independent protochlorophyllide reductase. The *chlN* gene product shows 19% identity to NIFD and NIFK (Fujita et al., 1993) and in particular, has cysteine residues at positions equivalent to Cys $\alpha62$, Cys $\alpha88$, Cys $\beta95$, Cys $\beta153$ and Cys $\alpha154$, which coordinate the P clusters in nitrogenase MoFe protein (Howard and Rees, 1994). By analogy with the role of the nitrogenase Fe protein in electron transfer to the MoFe protein, it has been suggested that CHLL may transfer electrons to the light-independent protochlorophyllide reductase for which CHLB and CHLN are probably structural subunits (Li et al., 1993). It therefore seems likely that the light-independent protochlorophyllide reductase is a three-subunit enzyme which may have evolved from an ancestral reductase such as nitrogenase (Burke et al., 1993).

Since the components of this reductase are likely to be oxygen sensitive and are required for chlorophyll biosynthesis in the dark, one might expect their synthesis to be regulated in response to light. Although Northern analysis indicates that low levels of *chlB* message are present in both light and dark grown cultures of *Chlamydomonas reinhardtii* (Li et al., 1993) and no light-dependent translational regulation has yet been demonstrated in this organism, in contrast *chlB* transcripts are light regulated in *C.eugametos* (Richard et al., 1994). It should however be borne in mind that at least seven nuclear loci are required for the light-independent protochlorophyllide reduction in *Chlamydomonas*, adding considerable complexity to analysis of biosynthesis and regulation of the reductase.

The potential structural similarities between NIFH, CHLL, FRXC (and the so called "chlorophyll iron proteins" BCHL and BCHX in the photosynthetic bacterium *Rhodobacter capsulatus* (Bauer et al., 1993; Burke et al., 1993) lead us to believe that it may be possible to express *nifH* as active Fe protein in the *Chlamydomonas* chloroplast in the absence of the *nifS*, *nifU* and *nifM* genes. This hypothesis is based on the assumption that CHLL contains an Fe_4S_4 cluster in a similar environment to that of the Fe protein and is therefore likely to be processed by proteins functionally equivalent to those required for Fe protein assembly. Moreover since CHLL, like NIFH, is most probably required for ATP-coupled electron transfer to the "catalytic" subunits of the enzyme, it is possible that NIFH could at least partially substitute for CHLL, leading to complementation of the $chlL^-$ (Gid^-) phenotype.

We have precisely replaced the coding sequence of *chlL* in the *Chlamydomonas reinhardtii* plastid genome with that of *nifH* from *Klebsiella pneumoniae* by chloroplast transformation using particle gun bombardment. This was achieved by first disrupt-

ing *petB*, which is adjacent to *chlL*, with the *aadA* gene (Goldschmidt-Clermont, 1991) which confers resistance to spectinomycin and streptomycin. The resulting *petB::aadA* transformants are photosynthetic mutants and will only grow on a medium containing a reduced carbon source such as acetate. A previously constructed *petB::aadA* mutant was also used for the next step (Kuras and Wollman, 1994). The *petB::aadA* photosynthetic mutants were transformed with a plasmid containing the wildtype *petB* gene flanked by a modified *chlL* gene in which the *nifH* coding sequence was inserted between the 5' and 3' regulatory regions of *chlL*. Homologous recombination between the plasmid and recipient chloroplast genome replaces the *petB::aadA* insertion with a functional copy of *petB* and this allows transformants to photosynthesize and grow on a medium lacking acetate. Since these transformants lack *aadA*, they are spectinomycin sensitive. The *nifH* coding region is introduced into the chloroplast genome of a proportion of *petB* transformants by a double recombination event in which one event flanks *petB* and the other occurs upstream of the 5' *chlL* regulatory region. A number of transformants gave positive DNA signals with a *nifH* probe and gave bands of the expected size for replacement of the *aadA* cassette with the wild-type *petB* gene. We are currently investigating the level of *nifH* expression in homoplasmic transformants. Hopefully sufficient levels of NIFH will be expressed in dark grown cultures to allow assay of Fe protein activity by in vitro complementation with purified MoFe protein.

The way forward

Although relatively little has been achieved so far, we believe that the feasibility of expressing nitrogenase component proteins in the chloroplast will be tested in the next few years, provided that appropriate funding is available and scientific interest in this area is maintained. It is rather surprising that, as far as we are aware, few laboratories have made a serious attempt to examine this problem. Given the interesting potential evolutionary relationships between nitrogenase Fe protein and the so-called "chlorophyll iron proteins" (Burke et al., 1993), we think that *Chlamydomonas* provides the best model for examining *nif* gene expression in chloroplasts at present. There are of course a number of potential bottlenecks along the way and, for example, we may have to introduce *nif*-specific processing components such as *nifS*, *nifU* and *nifM* in order to activate the Fe protein. Another potential problem is the relative inefficiency of nitrogenase; we may have to obtain very high levels of expression in order to achieve sufficient activity. We have sparse knowledge of the biochemistry of the light-independent protochlorophyllide reductases and further structure- function studies of these enzymes would no doubt provide useful pointers to the likely behaviour of nitrogenase in the plastid. For example are these enzymes as oxygen sensitive as nitrogenase and what co-factors are required for their activity?

If active Fe protein can be synthesised in the algal plastid, it will provide considerable impetus for further work on higher plants as well as the introduction of genes encoding Mo-Fe protein into chloroplasts. Although chloroplast transformation is well established in tobacco, all the currently available tobacco vectors make use of light-regulated promoters which are not useful in the context of *nif* gene expression. In order to circumvent the oxygen sensitivity of nitrogenase, we need to have efficient and regulated expression of the chloroplast- introduced *nif* genes in the dark, thus mimicking the temporal separation of photosynthesis and nitrogen fixation achieved by the non-heterocystous cyanobacteria. One way of approaching this would be to utilise nuclear encoded genes to regulate a complete *nif* gene set in the plastid, thus allowing modification of the temporal expression of the introduced *nif* genes without having to repeat the complete chloroplast transformation procedure. For example, nuclear encoded T7 RNA polymerase can be targeted to the plastid to give very high levels of gene expression (McBride et al., 1994) and synthesis of this polymerase could be controlled by a highly regulated nuclear promoter. In this way regulation of a single nuclear gene decides the expression of the complete set of *nif* genes.

As an alternative to controlling *nif* genes from the nucleus with a phage polymerase, one could consider the introduction of bacterial regulatory genes into the plant plastid. As mentioned above, transcription of *nif* genes in bacteria requires an alternative form of RNA polymerase holoenzyme containing the sigma factor, σ^N in association with a specific activator protein NIFA. If the bacterial sigma factor and activator protein could associate with the chloroplast polymerase to activate *nif* gene transcription, it would be possible to use one or two regulatory proteins to regulate the complete *nif* gene set without further modification of *nif* promoters. This is an attractive concept since a nuclear-encoded sigma factor with sequence

homology to a large number of bacterial and cyanobacterial σ factors is targeted to the chloroplast of the red alga *Cyanidium caldarium* and may control a subset of photoregulated genes in the plastid (Liu and Troxler, 1996; Tanaka et al., 1996). It would be a relatively simple task to test whether purified chloroplast RNA polymerase can function in vitro in the presence of σ^N to recognise *nif* promoters and respond to NIFA to activate transcription. Positive results would undoubtedly encourage the construction of transgenic plants in which these proteins are either targetted to the chloroplast or expressed directly in the plastid. This approach could have the advantage that the complete cluster would be expressed from unmodified promoters, hopefully allowing efficient expression and maintenance of the normal "balance" in the levels of each gene product. It would also allow nitrogenase synthesis to be tightly regulated in response to oxygen concentration, by introduction of either an intrinsically oxygen sensitive NIFA or the NIFL protein, a redox-sensitive flavoprotein which modulates NIFA activity (Hill et al.,1996).

We have made little mention in this review of the prospects for engineering nitrogen-fixing crop plants primarily because in our opinion all the feasibility experiments are most tractable in "model" lower and higher plants. Chloroplast transformation is now relatively routine in both algae and tobacco and in some cases it has been possible to obtain high level expression of foreign genes. However, no such system has so far been reported with respect to rice. Before we can even consider transferring nitrogen fixation to rice it is therefore necessary to develop vectors, methods and technology for rice plastid transformation. Since rice seeds can germinate and produce green shoots in the presence of spectinomycin, it appears that the *aadA* gene may not be a useful marker for rice chloroplast transformation. However, preliminary results indicate that the *bar* and *hph* genes encoding herbicide resistance and hygromycin resistance respectively are efficient selectable markers for rice (Datta et al., 1992; Lin et al., 1995). The sequence of the rice plastid genome is available which will facilitate the search for a suitable site within the genome for the insertion of foreign DNA. We are therefore reasonably confident that rice plastid transformation will be achieved within the next five years.

In considering these future prospects we have mainly discussed various biotechnological approaches to express active nitrogenase within the chloroplast. It is not possible to predict how many physiological bottlenecks we may encounter along the way. Although,

202

we are not aware of any absolute barrier to the expression of active nitrogenase in the plastid, even if we are able to construct plants in which the enzyme is fully active, there is no guarantee that nitrogen fixing plants will produce a satisfactory growth yield. But, in view of the tremendous benefits to mankind if such plants were effective in the field, can we afford not to try to construct them?

References

Allen R M, Homer M J, Chatterjee R, Ludden P W, Roberts G P 1993 Dinitrogenase and MgATP-dependent maturation of apo-dinitrogenase from *Azotobacter vinelandii* J. Biol. Chem. 268 23670–23674

Bauer C, Bollivar D and Suzuki J 1993 Genetic analysis of photopigment biosynthesis in eubacteria: a guiding light for algae and plants. J.Bacteriol 175, 3919–3925

Berman J, Gershoni J M and Zamir A 1985 Expression of nitrogen fixation genes in foreign hosts. assembly of nitrogenase Fe protein in *Escherichia coli* and in yeast. J. Biol. Chem. 260, 5240–5243

Burke D, Hearst J and Sidow A 1993 Early evolution of photosynthesis: clues from nitrogenase and chlorophyll iron proteins. Proc.Natl. Acad. Sci USA 90, 7134–7148

Choquet Y, Rahaire M, Girard-Bascou J, Erickson J and Rochaix J D 1992 A chloroplast gene is required for the light-independent accumulation of chlorophyll in *Chlamydomonas reinhardtii*. EMBO J. 11, 1697–1704

Datta S K, Datta K, Soltanifa N, Donn G and Potrykus I 1992 Herbicide-resistant Indica rice plants from IRRI breeding line IR72 after PEG mediated transformation of protoplasts. Plant Mol. Biol. 20, 619–629

Dean D, Bolin J and Zheng L 1993 Nitrogenase metalloclusters: structures,organisation and synthesis. J. Bacteriol. 175, 6731–6744

Dean D R and Jacobson M R 1992 Biochemical Genetics of Nitrogenase. *In* Biological Nitrogen Fixation. Eds. G Stacey, R H Burris and H J Evans. pp. 763–834. Chapman and Hall, New York.

Dixon R and Postgate JR 1972 Genetic transfer of nitrogen fixation from *Klebsiella pneumoniae* to *Escherichia coli*. Nature 237, 102–103

Dowson-Day M, Ashurst J, Mathias S, Watts J, Wilson T and Dixon R 1993 Plant viral leaders influence expression of a reporter gene in tobacco. Plant Mol. Biol. 23, 97–109

Dowson-Day M J, Ashurst J L, Watts J, Dixon R A and Merrick M J 1991 Studies of the potential for expression of nitrogenase Fe-protein in cells of higher plants. *In* Nitrogen Fixation: Proceedings of the 5th International Symposium on Nitrogen Fixation with Non-Legumes. Eds. M Polsinelli, R Materassi and M Vincenzini. pp 659–669. Kluwer, Dordrecht.

Eady R R 1995 The enzymology of biological nitrogen fixation. Science Progress 78, 1–17

Flint, D H 1996 *Escherichia coli* contains a protein that is homologous in function and N-terminal sequence to the protein encoded by the *nifS* gene of *Azotobacter vinelandii* and that can participate in the synthesis of the Fe-S cluster of dihydroxy-acid dehydratase. J. Biol. Chem. 271, 16068–16074.

Fu W, Jack R F, Morgan T, Dean D and Johnson M 1994 *nifU* product from *Azotobacter vinelandii* is a homodimer that contains two identical· [2Fe-2S] clusters. Biochemistry 33, 13455–13463

Fujita Y, Matsumoto H, Takahashi Y and Matsubara H 1993 Identification of a *nifDK*-like gene (ORF467) involved in the biosynthesis of chlorophyll in the cyanobacterium *Plectonema boryanum*. Plant Cell Physiol. 34, 305–314

Fujita Y, Takahashi Y, Chuganji M and Matsubara M 1992 The *nifH*-like (*frxC*) gene is inolved in the biosynthesis of chlorophyll in the filamentous cyanobacterium *Plectonema boryanum*. Plant Cell Physiol. 33, 81–92

Fujita Y, Takahashi Y, Kohchi T, Ozeki H, Ohyama K and Matsubara H 1989 Identification of a novel *nifH*- like (*frxC*) protein in chloroplasts of the liverwort *Marchantia polymorpha*. Plant Mol.Biol. 13, 551–561

Gallon J 1992 Tansley review no.44. Reconciling the incompatible: nitrogen fixation and oxygen. New Phytol. 122, 571–609

Gallon J and Chaplin A 1988 Recent studies of nitrogen fixation by non-heterocystous cyanobacteria. *In* Nitrogen fixation:hundred years after. Eds. H Bothe, F de Bruijn and W Newton. pp 183–203. Fischer, Stuttgart.

Gallon J, Reade J, Dougerty L, Pederson D and Rogers L 1995 Degradation of nitrogenase in Gleothece. *In* Nitrogen fixation: fundamentals and application. Eds. I A Tichonovich, N Provorov, V Romanov and W Newton. pp. 230. Kluwer Academic Publishers, Dordrecht.

Goldschmitt-Clermont M 1991 Transgenic expression of the aminoglycoside adenine transferase in the chloroplast: aselectable marker for site-directed transformation of *Chlamydomonas*. Nucleic Acids Res. 19, 4083–4089

Govezensky D, Greener T, Segal G and Zamir A 1991 Involvement of GroEL in nif gene regulation and assembly. J. Bacteriol. 173, 6339–6346

Hidalgo E and Demple B 1996 Activation of SoxR-dependent transcription in vitro by non catalytic or NifS-mediated assembly of [2Fe-2S] clusters into Apo-SoxR1. J. Biol. Chem. 271, 7269

Hill S 1992 Physiology of nitrogen fixation in free-living heterotrophs. *In* Biological nitrogen fixation. Eds. G Stacey, R H Burris and H J Evans. pp. 87–134. Chapman and Hall, New York.

Hill S, Austin S, Eydmann T, Jones T and Dixon R 1996 *Azotobacter vinelandii* NIFL is a flavoprotein that modulates transcriptional activation of nitrogen-fixation genes via a redox-sensitive switch. Proc. Natl. Acad. of Sci. USA 93, 2143–2148

Howard J and Rees D 1994 Nitrogenase: a nucleotide-dependent molecular switch. Ann.Rev. Biochem. 63, 235–264

Howard K S, McLean P A, Hansen F B, Lemley P V, Koblan K S and Orme-Johnson W J 1986 *Klebsiella pneumoniae nifM* gene product is required for stabilization and activation of nitrogenase iron protein in *Escherichia coli*. J. Biol. Chem. 261, 772–778

Khoroshilova N, Beinert H and Kiley P 1995 Association of a polynuclear Fe-S center with a mutant FNR protein enhances DNA binding. Proc. Natl. Acad. Sci. USA 92, 2499–2503

Kuras R and Wollman F-A 1994 The assembly of cytochrome b6/f complexes: an approach using genetic transformation of the green alga *Chlamydomonas reinhardtii*. EMBO J. 13, 1019–1027

Kustu S, Santero E, Keener J, Popham D and Weiss D 1989 Expression of σ^{54}(*ntrA*)-dependent genes is probably united by a common mechanism. Microbiol.Revs. 53, 367–376

Li J, Goldshmitt-Clermont M and Timko M 1993 Chloroplast-encoded *chlB* is required for light-independent protochlorophyllide reductase activity in *Chlamydomonas reinhardtii*. Plant Cell 5, 1817–1829

Lidholm J and Gustafsson P 1991 Homologues of the green alga *gidA* gene and the liverwort *frxC* gene are present on the chloroplast genome of conifers. Plant Mol. Biol. 17, 787–798

Lin W, Anuratha C S, Datta K, Potrykus I, .Muthukrishnan S and Datta SK 1995 Genetic engineering of rice for resistance to sheath blight. Bio/technology 13, 686–692

Liu B and Troxler R F 1996 Molecular characterization of a positively photoregulated nuclear gene for a chloroplast RNA polymerase sigma factor in *Cyanidium caldarium*. Proc. Natl. Acad. Sci. USA 93, 3313–3318

Liu X-Q, Hui X and Huang C 1993 Chloroplast *chlB* gene is required for light-independent chlorophyll accumulation in *Chlamydomonas reinhardtii*. Plant Mol. Biol. 23, 297–308

McBride K, Schaaf D, Daley M and Stalker D 1994 Controlled expression of plastid transgenes in plants based on nuclear DNA encoded and plastid-targeted T7 RNA polymerase. Proc. Natl. Acad. Sci USA 91, 7301–7305

Merrick M 1992 Regulation of nitrogen fixation genes in bacteria. *In* Biological nitrogen fixation. Eds. G Stacey, R H Burris and H J Evans. pp 835–876. Chapman and Hall, New York.

Merrick M J 1993 Organisation and regulation of nitrogen fixation genes. *In* New Horizons in Nitrogen Fixation, Proceedings of the 9th International Congress on Nitrogen Fixation. Eds. R Palacios, J Mora and W E Newton. pp. 43–54. Kluwer Academic, Dordrecht.

Merrick M and Dixon R 1984 Why don't plants fix nitrogen. Trends Biotech. 2, 162

Moshiri F, Kim J, Fu C and Maier R 1994 The FeSII protein of *Azotobacter vinelandii* is not essential for aerobic nitrogen fixation but confers significant protection to oxygen-mediated inactivation of nitrogenase in vitro and in vivo. Mol. Microbiol. 14, 101–114

Ohyama K, Fukuzawa H, Kohchi T, Shirai H, Sano T, Sano S, Umesono K, Shiki Y, Takeuchi M, Chang Z, Aota S, Inokuchi H and Ozeki H 1986 Chloroplast gene organisation deduced from complete sequence of liverwort *Marchantia polymorpha* chloroplast DNA. Nature 322, 572–574

Paul W and Merrick M 1989 The roles of the *nifW, nifZ* and *nifM* genes of *Klebsiella pneumoniae* in nitrogenase biosynthesis. Eur. J. Biochem. 178, 675–682

Postgate J 1981 Discussion. *In* The manipulation of genetic symbiosis in plant breeding. Eds. H Reese, R Riley, R Breese and C Law. pp 198–199. The Royal Society, London.

Postgate J 1992 The Leeuwhenhoek lecture 1992. Bacterial evolution and the nitrogen fixing plant. Phil. Trans. R. Soc. Lond. B 338, 409–416

Reddy P and Ladha J 1995 Can symbiotic nitrogen fixation be extended to rice ? *In* Nitrogen fixation: fundamentals and application. Eds. I A Tichonovich, N Provorov, V Romanov and W Newton. pp 629–633. Kluwer Academic Publishers, Dordrecht.

Richard M, Tremblay C and Bellemare G 1994 Chloroplast genomes of *Gingko biloba* and *Chlamydomonas moewusii* contain a *chlB* gene encoding one subunit of a light-independent protochlorophyllide reductase. Current Genet 26, 159–165

Suzuki J and Bauer C 1992 Light-independent chlorophyll biosynthesis: involvement of the chloroplast gene *chl* (*frxC*). Plant Cell 4, 929–940

Tanaka K, Oikawa K and Takahashi H 1996 Nuclear Encoding of a Chloroplast RNA Polymerase Sigma Subunit in a Red Alga. Science 272, 1932–1935

Thorneley R N F and Ashby G A 1989 Oxidation of nitrogenase iron protein by dioxygen without inactivation could contribute to high respiration rates of *Azotobacter* species and facilitate nitrogen fixation in other aerobic environments. Biochem. J. 261, 181–187

Thorneley R N F and Lowe D J 1983 Nitrogenase of *Klebsiella pneumoniae*. Kinetics of the dissociation of oxidised iron protein from molybdenum-iron protein: identification of the rate-limiting step for substrate reduction. Biochem. J. 215, 393–403

Viitanen P V, Schmidt M, Buchner J, Suzuki T, Vierling E, Dickson R, Lorimer G H, A G and Soll J 1995 Functional characterization of the higher plant chloroplast chaperonins. J. Biol. Chem. 270, 18158–18164

Yang Z and Bauer C 1990 *Rhodobacter capsulatus* genes involved in the early steps of the bacteriochlorophyll pathway. J. Bacteriol. 172, 5001–5010

Zheng L and Dean D 1994 Catalytic formation of a nitrogenase iron-sulphur cluster. J. Biol. Chem. 269, 18723–18726

Zheng L, White R, Cash V, Jack R and Dean D 1993 Cysteine desulfurase activity indicates a role for NIFS in metallocluster biosynthesis. Proc. Natl. Acad. Sci. USA 90, 2754–2758

Guest editors: J K Ladha, F J de Bruijn and K A Malik

Plant and Soil **194:** 205–216, 1997.

Enhancing biological nitrogen fixation: An appraisal of current and alternative technologies for N input into plants

Sivramiah Shantharam[1] and Autar K. Mattoo[2]

[1] *Biotechnology Evaluations, USDA/APHIS, BSS, Riverside, MD 20737–1237, USA* * *and* [2] *The USDA Vegetable Laboratory, ARS, BARC-W, Bldg. 010A, Beltsville, MD 20705-2350, USA*

Key words: carbon-nitrogen balance, ethylene, legumes, nitrogen economy, nitrogen fixing bacteria, non-legumes, rhizobium, senescence, signal transduction

Abstract

Biological nitrogen fixation (BNF) involves a highly specialized and intricately evolved interactions between soil microorganisms and higher plants for harnessing the atmospheric elemental nitrogen (N). This process has been researched for almost a century for efficient N input into plants. The basic mechanism and biochemical steps involved in BNF have been unraveled. It has become abundantly clear that the host plant (legumes) dominates in regulating the BNF process. Environmental factors as well influence this process. Perturbation or any manipulation of the interactions between the bacteria and the legumes seems to offset the critical balance, usually to the detriment of N fixation efficiency. Not much success has been obtained in either enhancing BNF in legumes or transferring important BNF traits to non-nitrogen fixing organisms. An appraisal is given for the lack of success in making the BNF process a popular and efficient agronomic practice. Alternative physiological approaches are presented for improving mobilization, redistribution and utilization of stored N reserves within the host plant.

Introduction

Nitrogen (N) is a key component of nutrition for plants. Plants derive nitrogen through nitrate and ammonia. In addition, atmospheric deposition of nitrous oxides and ammonia originating mainly from pollution also contribute to the N supply (Bockman, 1996). Leguminous plants fix atmospheric nitrogen in a process called Biological Nitrogen Fixation (BNF) which also provides N input into agricultural soils. Various estimates of the contribution of BNF to the soil N fall in the range of 44–200 Tg. N y^{-1}, with an average of about 140 Tg. N y^{-1} (Soderlund and Roswall, 1982). Globally, about 11% of arable land is used to cultivate legume pulses and oilseeds (Peoples et al., 1995). Legumes contribute an estimated 35 out of 75×10^{12} tonnes of fixed N annually (Elkan, 1992). Free-living bacteria contribute 15 kg N ha^{-1} y^{-1}, cyanobacteria 7–80 kg N ha^{-1} y^{-1}, associative bacteria 36 kg N ha^{-1} y^{-1}, *Azolla/Anabaena* 45–450 kg N ha^{-1} y^{-1}, *Frankia* 2-362 kg N ha^{-1} y^{-1}, and the highest by *Rhizobium*-legume

* FAX No: +13017348669.
E-mail: sshantharan@aphis.usda.gov

symbiosis being 24-584 kg N ha^{-1} y^{-1} (Elkan, 1992). In addition, clovers used in crop rotation, and the use of *Azolla* and *Sesbania* as green manure in parts of Africa and Asia have earned BNF an assured place in agriculture depending upon the environmental and socio-economic conditions (Ladha and Garrity, 1994; Lumpkin and Plucknett, 1982; Sarrantonio, 1991).

The success of green revolution was in part due to the the use of high yielding varieties and mineral fertilizers that kept pace with increase in population. The common limiting nutrient for plant productivity has been N, which has resulted in the increased use of fertilizer inputs. This runs counter to the principle of sustainable agriculture which demands low input with a high output. BNF is a fascinating biological phenomenon which has been extensively studied during the last hundred years with the sole objective of harnessing its potential to provide low-cost nitrogen to increase crop productivity. Although basic knowledge of the process has increased by quantum leaps, limited success has been achieved in manipulating the system to enhance nitrogen fixation in either the leguminous plants in which it naturally operates or in non-legumes.

There are two basic biological nitrogen fixation systems: symbiotic nitrogen fixation of the leguminous plants and asymbiotic or non-symbiotic nitrogen fixation also known as associative nitrogen fixation of non-leguminous C-4 plants. Of the two, only the legume-*Rhizobium* symbiosis held out a realistic chance for improving the rate of nitrogen fixation, as significant advances were achieved in the molecular genetics of *Rhizobium* and legumes in the 1980's. The main reason for optimism in the symbiotic nitrogen fixation system was that the root nodule symbiotic system is a highly specific, intricate, tightly controlled and complex metabolic system that showed the potential for manipulation. If we understand the complex nature of these interactions, it would then be possible to manipulate them using recombinant DNA techniques for improved agronomic purposes. Our current understanding of the complexity of symbiotic nitrogen fixation has undoubtedly increased considerably, unraveling some of the "key" factors that should help gain control over the process. However, it is still not clear exactly how many bacterial and plant genes control symbiosis, and how they are regulated.

The present paper briefly summarizes the recent developments in the study of BNF and provides an appraisal of the lack of wide scale adoption of BNF technologies, and success in transferring BNF trait(s) to non-legumes. More importantly, if the purpose of scientific effort in BNF is to improve N nutrition of plants, alternative approaches are presented which if tested might help achieve the same objective. The scope of this review is limited to a general appraisal of the use of inoculation technologies that so far seem to have not made a significant impact on agricultural development. Alternative strategies that use genetic engineering technologies to increase recycling N to plants are discussed. Readers are referred to recently published reviews for details on the current developments in the molecular biology, physiology, biochemistry, and genetics of BNF (Brewin, 1991; Cetano-Annolles and Gresshoff, 1992; Denaire et al., 1992; Devine and Kuykendall, 1996; Downie, 1997; Leigh and Coplin, 1992; Long, 1989a,b; Martinez et al., 1990; Triplett and Sadowsky, 1992).

Symbiotic nitrogen fixation

The host plants

Rates of N fixation are highly variable, and are dependent on the bacterial strain-legume cultivar used, soil and other environmental conditions. As high as 100-400 kg of N ha^{-1} are reported to be derived from symbiotic nitrogen fixation (Peoples et al., 1995). However, accurate estimates are lacking in a variety of legumes grown in developing countries in marginal soils. Consideration has to be given to environmental factors that limit BNF under uncontrolled field conditions. By all accounts, the legume host physiology and metabolism place strict limitations on the root nodule system.

Photosynthetic energy is a major factor that drives the system as it is the main source of energy to sustain the process. In this context, it is interesting to note that the highest amount of nitrogen fixed by the legumes takes place either at the flowering stage or during pod fill (Imsande, 1989; Imsande and Edwards, 1988; Imsande and Touraine, 1994). This implies that sufficient numbers of bacteria and root nodules should remain active in fixing nitrogen at least till the time of legume flowering and pod-fill stages if symbiotic nitrogen fixation is to be beneficial. Otherwise, the legume itself can switch over from its dependence on BNF to nitrate uptake from the soil (Imsande and Touraine, 1994). Leguminous plants have a very efficient NO_3^- uptake system which may be preferred to the energy-intensive BNF process, and therefore it is important to clearly distinguish between N fixed through BNF and that from other sources under variable agronomic conditions.

The highest demand for N fixation in legumes is during R1–R4 stages of plant development (Zapata et al., 1987). During these stages there is lowest gain in harvested biomass and lowest percentage gain in biomass (Imsande, 1997). Rapid nitrogen fixation during pod fill stages in the presence of low concentrations of nitrate that do not inhibit root nodulation and nitrogenase activity enhances net photosynthetic output of soybean which in turn results in increased biomass. Increase in biomass does not necessarily translate into increase in grain yield. Legumes consume approximately 10% of the plant's net photosynthetic output for N fixation. Root nodules become the largest sink of photosynthetic energy, depriving other, equally critical, plant metabolic processes. Fixing nitrogen for a legume is a physiological burden when it could use its highly efficient nitrate uptake system which does not depend on either the concentration of nitrate available, or N for its growth (Imsande, 1997). Another important aspect of plants N nutrition is that plants have evolved a very efficient N uptake system

Whatever the pattern of resource allocation in plants, the limited resource must be divided among plant parts for cellular functions. Therefore, trade-offs are always involved in resource allocation (Ronsheim, 1988). Although it is important to measure resource allocation patterns using the most limiting factor, this is not always possible because of diurnal fluctuations in photosynthesis and photorespiration. Reekie and Bazzaz (1987) have suggested measuring total carbon, which includes total biomass and respiratory carbon, as the appropriate index. They showed that respiratory costs of plant growth increase the concentrations of nitrogen and phosphorus in plant tissues. As N and P become more limiting, respiratory costs increase further. Thus, total carbon allocation reflects the true distribution of N and P that are limiting.

Legume host plants exert a substantial control over nodulation and nitrogen fixation (Vance et al., 1988), and it seems reasonable to suggest that the host plant mediates the success of root nodule symbiosis. Although substantial progress has been made in understanding the contributions of microbial genes and gene products to nodulation and N fixation, relatively little is known about key plant genes that control these processes (Nap and Bisseling, 1990; Vance et al., 1988). Through classical genetic studies some 45 genes across eight legume species seem to affect nodulation and N fixation (LaRue et al., 1985; Vance et al., 1988). In soybean alone at least eight genes control nodulation response (Devine and Kuykendall, 1996). The Rj4 and other nodulation restriction genes identified in soybeans are strain-specific that do not allow other strains, such as *Bradyrhizobium*, to nodulate. Most of the genes identified through classical genetic studies in soybean indicate that the relationship of soybeans carrying Rj4 allele with *Bradyrhizobium japonicum* strains have specialized during evolution, in that soybeans carrying Rj4 have adapted to select only those rhizobia which are symbiotically efficient. It appears that conditional restriction has evolved for survival of the most effective symbiotic partners in nature. In rhizobia, out of 17 operons 11 are critical (Downie, 1997; Freiberg et al., 1997). Many more genes are still being discovered. The isolated genes are being studied to understand their role and regulation during root nodu-

lation in order to realistically manipulate them in an ambient fashion to enhance the symbiotic process.

Another set of important plant coded genes, (nodule specific genes) are the "nodulins" and leghaemoglobins which are structural proteins. More than 30 nodulins have been reported but only about 10 of them have known functions (Horvath and Bisseling, 1993). Moreover, nodulation is autoregulated, mature root nodules inhibiting the growth of younger nodules. This aspect of nodule development is a complete "black box" at the biochemical and molecular levels. Very little is known about the regulation of these genes or about signal transduction pathways that control symbiosis, thereby limiting any genetic manipulation to enhance BNF (de Bruijn and Schell, 1993).

The goal of transferring BNF traits to non-legumes still remains elusive. Several publications have speculated on the possibilities of extending N fixation abilities to rice (Boivin et al., 1997; Cocking et al., 1994; Khush and Bennett, 1992; Reddy and Ladha, 1995). In a series of reports in late 1980's and early 1990's, research groups from the United Kingdom, Peoples Republic of China, and Australia reported exciting experimental results that suggested that under certain artificially created conditions (using hormones and cell-wall degrading enzymes) "nodule" like structures could be induced on rice and wheat roots with *Rhizobium* strains (Cocking et al., 1994). Based on these reports, optimistic predictions were made for achieving nodulation and BNF of rice plants by *Rhizobium* (Khush and Bennett, 1992). A critical examination of these results, however, revealed that it was possible to induce necrosis of epidermal tissue of the roots at junctions where lateral roots emerge. The bacteria accumulated at the site of injury due to the emergence of lateral roots, which gave the impression of the possibility of root nodule formation in non-legumes. In fact, such structures are not specific since they can be induced not only by *Rhizobium*, but by other bacteria in combination with hormones in legumes as well (Shantharam, unpublished results). Similarly, it is interesting that neoplastic growth of the root system can be induced merely by "signal" molecules without the physical presence of the bacteria (Denaire et al., 1992). None of these nodule-like structures could fix nitrogen as they lacked internal tissue differentiation of true nodules where the bacteroid is surrounded by peribacteroid membranes. Moreover, no evidence of any known nodule specific proteins, like leghaemoglobin or nodulins, was apparent. The suggestion that rice possesses certain complements of genetic make up to induce nodulation by rhizobia is untenable as it may simply be a hypertrophic response to general bacterial infestation. Extensive basic studies are needed to understand interactions between *Rhizobium* and rice with special attention to signal exchange mechanisms.

Atragalus and species of *Sesbania* are used as green manure both before and after rice cultivation in Africa and Asia. Excellent rates of N fixation by green manure have been reported with about 80% being contributed by BNF (Becker et al., 1995a). This technology holds great promise, but has not taken hold in rice cropping systems to a large extent. Green manure technology like other bacterial inoculation technology is also beset with high variability of its efficiency or performance. One important prognosis by Becker et al. (1995b) is that unless farmers are convinced of deriving economic benefit, green manure technology or any other microbial inoculant technology will remain an under or unexploited potential. Although green manure crops have the potential for high N supply for the rice crop, farmers are reluctant to use them for want of any economic advantage over chemical N fertilizer and reliability (Becker et al., 1995a,b; Ladha and Garrity, 1995).

Research efforts to breed high nitrogen-fixing soybeans for Africa and Australia, and groundnuts for India and USA, resulted in the release of a few lines with improved N fixation but little impact on legume cultivation (Herridge and Danso, 1995). The biggest obstacle in any breeding program seems to be the difficulty in manipulating multigenic traits which fluctuate under the vagaries of the environment. Maintenance and management of superior cultivars with desired traits in an agricultural system should be cost effective in order for a farmer to introduce a high N fixing cultivar.

Soybeans and peas are the few legumes that have been studied extensively for root nodule symbiosis both at the genetic and molecular levels (Herridge and Danso, 1995). The nature of nodulation restriction by the host plants was thought to be serogroup-specific (Cregan and van Berkum, 1984), but later studies suggested otherwise which means the system is more complicated. Host-plant-controlled nodulation restriction should be viewed as an extreme case of host-specificity. If one can use conventional breeding for increased competitiveness by the host plant, then it should be done with appropriate rhizobia as a challenge organism for selection (Cregan and van Berkum, 1984). This approach has not been taken by the legume breeders who have usually in the past scored for desirable agronomic traits other than N fixation. Herridge

and Danso (1995) have critically appraised the failures of legume screening and breeding program designed solely to enhance N fixation. If increased N fixation is used as a scorable trait for selecting legume varieties through breeding, sufficient attention must also be paid that other aspects of plant productivity are not compromised. Cregan and Van Berkum (1984) suggested a number of physiological and biochemical parameters of nitrogen metabolism as selection criteria for identifying germplasm with enhanced nitrogen metabolism and grain crop productivity. A properly designed physiological/biochemical selection program integrating the effects of plant genotype, environment, and their interactions should enable identification of factors limiting productivity of particular genotypes as well as provide an estimate of the end product. According to Cregan and van Berkum (1984) such a selection program should include NO_3^- uptake, N fixation, N accumulation, N remobilization, seed protein synthesis, and Nitrogen Harvest Index. This suggestion needs to be incorporated in breeding programs to identify the needed germplasm.

Whether increasing the number of root nodules would correspondingly increase the capacity to fix more nitrogen was tested with several hypernodulating mutants of soybean. Such mutants produced 100 times more nodules than the parent plant (Betts and Herridge, 1987; Carroll et al., 1985). Unfortunately, these mutants turned out to be poor agronomic performers (Pracht et al., 1994). The apparent reason for the failure of this approach was that plants expended considerable amount of energy in bearing root nodules, limiting energy needed for the nitrogen fixation process. Symbiotic nitrogen fixation is a highly energy intensive process consuming 6.5 g C g^{-1} of N fixed (Kennedy, 1997). Nitrogenase requires 36 ATP molecules for every N molecule reduced. Thus, bearing root nodules and fixing nitrogen for legumes seems to be an extra burden which has been aptly described as a "diseased" state. There are certain schools of thought which define root nodule symbiosis as refined parasitism (Djordjevic et al., 1987; Vance, 1983). Clearly, there is a delicate balance of physiological limitations placed on the root nodule symbiotic system which, when even slightly perturbed, results in inefficient nitrogen fixation.

Nitrogen fixation initiates with reduction of N_2 to NH_4^+, which, in turn, is utilized as a substrate in the ammonia assimilation pathways involving a series of plant enzymes. One of the key plant enzymes is the glutamine synthetase (GS) complex. This enzyme complex is also present in the nitrogen fixing bacteroids. Thus, both the bacteroid and plant enzyme can compete for this substrate, resulting in lesser amount of fixed N_2 available for the plant to assimilate. Apparently, this problem can be overcome by co-inoculating rhizobia with *Pseudomonas syringae* pv. *tabaci*. This plant pathogenic bacterium produces a root encoded, nodule specific toxin known as tabtoxinin-β-lactam. This toxin specifically inhibits the plant coded, root nodule specific GS, allowing only the bacterial GS to efficiently function, thereby enhancing N assimilation into plants (Knight and Langston-Unkefer, 1988). This approach has not been tested under field conditions to ascertain the BNF efficiency. Halvorson and Handelsman (1991) employing coinoculation with *Bacillus cereus* UW85 observed enhancement in soybean nodulation with little increase in nitrogen fixation. In another report dealing with time course of nitrogen fixation in field grown soybean using ^{15}N methodology, Zapata et al. (1987) confirmed earlier reports that soil N contributes more than BNF to the nutrition of soybean. Efforts should be made to increase the nitrogen fixation efficiency at the time of highest N uptake when the NO_3^- uptake system is not competing.

Mendoza et al. (1995) have shown that by enhancing NH_4^+ assimilating enzymes through genetic engineering in *Rhizobium etli* prevents nodulation of bean plants. Using recombinant DNA techniques, Mendoza et al. (1995) modified the ammonium assimilation pathway (GS-GOGAT), by adding an additional copy of glutamate dehydrogenase (GDH). Total inhibition of nodulation was observed. Bacterial cells that have high internal nitrogen down regulate *nod* genes; *NtrC* gene is involved in such regulation. This points to the fact that there is a delicate internal metabolite-regulated control of *nod* gene expression in bacteria. Similarly, pleotropic effects are commonly observed in transgenic plants when an introduced gene affects more than one trait in the plant (Flavell, 1994; Ingelbrecht et al., 1994; Kooter and Mol, 1993; Matzke and Matzke, 1995).

The bacteria

In a five year FAO/IAEA coordinated study on BNF in certain grain legumes (common bean and faba bean) highly variable amounts of nitrogen fixation ranging from 30–120 kg N ha^{-1} was recorded (Danso, 1992). As much as 163 kg N fixed ha^{-1} has been reported in Latin America involving certain common cultivars of common bean in combination with efficient strains of *Rhizobium*. On the other hand, soybeans were extremely well nodulated by the indigenous rhizobia of the cowpea miscellany group in African soils suggesting no need for inoculation (Danso, 1992). There are many such instances which clearly suggest that if suitable nodulating varieties of legumes are selected against indigenous rhizobia, it would be lot more productive in marginal soils where farmers can not afford costly chemical inputs. A perusal of literature in the area of BNF and sustainable agriculture, notably Mulongoy et al. (1992) and Ladha and Peoples (1995), reveals that by inoculating in large numbers strains isolated from cultivated plants during every seeding ensures certain marginal benefits of crop productivity in nitrogen deficient soils (Neves et al., 1990).

"Competition" is the single most critical factor in preventing successful inoculation of leguminous plants (Triplett and Sadowsky, 1992). This applies not only to *Rhizobium*, but also to all microbes, and is well known in microbial ecology (van Veen et al., 1997). It is now well known that any soil bacterium that is cultured in the laboratory either with or without genetic manipulation becomes unfit when reinoculated into the same environment from which it was isolated. Even spontaneous antibiotic resistant mutants of bacteria clearly show weakened competitive ability. One of the solutions to the problem of rhizobial "competition" suggested by Triplett and Sadowksy (1992) is to inoculate the soils with massive amounts of desired strains of *Rhizobium* over a long period of time to force the bacteria to establish itself in the soils. This is a lesson learned from the history of rhizobial inoculation in the Midwest with *Bradyrhizobium japonicum* USDA 123 serogroup strains. It has a reasonable chance to succeed in the long run provided the host plant variety is not changed. The caveat though is that introduction of new legume varieties or superior strains of bacteria as they become available subsequently may be out competed by the long established bacterial strains in the soil. This is the precise reason why repeatedly unsuccessful attempts have been made to inoculate superior strains into the mid-western soils.

The key to successful nodulation of legumes is that the inoculated bacteria should survive or persist in critical numbers for a long time in the rhizosphere of the host plant to outcompete the native microflora. The population of the inoculated bacteria declines rapidly within 48–64 hours of inoculation. A critical appraisal of the fate and activity of inoculated soil bacteria has been reviewed recently by van Veen et al. (1997). Microbiostasis is the term applied to the generalized phenomenon of population decline of inoculated microbes in the soil. A myriad of biotic and abiotic factors are responsible for the population decline of the introduced bacteria. The physiological status of the inoculated bacteria to a large extent regulates its own survival in the soil. Availability of substrates for the bacteria to multiply is also important for survival in critical numbers. Carbon is the key and is always found in limited supply in the soil. Soil type also plays a vital role in the establishment of inoculated microbes. On a larger scale, ecological selectivity of microorganisms takes place, precluding the survival of certain inoculated microorganisms. All these factors point to the well known fact of soil inoculation technology – that microorganisms chosen for the purpose of inoculation should be selected from local ecological niches and reinoculated into the same environment to ensure the desired benefits. This takes us back to the practical problem of developing a viable commercial operation for producing inoculants that will perform reliably and reproducibly in all agricultural environments. Perhaps, this is one of the major reasons as to why biofertilizer technologies have not prospered. Van Veen et al. (1997) suggest that instead of trying to inoculate single strains with a single trait, it may be more practical to inoculate a microbial consortia that have multiple benefits and can thrive together in unique ecological niches. Any successful association between the bacteria and the host must be achieved before survival in critical numbers for efficient nodulation becomes limiting. A refreshingly new approach using recombinant DNA technology has been reported by Mavingui et al. (1997). These investigators engineered *Rhizobium tropici* CFN299 with randomly chosen DNA sequences. After repeated cycling through various legumes, recombinant strains with increased competitiveness for nodule formation with *Macroptilium atropurpureum* were selected. The authors suggest that if this technique of random DNA amplification (RDA) is used for selecting strains with potential use in agriculture, defined greenhouse or field conditions should be employed during the selection cycles. This may be a novel means to overcome the problem of competition.

There has been no systematic and reliable investigation of survival rates of the inoculated bacteria during whole crop growth to establish a direct cause and effect relationship of the benefits of inoculation. Such investigations were limited by poor quality inoculants and inadequate methods for bacterial identification and enumeration. These barriers are now easily overcome with the availability of molecular markers and enumeration of bacteria by sophisticated methods like PCR and ELISA. These molecular epidemiological tools can now be employed to obtain reliable information on the survival of the inoculated bacteria.

Rhizobial genome is very complex, but more amenable to genetic studies than the host plant (Long, 1989b; Martinez et al., 1990). A recent study identifies a unique plasmid in a *Rhizobium* sp. NGR234 that harbors genes and signalling mechanisms necessary for the symbiotic process in addition to the symbiotic genes found on the megaplasmids (Freiberg et al., 1997; Long, 1989a,b). Due to millions of years of evolution these bacteria, like others, have developed intricate mechanisms for ecological "fitness" to survive. A very delicate balance exists between the microbe and the environment. Is this delicate balance altered and the "fitness" lost when microbial genome is manipulated for optimal performance?

Host specificity is another barrier for improving the nodulation capacity of rhizobial strains, especially in developing a "universal strain". Attempts to manipulate certain rhizobial genes in specific legume rhizosphere niches for improving competition have not produced impressive results (Nambiar et al., 1992; Ronson et al., 1990; Triplett, 1990). Some recombinant lines have not been subjected to rigorous field evaluation. Another set of recombinant constructs designed to increase nitrogen fixation are *R. meliloti* and *B. japonicum* strains which showed promise under greenhouse conditions by demonstrating 15% increase in the rate of nitrogen fixed via increased expression of NifA and dctA genes, and increased copy number of nifA genes (Ronson et al., 1990). But under field conditions, the same constructs were unsuccessful in showing any increase in N fixation or agronomically-significant yield improvement. Moreover, the strains did not show improved competitive ability to persist in the soil. Other recombinant constructs of pigeon pea Rhizobia engineered with δ-endotoxin gene of *Bacillus thuringiensis* have not been assessed for field performance (Nambiar et al., 1990).

Attempts to engineer hydrogen uptake (hup$^+$) ability by cloning hydrogenase genes into non-hydrogen uptake (hup$^-$) strains of *Rhizobium* resulted in experimental successes only in areas where soybeans are cultivated and where the photosynthetic energy is limited (Evans et al., 1987). Sunlight is therefore another limiting factor. In tropical areas where sunlight is not limited (2000 lux before 8:00 A.M.), the hup$^+$ system is not turned on as there is abundant photosynthetic energy available to drive the symbiotic nitrogen fixation. Hup⁻ strain of *B. japonicum* has proved to be agronomically inferior under field conditions (Hume and Shelp, 1990). No conclusive field experiments confirming the agronomic benefits of the utility of hydrogen uptake trait in rhizobia have been conducted with a combination of rhizobia and legumes other than the soybean.

Manipulation of common nodulation genes to improve the bacterial competition has usually resulted in either no nodulation, delayed nodulation, or inefficient nodulation (Devine and Kuykendall, 1996). Attempts to alter bacterial host specificity genes has also met with similar fate. Altering either the Genotype-Specific Nodulation (GSN) genes or the Cultivar-Specific Nodulation (CSN) genes simply alters the ability of the bacteria to either nodulate or not nodulate a given host legume with which it already has well defined symbiotic relationship. The GSN and CSN genes can affect only the nodulation ability, which is not the same thing as increasing the growth and plant productivity as measured by yield or biomass. The problem is further complicated by the interference of host plant genes that restrict nodulation.

Other physiological processes seem to be involved in regulating nodulation and N fixation. Nitrate fertilizer suppresses nodulation by *Rhizobium*. One of the consequences of this suppression is induction of the plant hormone ethylene (Mattoo and Suttle, 1991), ethylene production was shown to increase with increased concentrations of nitrate supplied to *Medicago sativa* (Ligero et al., 1987). Grobbelaar et al. (1970) had previously demonstrated that scrubbing ethylene increased nodulation in cultured root system, and applied ethylene (0.4 ppm) dramatically reduced nodulation. Jackson (1991) reviewed these observations and suggested that the role of ethylene on nodulation and N fixation has been neglected and should be investigated because it holds significance for practical farming. He further elaborated that since nodule porosity plays an important role in ensuring optimal oxygen tensions for the nitrogen-fixing bacteroids (Sprent, 1984), the role of ethylene in affecting "nodular aerenchyma" in legumes needs to be investigated.

Non-symbiotic nitrogen fixation

Plant associative bacteria

The other important bacterial inoculants that have been reported to increase yield are *Azotobacter chroococcum* and *Azospirillum braziliense*. In a comprehensive review, Jagnow (1987) summarized reports of yield increases due to these inoculants, which ranged from 20–50% depending upon the cereal crop used. Such increases are inconsistent with the low levels of N fixed by associative bacterial systems (Elkan, 1992), suggesting that factors other than N are involved. Variable results obtained with *Azospirillum* are attributed to the lack of good quality control and uncontrollable environmental factors, yield parameters, and yield components associated with responses. Therefore, no definitive evidence exists for the cause and effect of inoculation of associative diazotrophs.

Azospirillum is not only capable of nitrogen fixation but also codes for plant growth hormones auxins and cytokinins (Elmerich, 1984). It is believed that increase in yield is mostly due to the hormones produced by *Azospirillum* in establishing the early seedling vigor as opposed to nitrogen fixation *per se*. In associative bacterial symbiosis, there is inadequate N available for fixation, and the process is very inefficient (Ladha and Reddy, 1995). Jagnow (1987) has estimated that only 1–4% of the total bacterial flora are associated with roots (rhizoplane) when plants are grown without nitrogen fertilizer. Furthermore, most of the N will be only available to plants after death and lysis of bacterial cells. Fixed N usually inhibits bacterial colonization as well as N fixation. Analysis of all the available data and applying the basic knowledge about the organisms clearly suggests that yield responses are mostly due to factors other than nitrogen fixation (Jagnow, 1987). A precise understanding of how hormones produced by *Azospirillum* affect the plant under field conditions is lacking. As is the case with rhizobia, *Azospirillum* inoculation does not ensure persistence in sufficiently large numbers to overcome competition (Michiels et al., 1989). Overcoming this problem is equally daunting as in the case of *Rhizobium* inoculation. But the molecular, physiological, and biochemical bases are not very well understood.

If the objective is to provide enhanced N nutrition to the plants through a biological process, then *Azospirillum* or *Azotobacter* are not the best candidates. However, it is worth noting that these associative bacteria do contribute to general plant growth and productivity in establishing early seedling vigor. The best probable candidates for the associative N fixation are *Acetobacter diazotrophicus*, *Herbaspirillum* sp., and *Azoarcus* sp. (Ladha and Reddy, 1995). Boddey et al. (1995) have reported high rates of BNF in sugarcane fields in Brazil by using a combination of *Acetobacter diazotrophicus* and sugarcane varieties developed for high yield with low input. These results have not been replicated with other varieties of sugarcane and in other sugarcane areas of the world. This case also points to the fact that the host plant is the dominant partner in nitrogen fixation, and that soil and environmental factors play a critical role in regulating the process. With respect to N fixation in soil with heterotrophic bacteria and organic residues, no consistent estimates of N fixed are available (Roper and Ladha, 1995). Under controlled conditions, they show impressive nitrogen fixing efficiency, but poor performance under field conditions. In general, manipulation of the biological nitrogen fixation machinery seems to be an arduous task, and unless novel and new approaches can be devised, there is little hope for enhancing BNF.

The alternative physiological approach

The foregoing analysis of the problems highlights the strict limitations on our ability to genetically improve symbiotic nitrogen fixation, and to transfer nitrogen fixation abilities to non-legumes. That is not to say that we should abandon our research efforts in this area, but only that we should be prudent, selective, and innovative (Hardy, 1993). It is becoming apparent that researchers who are interested in increasing the nitrogen nutrition of crop plants would have to look for alternative avenues to use recombinant DNA technologies to tap N from other sources within the plant. One has to look at approaches other than biological nitrogen fixation for providing nitrogen to the plants. The answer might lie in mobilizing stored nitrogen resources in the plant itself using genetic engineering techniques.

Developmental phase of the plant and the need for the amino compounds determine the fate of individual nitrogenous solutes imported in the translocation streams involving assimilation via complementary transporters (Peoples and Dalling, 1988). Different plant organs contribute substantially to the mobilization and efficient utilization of nitrogenous compounds. For instance, nodules and roots are the source of newly assimilated nitrogen from nitrogen fixation and soil mineral nitrogen uptake. Their senescence or their longevity can influence nutrient mobilization or storage, respectively, from the remainder of the plant. Similarly, nitrogen remobilization from senescing leaves allows efficient use of nitrogen by crop plants. Nitrogen derived and exported from senescing tissues is an important source of protein nitrogen. A biological process that could be modulated to increase and recycle nitrogen in a plant would have a good potential to provide N nutrition to the plants. Induced or normal senescence of a dispensable plant organ is a sign that other plant parts need nitrogen (and other molecules). The developing leaves, pods or grains are the major sink for nitrogen accumulated as soluble and insoluble protein in the vegetative parts of the older leaves (source) during growth. Chloroplasts of mesophyll cells are a natural source of reduced nitrogen. Studies with ^{15}N have shown that a major proportion of remobilized and transported nitrogen is derived from the degradation of ribulose-1,5-bisphosphate carboxylase/oxygenase and other proteins in the chloroplast (Kamachi et al., 1991; Mae et al., 1985). These chloroplast proteins account for 64% of the total nitrogen remobilized and reutilized normally during the senescence of vegetative tissues. Regardless of any particular organ, or the plant as a whole, senescence is associated with recycling of soluble and insoluble protein in leaf chloroplasts (Dalling, 1987). Thus, leaf photosynthesis is linked to the recycling of nitrogen in the plant. In this context, it is important to evaluate both photosynthate limitation and nitrogen limitation for improving the efficiency of nitrogen fixation and plant yields. It would be more appropriate to direct research efforts in this area.

Breeding programs may have unwittingly selected for varieties that have high pod/grain number and size and enhanced synthesis and translocation of carbohydrates but with no corresponding improvement in the amount of amino acids transported to the developing pod/kernel (Feldman et al., 1990). Any new approach for more efficient N utilization, i.e. increased uptake, reduced excretion through roots or volatilization through leaves, and increased translocation from all plant organs to the sink tissue (pods, grains, fruit), depends on unravelling the key elements in the processes involved. Once that is achieved, it may be possible to use genetic engineering approaches to manipulate the corresponding genes and reintroduce them to produce transgenic plants with efficient nitrogen economy. Very little is, however, known about such processes or the genes involved.

One can use the information derived from mechanisms involved in nitrogen recycling/source-sink relationships during senescence to equip the plant with valuable genes that maximize nitrogen mobilization and utilization. In wheat, it is known that wild relatives that have higher grain protein yield have a different flux of nitrogen metabolism than the cultivated and common wheat which have lower grain protein yields (Feldman et al., 1990). The grain-filling period of var. *dicoccoides* lasts only about 25–30 days while in the cultivated lines this period is extended for an additional 10 days. Further, var. *dicoccoides* is characterized by a delayed anthesis, a significantly shorter grain-filling period (from anthesis to complete grain ripening), a very rapid leaf senescence, as well as a lower rate of photosynthesis per unit leaf area. Most starch and proteins begin to accumulate in the grain only 8 to 10 days after anthesis; the actual duration in the wild lines may be less than 2/3 of that period in the cultivated lines. In the wild lines, the coordination between carbon assimilation and nitrogen recycling results in a higher N/C ratio in the assimilates translocated to the grains. Increased mobilization of nitrogen at an early stage of grain-filling in high grain-protein wheat lines is correlated to an increase in proteolytic activity in the leaf (Dalling et al., 1976). Regulated proteolysis may therefore be an important process enabling plants to regenerate the amino compounds needed for maintaining a favorable N/C ratio. Some senescence-related proteases implicated in nitrogen mobilization have been identified and cloned (Jones et al., 1995; Mehta and Mattoo, 1996; Mehta et al., 1996; Minami and Fukuda, 1995; Pautot et al., 1993; Valpuesta et al., 1995).

As several facets of nitrogen economy in plants are being unraveled, scientists can incorporate this information into new models and genetic engineering approaches for enhancing the source-sink relationship to promote nitrogen economy and crop yield. It should be possible to genetically engineer plants such that they are able to depend upon renewable sources of nitrogen within. For instance, one can reconstruct genes encoding abundant proteins using regulatable (senescence) promoters and introduce these into crop plants. Their expression early in senescence should then meet the demands of the sink tissue for elevated nitrogen. Alternatively, if demand of the sink tissue for nitrogen occurs later in the senescence phase, one can use late-senescence promoters with the value-added gene(s) to modify nitrogen metabolism. The key to using these strategies is first to identify the "high qual-

ity" and abundantly expressed protein to act as the renewable source of nitrogen, and then to find the appropriate tissue specific and developmentally regulated promoters. Efforts are underway in various laboratories to use such an approach. At IRRI (Philippines), a concerted Working Group is addressing and assessing genetic engineering approaches to improve not only nitrogen acquisition and transport (Wang et al., 1993) but also to merge this area with opportunities for nitrogen mobilization (Miao et al., 1991) and utilization to improve senescence-dependent nitrogen recycling (Makino et al., 1984; Mehta et al., 1992) and its translocation to sink tissues such as the rice grain (Ladha et al., 1997a,b). Relationships between nitrate uptake rates and plant development are influenced by environmental perturbations, nitrate availability, and N fixation rates. Elucidation of these interrelationships will require a detailed identification of the signalling mechanisms. Identification and functional expression of a nitrate transporter gene in *Arabidopsis* (Tsay et al., 1993), together with other developments through molecular genetics and biochemistry, should make it possible to tackle the intricate problem of demand-driven control of nitrate uptake. The need of the day is to produce designer plants that can better utilize plant's capacity and ability to improve nitrogen economy, maintain a favorable C/N ratio and tolerate variable environmental perturbations.

Acknowledgments

We are grateful to Drs J K Ladha, Phil Chalk and P M Reddy at IRRI for their most useful and critical review of the manuscript.

References

Becker M, Ali M, Ladha J K and Ottow J C G 1995a Agronomic and economic evaluation of *Sesbania rostrata* green manure establishment in irrigated rice. Field Crops Res. 40, 135–141.

Becker M, Ladha J K and Ali M 1995b Green Manure technology; Potential , usage, and limitations. A case study for lowland rice. Plant and Soil 174, 181–194.

Bockman O C 1996 Fertilizers and biological nitrogen fixation as sources of plant nutrients: perspectives for future agriculture. Forskningssenteret, Porsgunn, Norsko Hydro, Norway.

Boddey R M, de Oliveira O C, Reis V M, de Olivares F L, Baldani V L D and Dobereiner J 1995 Biological nitrogen fixation associated with sugar cane and rice: Contributions and proposals for improvement. Plant and Soil 174, 195–209.

Boivin C, Ibrahima N, Molouba F, de Lajudre P, Dupuy N and Drey-fus B 1997 Stem nodulation in legumes: diversity, mechanisms, and unusual characteristics. Crit. Rev. Plant Sci. 16, 1–30.

Brewin N J 1991 Development of the legume root nodule. Annu. Rev. Cell Biol. 7, 191–226.

Caroll B J, Mcneil D L and Gresshoff P M 1985 A super-nodulation and nitrate tolerant symbiotic (nts) soybean mutant. Plant Physiol. 78, 34–40.

Cetano-Anolles G and Gresshoff P M 1992 Plant genetic control of nodulation. Annu. Rev. Microbiol. 45, 345–382.

Cocking E C, Webster G, Batchelor C A and Davey M R 1994 Nodulation of non-legume crops. A new look. Agro-Industry Hi-Tech. pp 21–24.

Cregan P B and van Berkum P 1984 Genetics of nitrogen metabolism and physiological/biochemical selection for increased grain crop productivity. Theoret. Gen. Genet. 67, 97–111.

Dalling M J, Boland G and Wilson J H 1976 Relation between acid proteinase activity and redistribution of nitrogen during grain development in wheat. Aust. J. Plant Physiol. 3, 721–730.

Dalling M J 1987 Proteolytic enzymes and leaf senescence. In Plant Senescence: Its Biochemistry and Physiology. Eds. W W Thomson, E A Nothnagel and R C Huffaker. pp 54–69. American Society of Plant Physiologists, Maryland, USA.

Danso S K A 1992 Biological nitrogen fixation in tropical agrosys-tems: twenty years of biological nitrogen fixation in Africa. In Biological Nitrogen Fixation and Sustainable Tropical Agricul-ture. Eds. K Mulongoy, M Gueye and D S C Spencer. pp. 3–13. John Wiley and Sons, Chichester, UK.

De Bruijn F and Schell J 1993 Regulation of plant genes specifically induced in developing and mature root nitrogen fixing nodules: cis and trans acting factors. In Control of Plant Gene Expression. Ed. D P S Verma. pp 241–258. CRC Press, Boca Raton, Florida, USA.

Denaire J, Debelle F and Rosenberg C 1992 Signaling and host-range variation in nodulation. Annu. Rev. Microbiol. 46, 497–531.

Devine T E and Kuykendall L D 1996 Host genetic control of sym-biosis in soybean (Glycine max L.) Plant Soil 186, 173–187.

Djordjevic M A, Gabriel D W and Rolfe B G 1987 Rhizobium, the refined parasite of legumes. Annu. Rev. Phytopathol. 25, 145–168.

Downie A 1997 Fixing symbiotic circle. Nature 387, 352–354.

Elkan G H 1992 Biological nitrogen fixation systems in tropical ecosystems: an overview. In Biological Nitrogen Fixation and Sustainability of Tropical Agriculture. Eds. K Mulongoy, M Gueye and D S C Spencer. pp 27–40. John Wiley and Sons, Chichester, UK.

Elmerich C 1984 Molecular biology and ecology of diazotrophs associated with non-leguminous plants. Biotechnology 11, 967–978.

Evans H J, Harker A R, Papen H, Rusell S A, Hanus F J and Zuber M 1987 Physiology, biochemistry, and genetics of the uptake hydrogenase in rhizobia. Annu. Rev. Microbiol. 41, 335–361.

Feldman M, Avivi L, Levy A A, Zaccai M, Avivi Y and Millet E 1990 High protein wheat. In Biotechnology in Agriculture and Forestry, Vol. 13 Wheat. Ed. Y P S Bajaj. pp 593–614. Springer-Verlag, Heidelberg, Germany.

Flavell R B 1994 Inactivation of gene expression in plants as a consequence of specific sequence duplication. Proc. Natl. Acad. Sci. USA 91, 3490–3496.

Freiberg C, Fellay R, Bairoch A, Broughton W J, Rosenthal A and Perret X 1997 Molecular basis of symbiosis between Rhizobium and legumes. Nature 387, 394–401.

Grobbelaar N Clark B and Hough M C 1970 The inhibition of root nodulation by ethylene. Agroplantae 2, 81–86.

Halvorson L and Handelsman J 1991 Enhancement of soybean mod-ulation by Bacillus cereus UW85 in the field and in a growth chamber. Appl. Environ. Microbiol.57: 2767–2770.

Hardy R W F 1993 Biological nitrogen fixation: present and future applications. In Agriculture and Environmental Challenges. Proc. Thirteenth Agricultural Sector Symposium. Eds. J P Srivastava and H Alderman. pp 109–117. The World Bank, Washington, DC.

Herridge D F and Danso S K A 1995 Enahancing crop legume N_2 fixation through selection and breeding. Plant Soil 174, 51–82.

Horvath B and Bisseling T 1993 Genes involved in early root nodule development. In Control of Plant Gene Expression. Ed. D P S Verma. pp 223-240. CRC Press, Boca Raton, Florida, USA.

Hume D J and Shelp B J 1990 Superior performance of the Hup⁻ Bradyrhizobium japonicum strain 532C in Ontario soybean field trials. Can. J. Plant Sci. 70, 661–666.

Imsande J 1989 Rapid dinitrogen fixation during soybean pod fill enhances net photosynthetic output and seed yield: a new per-spective. Agron. J. 81, 549–556.

Imsande J 1997 Nitrogen deficit during soybean pod fill and increased plant biomass by vigorous nitrogen fixation. Eur. J. Agron. (In press).

Imsande J and Edwards D G 1988 Decreased rates of nitrate uptake during pod fill by cowpea, green gram, and soybean. Agron. J. 80, 789–793.

Imsande J and Touraine B 1994 N demand and the regulation of nitrate uptake. Plant Physiol. 105, 3–7.

Ingelbrecht I, Houdt H V, Montagu M V and Depicker A 1994 Posttranscriptional silencing of reporter transgenes in tobacco correlates with DNA methylation. Proc. Natl. Acad. Sci. USA 91, 10520–10506.

Jackson M B 1991 Ethylene in root growth and development. In The Plant Hormone Ethylene. Eds. A K Mattoo and J C Suttle. pp 159–181. CRC Press, Boca Raton, Florida, USA.

Jagnow G 1987 Inoculation of cereal crops and forage grasses with nitrogen fixing rhizosphere bacteria: possible causes of success and failure with regard to yield response - a review. Z. Pflanzen-ernahr. Bodenkd. 150, 361–368.

Jones M L, Larsen P B and Woodson W R 1995 Ethylene-regulated expression of a carnation cysteine proteinase during flower petal senescence. Plant Mol. Biol. 28, 505–512.

Kamachi K, Yamaya T, Mae T and Ojima K 1991 A role for glu-tamine synthetase in the remobilization of leaf nitrogen during natural senescence in rice leaves. Plant Physiol. 96, 411–417.

Kennedy I R 1997 BNF: An energy costly process? In BNF: The Global Challenge and Future Needs. A position paper discussed at the Rockefeller Foundation Bellagio Conference Center, Lake Como, Italy, April 8–12, 1997.

Khush G S and Bennett J (Eds.) 1992 Nodulation and Nitrogen Fixation in Rice. Potential and Prospects. Int. Rice Res. Inst. Philippines.

Knight T J and Langston-Unkefer P J 1988 Enhancement of sym-biotic dinitrogen fixation by a toxin-releasing plant pathogen. Science 241, 951–954.

Kooter J M and Mol J N M 1993 Trans inactivations of gene expres-sion in plants. Curr. Opin. Biotech. 4, 166–171.

Ladha J K, de Bruijn F J and Malik K A 1997a Introduction: assessing opportunities for nitrogen fixation in rice - a frontier project. Plant Soil 194: 1–10.

Ladha J K and Garrity 1994 (Eds.) Green manure production systems for asian ricelands. IRRI, Los Banos, Philippines. 195 p.

Ladha J K, Kirk G J D, Bennett J, Peng S, Reddy C K, Reddy P M and Singh U 1997b Opportunities for increased nitrogen use efficiency from improved lowland rice germplasm. Field Crops Research (*In press*).

Ladha J K and Peoples M B (Eds.) 1995. Management of Biological Nitrogen Fixation for the Development of More Productive and Sustainable Agricultural Systems. Kluwer Academic Publishers, Dordrecht.

Ladha J K and Reddy P M 1995 Extension of nitrogen fixation to rice-necessity and possibilities. GeoJournal 35, 363–372.

LaRue T A, Kneen B E and Gartside E 1985 Plant mutants defective in symbiotic nitrogen fixation. *In* Analysis of Plant Genes Involved in Legume-*Rhizobium* Symbiosis. Ed. R Marticellin. pp 39–48. OECD Publication, Paris, France.

Leigh J A and Coplin D L 1992 Expolysaccharides in plant-bacterial interactions. Annu. Rev. Microbiol. 46, 307–346.

Ligero F Lluch C and Olivares J 1987 Evolution of ethylene from roots and nodulation rate of Alfalfa *Medicago-sativa* L. plants inoculated with *Rhizobium meliloti* as affected by the presence of nitrate. J. Plant Physiol. 129, 461–467.

Long S R 1989a *Rhizobium*-legume nodulation: life together in the underground. Cell 56, 203–214.

Long S R 1989b Rhizobium genetics. Annu. Rev. Genet. 23, 483–506.

Lumpkin T A and Plucknett D L 1982 *Azolla* as a green manure: use and management in crop production. Westview Tropical Agriculture Series, No. 5. Westview Press, Boulder, Colorado. 230 p.

Mae T, Hoshino T and Ohira K 1985 Protease activities and loss of nitrogen in the senescing leaves of field-grown rice (*Oryza sativa* L.). Soil Sci. Plant Nutr. 31, 589–600.

Makino A, Mae T and Ohira K 1984 Relation between nitrogen and ribulose-1,5-bisphosphate carboxylase in rice leaves from emergence through senescence. Plant Cell Physiol. 25, 429–437.

Martinez E, Romero D and Palacios R 1990. The *Rhizobium* genome. Critical Rev. Plant Sci. 9, 59–93.

Mattoo A K and Suttle J C (Eds.) 1991 The Plant Hormone Ethylene, CRC Press, Boca Raton, Florida, USA. 337 p.

Matzke M A and Matzke A J 1995 How and why do plants inactivate homologous (trans) genes? Plant Physiol. 107, 679–685.

Mavingui P, Flores M, Romero D, Martinez-Romero E and Palacios R 1997 Generation of *Rhizobium* strains with improved symbiotic properties by random DNA amplification (RDA). Nature Biotech 15, 564–569.

Mehta R A, Fawcett T W, Porath D and Mattoo A K 1992 Oxidative stress causes rapid membrane translocation and in vivo degradation of ribulose-1,5-bisphosphate carboxylase/oxygenase. J. Biol. Chem. 267, 2810–2816.

Mehta R A and Mattoo A K 1996 Isolation and identification of ripening-related tomato fruit carboxypeptidase. Plant Physiol. 110, 875–882.

Mehta R A, Warmbardt R D and Mattoo A K 1996 Tomato (*Lycopersicon esculentum* cv. Pik-Red) leaf carboxypeptidase: identification, N-terminal sequence, stress-regulation, and specific localization in the paraveinal mesophyll vacuoles. Plant Cell Physiol. 37, 806–815.

Mendoza A, Leija A, Martinex-Romero E, Hernandez G and Mora J 1995 The enhancement of ammonium assimilation in *Rhizobium etli* prevents nodulation of *Phaseolus vulgaris*. Mol. Plant-Microbe Interact. 8, 584–592.

Miao G H, Hirel B, Marsolier M C, Ridge R W and Verma D P S 1991 Ammonia-regulated expression of a soybean gene encoding glutamine synthetase in transgenic *Lotus corniculatus*. Plant Cell 3, 11–12.

Michiels K, Vanderleyden J and Van Gool A 1989 *Azospirillum*-plant root associations: A review. Biol. Fertil. Soils 8, 356–368.

Minami A and Fukuda H 1995 Transient and specific expression of a cysteine endopeptidase associated with autolysis during differentiation of *Zinnia* mesophyll cells into tracheary elements. Plant Cell Physiol. 36, 1599–1606.

Mulongoy K, Gueye M and Spencer D S C (Eds.) 1992 Biological Nitrogen Fixation and Sustainable Tropical Agriculture. John Wiley and Sons, Chichester, UK.

Nambiar P T C, Ma S W and Iyer V N 1990 Limiting insect infestation of nitrogen fixing root nodules of the pigeon pea (*Cajanus cajan* L.) by engineering an entomocidal gene in its root nodules. Appl. Env. Microbiol. 56, 2886–2869.

Nap J P and Bisseling T 1990 Developmental biology of a plant-prokaryote symbiosis: the legume root nodule. Science 250, 948–954.

Neves M C P, Ramos M L G, Martinazzo A F, Botelho G R and Dobereiner J 1990 Adaptation of more efficient soybean and cowpea rhizobia to replace established populations. lit In Biological Nitrogen Fixation and Sustainability of Tropical Agriculture: Proceedings 4th International Conference of the African Association for Biological Nitrogen Fixation, Int Inst of Tropical Agric. , Nigeria, Sept 24-28, 1990. pp 219–233.

Pautot V, Holzer F M, Reisch B and Walling L L 1993 Leucine aminopeptidase: an inducible component of the defense response in *Lycopersicon esculentum* (tomato). Proc. Natl. Acad. Sci. USA 76, 9906–9910.

Peoples M B and Dalling M J 1988 The interplay between proteolysis and amino acid metabolism during senescence and nitrogen reallocation. *In* Senescence and Aging in Plants. Eds. L D Nooden and A C Leopold. pp 181–217. Academic Press, NY. USA.

Peoples M B, Herridge D F and Ladha J K 1995. Biological nitrogen fixation: an efficient source of nitrogen for sustainable agricultural production. Plant and Soil 174, 3–28.

Pracht J E, Nickell C D, Harper J E and Bullock D G 1994 Agronomic evaluation of non-nodulating and hypernodulating mutants of soybean. 34, 738–740.

Reddy P M and Ladha J K 1995 Can symbiotic nitrogen fixation be extended to rice? 10th Int. Cong. Nitr. Fix., St. Petersburg, Russia.

Reekie E G and Bazzaz F A 1987 Reproductive efforts in plants. 2. Does carbon reflect the allocation of other resources. Am. Nat. 129, 897–906.

Ronsheim M L 1988 Determining the pattern of resource allocation in plants. Trends Ecol. Evol. 5, 30–31.

Ronson C W, Bosworth A, Genova M Gudbrandsen S and Hankinson T 1990 Field release of genetically engineered *Rhizobium meliloti* and *Bradyrhizobium japonicum* strains. *In* Nitrogen Fixation: Achievements and Objectives. Eds. P M Gresshoff, L E Roth, G Stacey and W E Newton. pp 397–403. Chapman and Hall, New York, USA.

Roper M M and Ladha J K 1995 Biological nitrogen fixation by heterotrophic and phototrophic bacteria in association with straw. Plant Soil. 174, 211–224.

Sarrantonia M 1995 Soil-improving legumes. Rodale Institute, Kutztown, USA. 312 p.

Soderlund R and Roswall T 1982 The nitrogen cycle. *In* The Handbook of Environmental Chemistry. The Natural Environment and the Biogeochemical Cycles. Ed. J Hutzinger. Vol. 1B. pp 60–81. Springer Verlag, Berlin, Germany.

Sprent J I 1984 Nitrogen fixation. *In* Advanced Plant Physiology. Ed. M B Wilkins. pp 249–262. Pitman, London, UK.

Triplett E W 1990 Construction of a symbiotically effective strain of *Rhizobium leguminosarum* bv. trifolii with increased nodulation competitiveness. Appl. Environ. Microbiol. 56, 98–103.

216

Triplett E W and Sadowsky M J 1992 Genetics of competition for nodulation of legumes. Annu. Rev. Microbiol. 46, 399–428.

Tsay Y-F, Schroeder J I, Feldmann K A and Crawford N M 1993 The herbicide sensitivity gene CHL1 of *Arabidopsis* encodes a nitrate inducible nitrate transporter. Cell 72, 705–713.

Valpuesta V, Lange N E, Guerrero C and Reid M S 1995 Up-regulation of a cysteine protease accompanies the ethylene-insensitive senescence of day lily (*Hemerocallis*) flowers. Plant Mol. Biol. 28, 575–582.

Vance C P 1983 *Rhizobium* infection and nodulation: a beneficial plant disease? Annu. Rev. Microbiol. 37, 399–424.

Vance C P, Egli M A, Griffith S M and Miller S S 1988 Plant regulated aspects of nodulation and N_2 fixation. Plant Cell Environ. 11, 413–427.

van Veen A J, van Overbeek L S and van Elsas J D 1997 Fate and activity of microorganisms introduced into soil. Microbiol. Mol. Biol. Rev. 61, 121–135.

Wang Y M, Siddiqui M Y, Ruth T J and Glass A D M 1993 Ammonium uptake by rice roots. II. Kinetics of $^{13}NH_4$ influx across the plasmalemma. Plant Physiol. 103, 1259–1267.

Zapata F, Danso S K A, Hardarson G and Fried M 1987 Time course of nitrogen fixation in field grown soybean using Nitrogen-15 methodology. Agron. J. 79, 172–176.

Guest editors: J K Ladha, F J de Bruijn and K A Malik